M. Matsuura S. Yamane

Biology of the Vespine Wasps

Foreword by Shôichi F. Sakagami

With 104 Figures

Springer-Verlag
Berlin Heidelberg New York
London Paris Tokyo
Hong Kong

Dr. Makoto Matsuura
Laboratory of Entomology
Faculty of Bioresources
Mie University
Kamihama-chô
Tsu 514, Japan

Seiki Yamane
Department of Biology
Faculty of Science
Kagoshima University
Kôrimoto
Kagoshima 890, Japan

English edition arranged with Hokkaido University Press through Japan Foreign-Rights Centre

ISBN-13:978-3-642-75232-2 e-ISBN-13:978-3-642-75230-8
DOI: 10.1007/978-3-642-75230-8

Library of Congress Cataloging-in-Publication Data. Matsuura, Makoto, 1941 – [Suzumebachirui no hikaku kōdōgaku. English] Biology of the vespine wasps / Makoto Matsuura and Seiki Yamane. p. cm. Translation of: Suzumebachirui no hikaku kōdōgaku. Includes index. 1. Vespidae--Behavior. 2. Social behavior in animals. I. Yamane, Seiki, 1948 – II. Title. QL568.V5M3813 1990 595.79′8--dc20 89–26314

Softcover reprint of the hardcover 1st edition 1990

Typesetting: International Typesetters Inc. Makati, Philippines

2131/3145-543210 – Printed on acid-free paper

Foreword

It is a great privilege and pleasure for me to write the foreword to this book, which is the result of considerable effort on the part of Dr. Makoto Matsuura and Mr. Seiki Yamane. Such an introduction might sound like a cliché, but in this case the sentiments are real, since the preparation of this book really has been a laborious task. In fact, I think it would not be an exaggeration to say that such a book could only be published once every 100 years. I would like to explain why I regard this book as unique.

It is well known that as social insects *„hachi“* (as bees and wasps are called in Japanese), as well as ants and termites, have a very different way of life from that of other insects.

Each of these three groups has its own characteristics. Ants and hachi are classified in the same category; in other words, an ant is simply a wasp which has lost its wings and its sting, and has adapted itself to an underground life. Termites, on the other hand, are an aberrant cockroach, descended from a phyletic line which is completely different from that of ants and hachi, which undergo complete metamorphosis. On the other hand, ants and termites have a common feature: all members have adopted a social life. There are no solitary species of ant or termite, whereas most *„hachi“*, with the exception of a very few social groups, lead a solitary existence. These social *„hachi”* are descendants of two phyletic lines: (1) those which originate from bees, and which are exclusively dependent on flowers (pollen and nectar being protein and energy sources respectively), and (2) those which originate from wasps, and which use other animals (mainly insects) as their food source. The highest level of the former type is the honeybee, while the latter type is represented by the vespid wasp, whose life cycle is the main subject of this book.

Among the many types of vespid wasp, it is the polistine wasp whose life cycle has been studied in most detail. Their exposed nests, relatively small community size, and relatively weak aggression and virulence permit both direct observation of their behavior and analysis of their social relationships by means of individual identification. One of the most important discoveries in animal sociology, that social dominance also exists in invertebrate animals, was proved by Pardi, using *Polistes*. In contrast, far less information has been obtained on other groups of Vespidae. These include not only the Vespinae, but also many species of Polybiini, which belong to the Polistinae but form large communities and are endemic mainly in Central and South America.

As described in detail in this book, the Vespinae consist of three groups: a small-sized *Vespula* group (*Vespula* and *Dolichovespula*), which mainly inhabit the temperate zone in the Northern Hemisphere, the genus *Vespa* (including large-sized species), which is native to Asian subtropical and temperate zones and whose habitat extends to tropical zones, and the genus *Provespa,* which is endemic in Southeast Asia. A comparison of the present state of studies on each of these three groups and the contribution made in this book can be summarized as follows:

(1) *Vespula* group: The studies made on these wasps are the most advanced, since they inhabit not only Asia but also Europe and North America. In this book, new information obtained in Asia, mainly in Japan, is provided, in addition to previously extant knowledge. However, it cannot be said that this book makes a major contribution to knowledge on the *Vespula* group;

(2) *Vespa:* Most descriptions of this genus in technical journals published in Europe and North America are fragmentary, and the data can be traced back to the studies published by Janet in France in 1895 – 1903. This is partly due to the distributional characteristics of the genus. In Europe only one species (*V. crabro*) exists, except in the southern area, and the population density does not seem to be very high. In North America, this genus is not native, although the above-mentioned *V. crabro* is establishing itself in the eastern area of the United States as a secondary immigrant. For these reasons, little research has been done on this genus, even since World War II, except for a series of studies which tended to stress physiological aspects, and which were performed by Ishay and his co-workers in Israel. Consequently, with regard to descriptive work on this genus, the present book would seem to be unrivaled, and its publication should greatly increase the amount of information available on Vespinae;

(3) *Provespa:* Almost nothing has been reported on the ecology of this genus or even on its nest, except for the assumption that this genus is nocturnal, since few individuals have been collected in the daytime but are found occasionally around lamps at night.

Therefore, it could be said that the life history of this genus has come to light for the first time only with the publication of recent data which have been obtained in Sumatra and which are cited in this book.

Having described the contribution made by this book, I would like to elaborate on the reasons why its compilation was such a laborious task. At least the *Vespula* group is common in Europe and America, and *Vespa* is abundant in Japan. Why, then, have so few researchers studied the life cycle of these wasps? One of the reasons might be the formidable aggression and virulence of most of the wasps of these groups.

As detailed in this book, one sting from a large vespine wasp is often capable of endangering a human life and this book is a record of investigations performed in defiance of this danger and despite many accidents caused by the attack-defense behavior of these wasps. I myself have been stung by various kinds of bees on numerous occasions during my work, but I could not bring myself to do any work on vespine wasps. That is why I consider this to be a „once in a 100 years" book, as I stressed at the beginning of this foreword. The life history of *Vespa* in the tropics has been described for the first time in this book, overturning the formerly held hypothesis that this genus was a group showing strict monogyny. These results were obtained despite having to face numerous wasp stings.

In addition, the following points are also outstanding aspects of this book: first, 70 - 80 % of the contents are based on published and unpublished data obtained by the authors themselves, whereas less than one tenth of the photographs, figures and tables are cited from other authors; second, the work refers to almost all of the important references up to 1983; finally, a fault occasionally associated with some researchers who are concerned with evolution but not with systematics is avoided, since the book has been jointly produced by Dr. Matsuura, who is a professional ecoethologist, and Mr. Yamane, a systematist with rich work experience in ecology and ethology.

Since a foreword couched totally in words of praise seems a bit improper, I should perhaps point out one aspect which has not been dealt with in this book, although it may not necessarily be a weak point. There is no discussion of the latest sociobiological theories and ideas. Presumably the authors are of the opinion that studies on the Vespinae have not yet reached the stage for such discussion. To put it somewhat ironically, this book may be expected to have quite a long life because of this „fault".

Finally, there is another point which I feel I must add; it is not a „fault" as such, but one which occupied my thoughts after I had read this book. Reproducibility of results is a requisite for any scientific research, and this rule, as a matter of course, must also be applied to the content of this book. When I read it, however, I often wondered who would wish to reaffirm some parts of the descriptions. I, at least, would be anxious to be excused from carrying out such a dangerous study, and my somewhat grandiose expression, „once in a 100 years", stemmed in part from the impression which this book made on me. In any event, let us congratulate the authors on the publication of their book, which is far superior to those currently available anywhere in the world, including those published in Western countries.

SHÔICHI F. SAKAGAMI

Zoology Section
The Institute of Low Temperature Science
Hokkaido Unversity, Japan

Preface to the Japanese Edition

This book presents observations on the natural history of the Vespinae, a wasp group that has evolved one of the most highly social lives among the Vespidae. Accordingly, this work deals especially with their social behavior and related aspects of their biology. In most chapters we attempt comparisons between species, species groups, or genera, and for this purpose two chapters providing fundamental comparative knowledge are devoted to the systematics and distribution of vespine wasps.

Recently, vespines have been given increasing attention as significant pests of honeybees and fruits, and as nuisances in urban areas. To supply apiculturists, specialists concerned with agriculture and forestry, and the staff of public health centers with salient information, both the social and economic aspects of the relation of vespines to human life will be discussed. In this regard, Chapter 13, which presents a summary of the biology of the Japanese species, will help specialists in other fields of biology, as well as non-specialists.

Readers may begin with any chapter. Chapters 1 and 2 present the preliminary material necessary to grasp the basic features of the life cycle of Vespinae and their behavior as social insects. Topics that are briefly touched upon in those chapters are discussed later at greater length. Some topics may appear in several sections of chapters; for example, the visit of conspecific queens to established nests and the interaction between queens during the preemergence and cooperative periods are mentioned or discussed both in Chapters 3 (Sects. 2, 3, 6), 6 (Sect. 6), and 8 (Sect. 2). For a deeper understanding of this behavior of the queen, readers should consult the relevant sections of these chapters.

This book was written chiefly from the results of our surveys made in Japan. While using a common Japanese name for each species or genus seemed more convenient for Japanese readers, we decided to use scientific or Latin names. In contrast to butterflies and birds, only a few foreign wasps have Japanese names. For the collation of the Japanese names with the scientific names of the Japanese species, see Chapter 13 (Sect. 2). From Table 10.3 readers can judge the taxonomic position of any species (i.e., the species groups, subgenera or genera to which it belongs) mentioned in the text. The families and subfamilies

Preface

This English edition is a complete translation of the 1984 Japanese edition, with slight modifications. Several important papers on the biology and systematics of vespines appeared after the original publication of our book, and are accordingly not cited in the present edition. Therefore, our original notions, some of which would be revised in the future in the light of new information, are preserved. The publication of this edition was arranged through the courtesy of Prof. Yosiaki Itô (Nagoya University) and Prof. Ulrich Maschwitz (Fachbereich Biologie der Universität Frankfurt), to whom we are greatly indebted.

The earlier draft of our manuscript was revised by the following specialists, to whom we would like to express our sincere gratitude: Dr. C.K. Starr of the University of Georgia (Chaps. 8, 10), Dr. S.W.T. Batra of USDA [Chaps. 1 (Sect. 2), 3 (Sects. 3, 4), 4 (Sects. 3, 4, 6), 9 (Sect. 1), 11 and 12 (Sects. 1, 5)].

Tsu/Kagoshima, Spring 1990

MAKOTO MATSUURA
SEIKI YAMANE

are cited with Japanese names alone, or Japanese and scientific names together.

The genus and subgenus names are often abbreviated as: *Vespa* → *V.*, *Provespa* → *Pv.*, *Vespula* → *Vl.*, *Dolichovespula* → *D.*, *Paravespula* → *Pl.*, etc. Subspecies names are often omitted. For example, strictly speaking the population of *Vespa analis* in the Japanese mainlands has the trinomen *V. analis insularis*. But in cases where it is clear that the data cited were obtained on the Japanese archipelago, the subspecies name is omitted. When the precise locality is not given the full trinomen is generally used.

There are national and entomological disagreements as to defining and ranking supraspecific taxa in the Vespinae. We treat *Paravespula* as a subgenus of *Vespula*, while European entomologists (who have been influenced by P. Blüthgen) often consider it as a genus. One of the leading vespidologists in the USA, R.D. Akre of Washington State University, appears to be in the process of reforming his views and adopting *Paravespula* as a genus name, as is seen in his contribution to a new book, *Economic Impact and Control of Social Insects*. *Dolichovespula*, which is treated in our book as a genus, has long been considered a subgenus of *Vespula* by most entomologists in both the USA and Japan. Though no one view (including our own) can be considered correct, agreement among entomologists is to be hoped for.

Most photos used in this book, except in Chapter 7, were taken by us in Japan, chiefly in Hokkaido and southwestern parts of the Japanese archipelago. These photos cover 13 of the 16 Japanese species, and therefore may help readers to determine specimens (both wasps and nests). We cite many, if not all, articles written in Japanese that are difficult for researchers outside Japan to discover, in the belief that this will be of use to both Japanese and foreign researchers.

In the original version of the manuscript, Chapters 1 (Sect. 1), 2, 3 (Sects. 2, 5-7), 4 (Sects. 1, 2, 5), 5-7, 9 (Sect. 2) and 12 (Sects. 2-4) were contributed by M. Matsuura, and Chapters 1 (Sect. 2), 3 (Sects. 3, 4), 4 (Sects. 3, 4, 6), 8, 9 (Sect. 1), 10, 11, 12 (Sects. 1, 5) and 13 by Sk. Yamane, but the final version was produced through mutual revisions.

December 1983

Acknowledgments

We are sincerely grateful to Prof. Shôichi Sakagami of the Institute of Low-Temperature Science, Hokkaido University, for his continuous guidance and encouragement throughout the present enterprise and for writing the foreword to this book. Cordial thanks are also extended to Prof. emeritus Kunio Iwata of Kobe University; Prof. emeritus Zenpei Yamashita of Mie University; and Prof. emeritus Toshio Nakashima and Prof. Sadao Takagi of Hokkaido University, for their immeasurable erudition and guidance which inspired us in the study of wasps and bees.

We are also grateful to Profs. Toshitaka Hidaka of Kyoto University; Yosiaki Itô of Nagoya University; and Takehiko Nakane of Kagoshima University, for their kind encouragement during the preparation of this book.

We are indebted to many people for their staunch support during our investigations in Indonesia, including Prof. Shunzo Kawamura of the Primate Research Institute, Kyoto University; Prof. Roichi Ogushi of Kanazawa University and many other Japanese and Indonesian researchers who participated in the Sumatra Nature Study Project; also to Prof. Hideo Tagawa and Dr. Junichi Yukawa, both of Kagoshima University, who were investigating the ecosystem of Krakatau.

Many published and unpublished data and photographs were given to us by Mr. Toshiki Aoi of Hokkaido University; Dr. Shigeyuki Aoki of Rissho University; Mr. Teruhiko Ban of Otaru Public Health Center; Mr. Akio Kawazoe of Kuwana City; Dr. Tosio Kumata and Mr. Shun'ichi Makino of Hokkaido University; Mr. Kazumi Matsumura of Ishikawa Prefecture; Mr. Kenji Ohara of Kyushu University; Mr. Robert E. Wagner of the University of California, Riverside, USA; Dr. Sôichi Yamane of Ibaraki University and also of Asahi Graph Co. Ltd. Our particular thanks go to Messrs. Teruhiko Ban and Shun'ichi Makino.

Some invaluable advice was given to us by Mr. Robin Edwards Esg., of the Research and Development Department of Rentokil Ltd., England. Many valuable suggestions, in addition to a great number of reprints and copies of unpublished manuscripts, were provided by Dr. Michael E. Archer of the College of Ripon and York St. John; Dr. J. van der Vecht, a former curator of the Leyden Museum; Prof. Roger

D. Akre of Washington State University; Dr. Jacob Ishay of the University of Tel Aviv and Prof. Robert L. Jeanne of the University of Wisconsin.

Many specimens and data were given to us by Dr. Regine Eck of the Zoological Museum, Dresden; Dr. Josef Gusenleitner of Linz Agricultural Chemistry Institute; Dr. Robert S. Jacobsen of East Carolina University; Dr. N.V. Kurzenko of the Institute of Biology and Pedology, Vladivostok and Dr. Tie-Shéng Li of the Institute of Zoology, China Academy of Sciences.

We also extend our thanks to the people who helped us in discovering and collecting nests of vespine wasps and in rearing the wasps. They are (in random order): Messrs. Shigeaki Miyata, Mamoru Egawa, Teruhiko Ban, Shun'ichi Makino, Yasuo Maeta, Kazuaki Kamijo, Tsukasa Sunose, Noboru Hayashi, Miyuki Hayashi, Satoru Yamazaki, Masato Okubo, Sei Kitazawa, Akio Kawazoe, R.P. MacFarlane, Masanori Toda, Takakuni Sasai, Kazumi Matsumura, Katsumi Yabumoto, Fumitaka Ikeda, Kenzo Hiratani, Nagaaki Sekiya, Chiyoji Nakazawa, Yukihiro Tachi, Takanobu Minami and Yoshihisa Hanada.

The Japanese publication of this book was supported by a grant from the fourth academy publication subsidy of the Nihon-Seimei Foundation. Publication of such a specialized book would not have been possible without their help. The study was also partly supported by a Grant-in-aid for Overseas Scientific Surveys (Nos. 56041027 and 57043024) and a Grant-in-aid for Special Project Research (No. 58121001) from the Ministry of Education, Science and Culture, Japan, and by a grant from the Japan Society for the Promotion of Science for JSPS-DGHE Scientific Cooperation (1982). It was also facilitated by research subsidies from the Japan Society for the Promotion of Science (a scientific cooperation project for underdeveloped countries) for the fiscal year 1982.

Our special thanks go to the staff of Hokkaido University Press, including Mr. Haruo Tamiya, who encouraged us to write this book, waited patiently for completion of the manuscript, and took the trouble of editing.

Finally, we would like to thank our wives, Chieko and Etsuko. Our foolhardiness in choosing vespine wasps as the subject of our study caused them considerable anxiety during every period of our investigations. They have helped us in many ways, including the preparation of a fair copy of the manuscript.

March 1984

MAKOTO MATSUURA
SEIKI YAMANE

Contents

Chapter 1

The Vespinae in Insect Sociology

1 Historical Review

The northern hemisphere is the main distribution area of the subfamily Vespinae, and Japan, possessing 16 species of 3 genera of this subfamily, is an area particularly rich in these insects. The Japanese seem to have quite a long history of acquaintance with vespine wasps, starting in prehistoric times, since the wasps' habitats are the natural ecosystems of hills and fields which are closely related to human habitation. Although people in Europe and America have usually detested and avoided vespine wasps as an insect pest with a highly poisonous sting, in Japan there have been cases where the relationship has been a positive one. For example, many attitudes to these wasps can be seen to be based on the Oriental view of nature, as evidenced by the use of hornet nests in Chinese herbal medicine, or the custom of using the young from the nests as a source of food. The fact that they have been an object of folk belief is evidenced by the name "*bugensha bachi*" (wealthy wasps), which they were given because of their way of constructing huge nests.

Serious studies on vespine wasps in Japan have been started only quite recently. Only a few simple descriptions of their nests and the like can be found, even in the compilations on natural history published in the Edo Period by Kaibara ("*Yamato Honzô*", 1709) and Ono ("*Honzô Kômoku Keimô*", 1805). The great difference in the extent of knowledge on these wasps can be seen by comparison with that available on the Japanese honeybee, *Apis cerana*, which was already being used for breeding during the same period and which was described in several apiculture books such as "*Kahô Chikuyô-ki*" by Kuse (1791). From the Meiji to Taisho periods, reports on vespine wasps were published frequently in journals such as *Dobutsu-gaku Zasshi* (Zoological Magazine) and *Konchu-Sekai* (Insect World), but dealt with them mainly as edible insects. A report entitled *Studies on Vespa in Taiwan* (1), published by Sonan (1929) in *Taiwan Hakubutsu Gakkaihô* (Trans. Nat. Hist. Soc. Formosa) when Taiwan was a dominion of Japan, is considered to be the first real research work in this field. Sonan subsequently contributed to the same journal a series of reports on the taxonomy and ecology of species of Vespidae in Taiwan. As for Japanese vespine wasps, a great contribution was made by Takamatsu when he was at the Nagano Agricultural High School. He has published many reports since 1947 on the taxonomy, morphology, physiology and ecology of Vespinae, mainly *Vespula flaviceps*. The results of his histological studies on Japanese vespine wasps are included in "*Konchû Soshiki Zusetsu* (Illustrated Insect Tissues)" published in 1958. In the 1950s, detailed studies were performed by Shida on the ecology of *Vl. flaviceps* and the results were compiled as "Wasps of Musashino" in "*Nippon Konchû-ki* (1959)". This work became widely known to foreign scholars after its introduction by Iwata (1971a), a leading Japanese specialist on wasp behavior.

With regard to *Vespa*, on the other hand, the habits of *V. tropica* and *V. dybowskii* have been successively reported by Fukushima of Nagano Prefecture (Sakagami and Fukushima 1957 a, b) and these studies are considered to be the advent of behavioral studies on *Vespa* in Japan. Attention has also been paid to *Vespa* as a natural enemy in apiculture since the introduction of the European honeybee, *Apis mellifera*. Okada (1961) was the first to report a series of studies on the habits of *V. mandarinia*, which is the bee's worst enemy in this respect. Since its inception in 1946, an apiculture journal, *Gekkan Mitsubachi* (Jap. Bee J. Mon.), has published the records of many studies on the ecology of and protection against *Vespa* as an enemy of the honeybee. A review published serially in this journal [*Life of Hornets – from the viewpoint of protection of honeybees* (1) – (17); Matsuura 1967 a,c, 1968 a,b] introduces a considerable volume of new information on the ecology of *Vespa* in Japan and gives guidelines for protection against it. In the 1970s, great progress was made worldwide on the study of vespine wasps. Also in Japan, considerable strides were made, mainly in taxonomy and ecology, by Ishikawa, Matsuura, Sôichi Yamane, Seiki Yamane, Makino and many amateur naturalists. A book, *The Wonder of the Hornet* (in Japanese; published by the author in 1979), written by Adachi of Hyogo Prefecture, is a unique publication containing much valuable data, and clearly demonstrates the passion with which he has devoted many years to the study of hornets. Late in the 1970s, a growing interest was taken in vespine wasps as a "nuisance", mainly by people in residential areas on the outskirts of towns. In response to this pressing situation, Ban at the Otaru Public Health Center has been energetically studying means of exterminating these wasps (Ban 1980a-c). In the 1980s, studies on vespine wasps have progressed at a rapid rate, along with studies on other social insects, performed by researchers in various fields, including amateur naturalists, health center-related personnel and researchers in apiculture, toxicology, insect sociology and behavioral ecology.

Although the studies given most priority in this book are those conducted in Japan, an effort has been made to present as much recent information as possible from other countries, including Europe and America. Especially over the past 10 years, foreign reports, reviews and technical books on social insects including vespine wasps have been accumulating in large numbers. In the latest technical book on vespine wasps (*Social Wasps*, 1980), Edwards describes how the total number of reports on vespine wasps published worldwide during the 5-year period from 1973 to 1977 reached more than 340, which is equivalent to 12% of the total number of reports published over the last 300 years. Therefore, it is clearly not easy to review all the recent reports, which are being published at an increasing rate. As for the history of studies on vespine wasps in other countries, the above-mentioned book by Edwards (1980) is the most detailed and should be consulted.

2 The Vespine Wasps as Social Insects

Most species of the family Vespidae lead a social life (Table 1.1). Their social lives involve various types of societies that are sorted into three main categories by contemporary insect sociologists. The first is subsociality, in which parent(s) cares for its (or their) offspring. Earwigs, some pentatomid bugs, some bark beetles, and some aculeate wasps and bees fall into this category. Semisociality is characterized by the living together in a nest of more than one individual (normally female) of the same generation, among which polyethism is established to produce reproductive and more or less sterile individuals. This type of society has been observed in some bees (Michener 1958, 1974). The third category, which is called eusociality, must satisfy the following three conditions (E.O. Wilson 1971; Oster and E.O. Wilson 1978): (1) Cooperation among adult insects in caring for young (eggs, larvae and pupae), (2) reproductive division of labor among colony members – differentiation into individual(s) engaged in reproduction (queens)

Table 1.1. Classification, distribution, and social pattern of vespoid wasps

Groups	Distribution	Social structure
Masaridae	The Americas, Africa, Eurasia, Australia	Solitary. A few are communal
Eumenidae	Cosmopolitan	Solitary. Some are subsocial or communal
Vespidae		
Stenogastrinae	Tropical Asia, New Guinea	Subsocial, communal or eusocial(?)[a]
Polistinae		
Polistini	Cosmopolitan	Eusocial. Some are social parasites
Ropalidiini[b]	Tropical Asia, Australia, New Guinea	Eusocial
Polybiini	The Americas (mainly Central and South America), tropical Asia, Africa	Eusocial[c]
Vespinae	Eurasia, islands of South Pacific, North America.	Eusocial. Some are social parasites

[a] Not much is known on the social structure of this subfamily. S. Yamane et al. (1983) showed that the Sumatran *Parischnogaster mellyi* often has a society which comprises stepmother(s) and stepdaughters.

[b] Carpenter (1982) opposed to the taxonomic treatment in which the single genus *Ropalidia* constitutes a tribe. However, we tentatively follow Richards (1962).

[c] The wasps of the genus *Belonogaster* (dominant in Africa) have been considered as subsocial (cf. E.O. Wilson 1971), though recent investigations strongly suggest their eusociality (e.g., Pardi and Marino Piccioli 1981).

and more or less sterile individuals (workers that are usually daughters) occurs –, (3) coexistence of at least two adult generations that can contribute to the colony labor. The honeybees, stingless bees, bumblebees, some halictid bees, all the ants, most vespid wasps (Hymenoptera), some aphids (Homoptera)[1] and all the termites (Isoptera) are eusocial insects. Eusociality in the Hymenoptera may have evolved mainly from subsocial societies, but some apoid sociality is claimed to have originated through semisocial conditions (Michener 1958, 1969; also see Sakagami 1968, 1970). Varieties and subdivisions of these three basic social patterns are discussed in Sakagami (1970, 1975b), Michener (1974), and Eickwort (1981).

The species in the subfamily Vespinae, except for a few social parasites (see Chap. 8), are eusocial insects, in the strict sense of the term. In general, one female (foundress) initiates a nest (haplometrosis), cares for her young, and becomes absorbed in oviposition after the eclosion of her first brood, which are sterile workers. This polyethic family finally produces the next generation of reproductives (males and new queens) in the fall, then the colony dissolves. Queens differ from workers in their physical make-up as well as behavior. One colony normally contains only one queen (monogyny). The Polistinae, also belonging to the family Vespidae, have a far richer array of social patterns (Fig. 1.1). In this subfamily, pleometrotic nest foundation is commonly observed in tropical and subtropical regions where permanent polygyny and colony multiplication by swarming are known also to occur (Richards and Richards 1951; Jeanne 1980; Akre 1982). The polistine wasps afford ample material for comparative sociology to infer the evolutionary processes leading to eusociality. The vespine societies, most of which present an ideal model of eusociality, are less informative as material for comparative sociology. However, a detailed comparison of their seemingly monotonous societies have revealed that in fact they are rich in variety. Recent investigations on the vespines, made in the tropics mainly by Matsuura, show that the generally accepted idea about vespine life as mentioned above is no longer supported. All the vespine species (populations) are not haplometrotic or monogynous, and all do not adopt independent founding. At least one tropical hornet (*V. affinis)* is polygynous even after the eclosion of the first workers (Matsuura 1983a), and *Provespa* species probably found their nests by swarming (Chap. 7). The sociology of Vespinae is now at a turning point.

Impact on other social species is another characteristic of vespines. Some large species of *Vespa* attack congeners as well as polistines, yellowjackets and honeybees, and plunder larvae, pupae and honey from their nests. Extraordinary aggressiveness toward larger animals such as mammals is also worthy of special mention. Their unique position in ecosystem, as suggested by their aggressiveness and tyranny, will become an important aspect of future behavioral study.

---

[1]In aphids with a soldier caste (Aoki 1979) parental care is not conspicuous, but they can be regarded as eusocial insects because their societies satisfy the most important condition: the presence of a sterile caste usually consisting of foundress' daughters. Krebs and Davies (1981) hold a similar opinion. Aoki (1982) correctly argued that it is a matter of definition whether one counts these aphids among eusocial insects or not.

Fig. 1.1. Foundress association in a Japanese polistine, *Polistes jadwigae* (*top*), and polygynous nest of a South Indian polistine, *Ropalidia montana* (*bottom*)

Insect sociologists have sought the route along which eusociality evolved, and the approach involving the seeking of genetic reasons for the evolution of a sterile caste as the core of eusociality has made rapid and considerable progress since Hamilton (1964a,b) advocated the kin selection theory. This theory interprets the seemingly self-sacrificing act of an individual as behavior that efficiently passes copies of its genes to the next generations by helping its close relatives. In other words an individual organism may behave as if in order to maximize its inclusive fitness. Hamilton further proposed the "3/4-hypothesis" that was said to explain the evolution of the sterile worker caste in Hymenoptera where the relatedness among sisters (3/4) exceeds that between a female and her daughters (1/2), so that female offspring may benefit by helping their sisters (altruism) rather than by producing their own offspring (selfish act). Although this subhypothesis was later criticized by Craig (1979) and Kasuya (1981, 1982a), based on population genetics, kin selection theory itself has remained a powerful principle to explain the occurrence of eusociality (cf. Crozier 1982). Alexander's (1974) "parental manipulation of offspring" can be regarded as a subhypothesis within the frame of kin selection theory.

As mentioned above, insect sociology has entered an age of controversy over the origin of eusociality, but this book does not attempt to deal with this problem. There are two reasons for this, in addition to our inability to fully discuss this difficult theory: first, numerous excellent reviews have been published concerning the evolution of eusociality (sterile caste), and this situation does not require an additional repetition. Readers interested in this subject can consult E.O. Wilson (1975) and Starr (1979) as well as the above-mentioned articles by Hamilton and Alexander. Reviews and books written by Kasuya (1981, 1982a), Sakagami (1975a), Iwasa (1981) and Itô (1982) in the Japanese language are also useful. From references in these pages, readers can learn of many of the important articles on the evolution of insect eusociality; second, many hypotheses have appeared to explain the origin and structure of sociality, though few data to test them are available. As to the vespine wasps, no monograph has been published on the biology of the species inhabiting eastern Asia, where all the four genera occur, including most species of hornets. For these reasons, we will devote most of this book to the description of vespine natural history, rather than present theories. At the onset we planned to make an extensive comparison of the world species, though it cannot be said that this purpose was fully accomplished, mainly due to our inability and time limitation. Therefore, primarily Asian (especially Japanese) species receive intensive attention. At any rate we are confident that there has been no other book that gives such a comprehensive comparison of the species and genera. We hold the conviction that an understanding of the evolution of insect societies should ultimately be based upon a thorough understanding of phylogeny, a view also held by our teacher, Dr. Sakagami. For this reason, two chapters are devoted to classification, phylogeny and distribution.

Chapter 2

Life Cycle

1 Features of the Life Cycle

Three genera of vespine wasps in temperate regions (*Vespa, Vespula* and *Dolichovespula*) nest annually. A new colony is founded by a single queen, which has copulated the previous fall and completed hibernation, and the new colony life begins after the sterile females (called workers) have been raised. The colony develops rapidly with the establishment of a division of labor between the queen (foundress), whose job is oviposition, and workers, which perform all other duties. Both groups are all females. Rapid decline of the colony starts with the death of the foundress some time between late summer and fall, a period during which the number of workers reaches a maximum for the year and new queens and males are bred as reproductive individuals for the next generation. The new queens and males leave their nest and copulate in the field, and only the queens hibernate and spend a long time dormant until the following spring.

With regard to the polistine wasps, relatives of the vespine wasps, Yoshikawa (1959) has proposed comparing differences between their life styles by dividing their life cycles in a stepwise manner. The qualitative changes in the life cycle of vespine wasps, presented in this chapter, are divided into six steps based on the social relationship between queen and workers: (1) prenesting, (2) preemergence (= solitary), (3) cooperative, (4) polyethic, (5) reproductive and (6) hibernating periods. The sequence does not essentially differ from those in other temperate genera which produce annual colonies, including *Polistes* (Yoshikawa 1959; S. Yamane 1969; Matsuura 1977a) and *Parapolybia* (Sugiura et al. 1983) of the Polistinae and *Bombus* of the Apidae (Sakagami 1976).

The typical life cycle of five species of *Vespa* common in the warm southwest region of Japan is shown in Fig. 2.1, based on the above-mentioned division of life cycle. The nesting period from nest foundation to colony disorganization varies even among species of the same genus. Among the genus *Vespa, V. tropica* is a short-cycle species (four months), *V. simillima* is a long-cycle species (ca. 7 months) and the other species can be placed somewhere between the above two species. In the case of *Vespula, Vl. schrenckii* found in the highlands of central Honshu is a type showing short-cycle nesting (ca. 4 months; Takamizawa 1981a) and *Vl. flaviceps* found in southern Honshu is a long-cycle nesting species (7–8 months; Matsuura 1976b). *Vl. shidai* is considered to belong to the latter type. Among *Dolichovespula* species, *D. media* in Hokkaido is a short-cycle type with a nesting period of about three months (Fig. 2.2; Makino et al. 1981) and *D. norvegicoides* and *D. saxonica* are also considered to have the same type of life

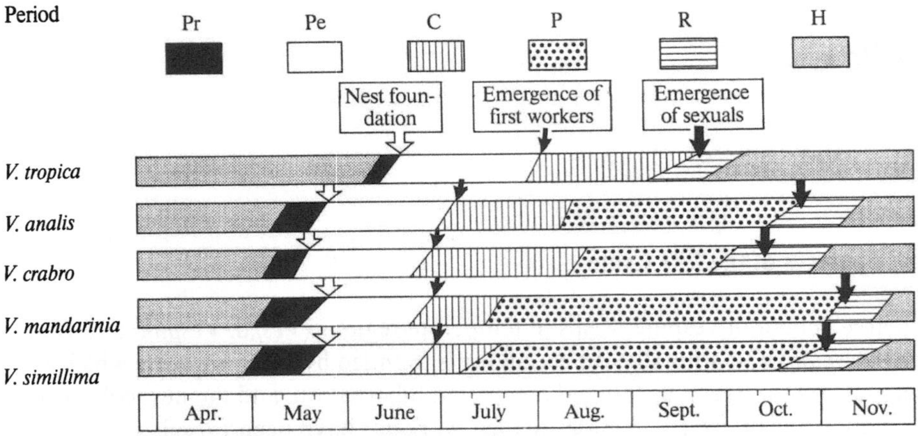

Fig. 2.1. Life history of five *Vespa* species in southwestern parts of Honshu, Japan. *Pr* Prenesting period; *Pe* preemergence period; *C* cooperative period; *P* polyethic period; *R* reproductive period; *H* hibernating period (After Matsuura 1984)

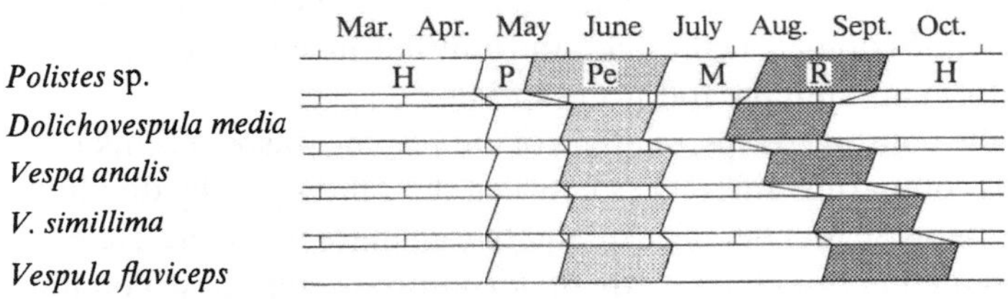

Fig. 2.2. Life history of vespine wasps in Hokkaido, northern Japan. *H* Hibernation period; *P* prenesting period; *Pe* preemergence period; *M* matrifilial period; *R* reproductive period (After Makino et al. 1981)

style, since the emergence of their new queens can be observed early in mid-August (Matsuura 1984).

In the case of vespine wasps, a relationship can be seen between this variation in nesting period and species-specific colony sizes. In general, a species with a longer nesting period has higher numbers of workers and cells and, in fact, the nesting periods of *V. similima* and *Vl. flaviceps,* whose colony sizes are the biggest among Japanese vespine wasps, are remarkably long. *Polistes* and *Parapolybia,* on the other hand, show almost the same nesting period of about five months (late April to mid-August), despite the remarkably different colony sizes, at least among the seven species which occur in Honshu. Vespine wasps improve the thermoregulation function of their nests by developing an envelope structure, which does not exist in the nests of polistine wasps, thus making it possible to maintain a high internal temperature of around 30°C, which is essential for breeding new queens during the cold fall season (Himmer 1932). In contrast to the Polistinae, the Vespinae have presumably differentiated, at the level of species or genus, into various types of life cycle, from short-cycle to long-cycle types, by greatly extending the possible nesting period and thus enlarging the degree of freedom of the nesting period even in temperate regions (cf. Chap. 6).

2 Behavior of Queens Before Nest Foundation

2.1 Behavior of Queens Emerging from Hibernation

The time of appearance of queens in the field after a long dormancy of nearly half a year in their hibernacula differs according to the species. In the warm southwest region of Japan, *Vespa* queens are first encountered in early to late April (*V. simillima*), mid-April to early May (*V. mandarinia* and *V. crabro*), early to mid-May (*V. analis*) and after the end of May (*V. tropica*), showing a disparity of 1–2 months, even among species of the same genus (Table 2.1). As for *Vespula* in the same regions, *Vl. flaviceps* first appears in late March to early April (Matsuura 1976b). In Hokkaido, however, *Dolichovespula media* and *Vl. flaviceps* start their activities all at once early in May (Makino et al. 1981).

Table 2.1. Date of first encounter of hibernated queens of *Vespa* species in and near Kibi, Wakayama (Matsuura 1984)

Species	Year									
	1966	1967	1968	1969	1970	1971	1972	1973	1974	1975
V. simillima	Apr. 12	Apr. 5	Apr. 10	Apr. 29	Apr. 15	Apr. 20	Apr. 17	Apr. 20	Arp. 14	Apr. 18
V. mandarinia	Apr. 8	Apr. 22	Apr. 6	Apr. 11	Apr. 26	Apr. 12	Apr. 24	Apr. 30	Apr. 18	Apr. 27
V. crabro	Apr. 28	Apr. 25	May 2	Apr. 21	May 5	May 3	Apr. 24	May 2	Apr. 25	May 2
V. analis	May 3	May 5	May 2	May 9	May 3	May 5	May 9	May 7	May 8	May 2
V. tropica	Jun. 8	May 22	May 22	Jun. 4	Jun. 8	May 25	May 25	May 29	May 24	May 26

After hibernation the queens do not immediately start to found their nests, but fly about for a while, seeking tree sap and nectar. *V. simillima* in warm regions occasionally visits flowers of *Camellia japonica*, but other *Vespa* species seldom frequent flowers. *Vespula* and *Dolichovespula,* on the other hand, use the nectar of raspberries and the honeydew of aphids as food sources, but seldom utilize tree sap.

The most important food source for large-sized *Vespa* species in spring is the sap secreted from the trunks of *Quercus* spp., in addition to those of *Betula* spp., and *Acer mono* and *Salix* spp. in some districts. A large number of various *Vespa* species visit these sap-rich trees for a while just after hibernation. Post-hibernating *V. mandarinia* in particular depends almost solely on the sap as food, and 20–30 queens may gather around one tree of *Quercus* spp. to feed on the sap from mid-April to early May (Fig. 2.3). These individuals take tree sap, not only during the daytime, but also in the evening, for several days. Because of the limited number of sap-secreting apertures, however, a linear dominance relationship is created, to form a specific order of sap-taking among the intra- and interspecific

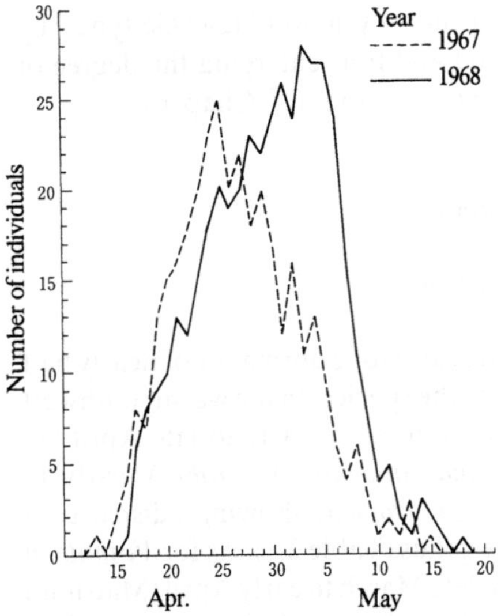

Fig. 2.3. Daily change of number of overwintered *Vespa mandarinia* queens visiting *Quercus acutissima* tree sap (Matsuura 1984)

individuals of *Vespa.* Thus, one aperture is always monopolized by one top-ranked queen, while the other individuals wait by her side for her to complete sap-taking or leave the aperture in search of other sap. Such behavior, which will be described in detail in Chapter 5, indicates that a certain qualitative difference already exists before the period of nest foundation, even among queens of the same species.

The fresh weight of post-hibernating queens is very much lower than that of the pre-hibernating ones. In *V. mandarinia,* the mean fresh weight of queens has been shown to be 3.46 g (s.d. ±0.21) just before hibernation, whereas the weight of queens just after hibernation is reduced by about 40%, with a mean value of 2.14 g (s.d. ±0.22) (Matsuura 1966b). This decrease in fresh weight is related to the fact that the fat body which accumulates in great quantity within the gasters of pre-hibernating queens can barely be observed after hibernation. According to Spradbery (1965), fat content as a percentage of dry weight in new queens of *Vespula* is over 40%, of which 30% is utilized during hibernation. The quantity of fat body contained in post-hibernating queens does not increase in spite of their feeding activity.

The post-hibernating queens feed in order to recover their strength, which has been reduced by the long dormancy. They then begin a searching activity in order to find sites for construction of their new nests. *V. mandarinia* disappears completely from tree sap sources in mid- to late May (Fig. 2.3) which is the period when it begins nest foundation. The behavior of the queens during this period, however, is mostly unknown. In the case of *Vl. flaviceps,* it has been found that the post-hibernating queens select certain sites where dead leaves lie on the ground and use them as a place to rest during the day and at night (Shida 1959).

2.2 Migration Before Nesting

When post-hibernating queens of *V. mandarinia* gathering around the sap of a *Quercus* tree were marked, they were found to have disappeared completely from the area (within several hundred meters from the tree) a few days after marking. During five years from 1965 to 1969, a total of 640 queens were marked and recaptured at the same place in Kibi-chô, Wakayama Prefecture, but marked individuals could be found after one week of marking, neither in the vicinity nor in other districts (Matsuura unpubl.). The same type of investigation was also made of 120 queens of *V. simillima,* but again, no marked individuals could be recaptured. This therefore indicates the possibility that post-hibernating queens of vespine wasps move out of their hibernacula and feeding sites prior to the initiation of nest foundation activity.

Several observations have been reported in Europe on the migration of queens in spring. In England, it has been reported that post-hibernating queens of *V. crabro* and *Vl. vulgaris* (both occur also in Japan) and *D. sylvestris* flew out to a lightship situated far from the shore (Rudebeck 1965). There has also been a remarkable example of group migration in the Scandinavian Peninsula, in which queens of *Vl. rufa*, just after hibernation, flew in a group in a west-south-westerly direction along the shore line from June 16 to 20, 1957. The width and height of the swarm were 20 and 1–3 m respectively, and the maximum number of individuals per hour reached 4500 (Rudebeck 1965). A swarm of the same species on the same scale had been observed in a north-eastern seaside district 50 km distant from the above-mentioned place from June 9 to 15, and the two groups were thought to have been the same. These queens reached an island at the south-western edge of the North Sea and, although part of the group made a turn of 150°, most of them flew away in a west-south-westerly direction. It was assumed that the latter major group arrived in Denmark, 25 km distant from the island. A similar phenomenon over the space of a few days was observed successively for 3 years in a southern seaside district in Finland (Mikkola 1978). The individuals involved were post-hibernating queens of *Vl. rufa* and *D. saxonica*, the maximum number of individuals reaching 2300 per hour and the width of the moving mass being over 800 m. Mikkola has estimated that at least 200 000 queens moved from west to east, passing over Helsinki during the course of three days, based on the assumption that the numbers of moving individuals during the day on which they were most frequent would be 120 000. It was assumed that these vespine queens had gathered from south-western districts of Finland and that the total area of the districts would be 10 000 to 20 000 km^2.

Since the origin of such group migration in spring is not yet clear, it is necessary to confirm whether it is induced by the lack of nesting sites and food sources caused by an outbreak the previous year, or whether it is repeated every year simply as a form of dispersal from the hibernacula. In any case, such group migration plays a role in reducing intra- and interspecific competition for nesting sites and food sources in the early stage after hibernation, since the migration is performed by prenesting queens. Consequently, there is a possibility that group migration is concerned with the movement of vespine population as an effective density regulation mechanism.

3 Nesting Periods

3.1 Preemergence (Solitary) Period

This is the period from the time when queens start nest foundation to the time when emergence of workers begins. Vespine wasps in temperate regions found their colonies with a single fecundated female (queen). The proportion of fecundated queens has been shown to be about 30% in *V. mandarinia* gathering around tree sap just after hibernation (Matsuura 1969), while those of hibernating *V. analis* and *V. simillima* are 96.6 and 93.3% respectively (Matsuura unpubl.). As for *Vespula*, it has been reported that in North America almost all queens of *Vl. pensylvanica* and *Vl. atropilosa* are fecundated in spring (MacDonald et al. 1974). On the other hand, queens of *V. mandarinia* which remain unfecundated in the fall of the previous year continue a solitary life until late July, visiting tree sap and then disappearing without showing any nest-foundation activity (Matsuura 1969).

The first nest-founding task of a queen is to search for a nesting site and, in southern Honshu, the times of the tasks necessary in *Vespa* species occur in the following order: late April to early May (*V. simillima*) / early May (*V. analis* and *V. crabro*) / mid-May (*V. mandarinia*) / early June (*V. tropica*). A queen of *V. simillima* makes a close examination of possible places for her nesting site by creeping into stone walls and underground spaces, but the same individual sometimes visits and inspects open spaces such as areas under the eaves of houses. Since, as will be described later in detail, in most cases this species first builds its nest in an enclosed space such as an underground gap and then moves to a wide space such as an area under the eaves of a house, it is possible that the queen decides her first nesting site simultaneously with consideration of her second move. In *V. mandarinia* which builds underground nests, a queen repeatedly enters and inspects burrows excavated by snakes or moles. When one queen which had crept into the ground was marked in mid-May, she was still searching for her nesting site 5 days after marking, at a place about 200 m distant from the marked ground. A queen may therefore take several days of cautious inspection work before she decides upon her nesting site.

As soon as a definite nest site is found, queens start minute inspection activity, searching for the nest attachment point. In *V. mandarinia*, a queen chooses a plant root hanging down in an underground space and removes any mud from the surface of the root with her mandibles. The first load of building materials is then moved in. Some of this material is first chewed gently and slowly to loosen it, then attached to the root and stretched out. The base of the pole thus attached is again chewed and loosened to repeat the same process, and attachment of the first load is then completed, taking about 20 minutes to make a club-shaped pedicel. The moving and processing of building materials is repeated and by evening, two to three cells, circular in cross-section, are completed on the tip of the pedicel. The queen's task for the first day of nest construction finishes after an umbrella-like envelope is constructed, covering the pedicel. In the case of the aforementioned observation, the *V. mandarinia* queen did not stay in the nest during the first night, and oviposition was carried out the following day (Matsuura unpubl.). Nest foundation in *Vespa* starts with the construction of a club-shaped pedicel, as described above. In the case of *Vespula* and *Dolichovespula*, however, the pedicels have the shape of a triangular sheet with a spiral twist. Consequently, and as will be described in Chapter 3, the nests founded by queens have different characteristics, not only among the three genera (*Vespa, Vespula* and *Dolichovespula*) but also among species of the same genus.

The number of cells constructed by a single queen up to the time of emergence of workers is 35–60 in *Vespa* and *Vespula* and 35–40 in *Dolichovespula*, and the number varies from one nest to another, even in the same species (cf. Chap. 3). Of these, the number of individuals reared by a single queen herself is 10–15. During a period of 30–50 days until the emergence of the first worker, the queen has to carry out all tasks herself, including food collection, nest construction and preservation of the nest. Consequently, disintegration of colonies occurs more frequently during this period than in any other throughout the whole nesting period; for example, because of the accidental death of the queen (cf. Chap. 6).

3.2 Cooperative Period

This is the period from the time when the first worker emerges to the time when the queen ceases her extranidal activities. As larvae reared by a queen mature, they construct solid white cocoons on the surface of the cells by secreting silk threads from silk glands near their mouths, rather like a silkworm. Rapid changes (larva -prepupa -pupa -adult) occur in the cocoon, and a worker emerges about two weeks after pupation.

Workers, having emerged and grown their wings inside the cocoon, secrete salivary fluid and daub it on one point of the cocoon wall in order to soften it. They then give the point a firm push with their mandibles. Thus one of the pair of mandibles pierces the softened and swollen point and gnaws away the wall step by step; in all, it takes about 1 h for the worker to get out of the cocoon solely by its own efforts.

The queen continues her tasks, having no interest in the first worker to emerge from the cocoon, and there is little contact between them. The workers move very little and spend most of the first day thrusting their heads into cells, except for an occasional visit to full grown larvae to ingest liquid secreted from the larval salivary gland. On the second day, their body color becomes lustrous, their once-opened wings are folded up and they start to walk around on the comb, seeking appropriate jobs. On and after the third day, as soon as they are able to fly, they make an orientation flight and then leave the nest to collect nest materials, food and water (cf. Chap. 4).

The appearance of the workers produces psychological and physiological effects on the queen, who then shows changes in her behavior. The queen starts to spend more time standing still on the comb as the emergence of workers proceeds, retires from extranidal activities, and then gradually ceases her intranidal activities except for oviposition. The behavior of queens in this period will be described in detail in Chapter 4. In general, the period for cooperative work between workers and a queen is short in species showing large colony size. In the case of *Vespa*, a queen of *V. tropica*, the colony size of which is the smallest in Japan, performs various intranidal activities throughout her life, whereas queens of large colony size species like *V. simillima* and *V. mandarinia* concentrate on oviposition 2–3 weeks after the emergence of workers. Even in *V. tropica*, however, the construction of new cells is performed by workers. In contrast, in polistine wasps which have been considered to have a society showing a lower degree of differentiation than vespine wasps, it has been revealed that queens of many species perform cell construction for oviposition and also go on outings to collect materials for the construction, even after the emergence of workers (West-Eberhard 1969; S. Yamane 1976; Sugiura et al. 1983).

In the cooperative period, development of colonies is slow, and the number of workers is small, amounting only to about ten. The gaster of a queen contracts and the number of mature eggs in one ovary is also small. Thus there is no apparent external difference between the profiles of the preemergence and cooperative periods.

3.3 Polyethic Period

This period includes the progress of activities which occur in the following order: increase in the number of workers, attainment of division of labor between the queen and the workers, rapid development of the colony and the emergence of males and new queens.

In the latter half of this period, *Vespa* workers start to display an obvious attraction to their queen. Workers who come across the queen orient their head towards her and antennate and lick her body. A few (up to about ten) workers continually cluster around the queen, forming the royal court which is familiar in honeybees (Fig. 2.4A). This phenomenon is explained by the secretion of a worker attractant, called queen substance, from the body of the queen. The chemical structure of the substance has been determined in *V. orientalis* (Ikan et al. 1969). The queen substance is considered to repress the development of ovaries in workers. Even in *Vespa*, however, *V. tropica* does not form a royal court throughout the whole nesting period (Fig. 2.4B). As for *Vespula*, no clustering

Fig. 2.4. Interaction between the queen and workers in some Japanese vespine species. **A** Royal court in *Vespa analis*; **B** *Vespa tropica* (no royal court observed); **C** *Dolichovespula saxonica* (workers occasionally attracted to the queen)

A
B
C

behavior of workers around their queen has been found in two Japanese species, *Vl. flaviceps* and *Vl. vulgaris*, (Matsuura unpubl.). The same result has been found in England for *Vl. vulgaris* and *Vl. germanica* (Archer 1972). In this genus, there is a possibility that a queen spreads the queen substance on the comb by rubbing her gaster against it, and controls her workers by letting them feed on the substance (Akre & Reed 1983). In *Dolichovespula*, attraction behavior between queen and workers is generally rare in the Japanese species, *D. saxonica* and *D. media*, except for an occasional attraction by the queen of the former species during the night (Sk. Yamane unpubl.). Workers of *D. maculata* in North America sometimes antennate their queen's body (Greene 1979).

In the case of royal court-forming species, the behavior of workers in the presence of their queen becomes more active as the nesting activities proceed; not only by orienting towards the queen and touching her body with their antennae, but also biting her gaster severely with their mandibles and licking all over her body while sitting astride her. The hairless body of the *Vespa* queen during the polyethic period is explained by this type of violent behavior of the workers, since the degree of hair loss is quite small in the queen of *V. tropica*, which does not form a royal court. In *Vespula* queens in the polyethic period, on the other hand, their bodies become naked and polished, the hairs having been worn away as in the case of the royal court-forming *Vespa*, in spite of the lack of any queen-licking behavior by *Vespula* workers. The reason for this phenomenon is not yet known.

As the end of this period approaches, construction of large-sized cells starts in the nests of vespine wasps to accommodate the production of new queens. In *Vespa* species, except for *V. tropica*, the time of initiation of large cell construction coincides with the time of royal court formation. It has been indicated that, in *V. orientalis*, the construction of large cells for new queens is regulated by the queen substance (Ikan et al. 1969). This mechanism is considered to be common to all royal court-forming species of *Vespa*. In *V. tropica* and *Vespula*, however, the mechanisms involved in the control of workers by the queen and in the transfer of queen substance to the workers are still unclear.

3.4 Reproductive Period

The new queens reared in the large cells emerge at almost the same time as the males, and leave the nest as they mature. Production of the new queens and males indicates the final step in nesting activity since, in general, the emergence of workers ceases when that of the new queens starts. For their remaining days, workers are engaged in the rearing of reproductives of the next generation. The founding queen, on the other hand, survives until about the time of emergence of the new queens. After these events, the colony goes into rapid decline, and disintegration of the colony proceeds, along with the additional effects of lack of food and lowered temperature in the field.

The new queens are reared in the large cells built in lower comb(s), but the males emerge from most of the combs and are reared even in the cells left by emerged workers. Emerged males and new queens remain quiet for the first 1–2 days, thrusting their heads into empty cells then gathering in groups and resting at the back of the comb or on the inside wall of the envelope, having nothing to do with the division of labor in the colony. They walk about inside the nest only at the time of feeding, receiving secretion from larvae at the final instar stage, or else liquid food such as tree sap collected by workers. The males and new queens stay in the nest for 1–2 weeks. As the population in the nest increases, the foraging activity of workers changes principally from searching for protein food such as prey to carbohydrates such as tree sap, which may be fed to the adult males and new queens. This results in a decrease in attention given to the larvae, which fail to be fed and remain at the same instar stage before finally dying due to rapid prostration. The prostration is also accelerated by exploitation of their salivary secretion by the adult males and new queens. Workers pull out the dead bodies of larvae from the cells and dispose of them outside the nest. Countless corpses of larvae can often be seen just below the nest of *V. crabro* and *V. simillima.*

Fully matured new queens, having had sufficient food in the nest, accumulate a large amount of fat body in their gaster before their departure from the nest. In *V. mandarinia*, the mean fresh weight of new queens has been found to be 2.88 g (s.d. ±0.25) just after emergence and 3.46 g (s.d. ±0.21) just before their departure, showing a weight increase of about 20% (Matsuura 1966b). In contrast, emerged males and new queens sometimes fail to get a nutritionally adequate amount of food in the last period of nesting, due to the rapid decrease in the number of workers. In *V. simillima, Vl. flaviceps* and *D. media,* some of the individuals emerging as new queens show tree sap and fruit juice-collecting behavior and give these materials to other individuals (Matsuura unpubl.). It is doubtful whether these new queens can found nests in the following spring, even if they are able to hibernate, because the amount of accumulated fat body in their gaster is quite small, and males show no copulation reaction to them. There are some reports on the extranidal activities of pre-departing new queens of *Vespula* and *Dolichovespula* (Spradbery 1973a; Makino 1982). These individuals can also be explained as "new queens who became workers" due to an insufficient supply of nutrients after emergence.

Production of males and new queens continues for 1–2 months, after which the nesting activities start to become disorganized. In *Vespa*, the disintegration of colonies takes place from early September to late November in the following order: *tropica – crabro – analis – simillima – mandarinia.* As for *Vespula*, the males and new queens of *Vl. schrenckii* in cold districts complete nest-leaving by the end of September, but each species of the subgenus *Paravespula* continues nesting activities until October or November. Especially in the case of *Vl. flaviceps*, which inhabits level land, a nest showing continued emergence of males and new queens can occasionally be found until January or February. In *Dolichovespula,* on the other hand, the reproduction period of males and new queens of *D. saxonica* and *D. media* is earlier (August to September) than those of other vespine wasps, and their nesting activities last only for a short period.

4 Hibernating Habits

4.1 Departure of New Queens and Males from Nest

Males and new queens are fed by workers and ingest the secretions of larvae. One after another they reach full maturity and then leave the nest. Since they do not return to the nest after leaving it, they do not perform an orientation flight (a necessary process for workers when they leave the nest for the first time), but leave the nest straight away and fly off in various directions. Workers continue their labor, having no interest in the departing males and new queens. In *V. simillima*, however, workers bite at a male with their mandibles to urge the male to leave if he does not fly away but walks around on the surface of the envelope instead, although they never attack the new queens (Matsuura unpubl.). The departure of males and new queens can frequently be observed on the morning of a clear day, but many of them leave the nest together after a spell of rainy weather, even on a cloudy day. Vespine wasps hardly ever leave their nests on a rainy day.

4.2 Copulation

The actualities of copulation in the field remain practically unexplained except for that of *V. mandarinia*. Males of *V. mandarinia* wait for departing new queens around the nest entrance and copulate on the spot (Matsuura 1968b), but males of the other species of *Vespa* and those of *Vespula* and *Dolichovespula* seem to await the new queens at other sites away from the nest. Since males of *V. simillima* are often found flying a certain course along a road or line of trees, it seems that they wait for new queens at such places in order to copulate. Males of *Vl. flaviceps* fly along a road or line of trees (Sk. Yamane et al. 1980b), but also fly in groups of several tens over the tops of large trees ten-odd meters in height, wait for new queens and copulate with them in the tree crown after catching them in the sky (Matsuura 1984; Fig. 2.5c). There are no reports on the copulating sites of *Dolichovespula*.

Mating in *V. mandarinia* occurs between 9.00 and 12.00 in the morning when new queens leave the nest. During the departing season for new queens of this species (October to November), males search for nests of the same species, fly around the nest entrances in groups and wait for the departing new queens. The number of males around one nest becomes several tens, sometimes over 100, and they fly about in front of the entrance, buzzing loudly within a space 3–5 m in depth and length and 1–3 m in height. New queens leave the nest mostly from 9.00 to 12.00 in the morning, but are pursued by the males as soon as they leave the nest, are captured in the air and copulate on the ground (Fig. 2.5A). A new queen bites the male's gaster or escapes from the site before insertion of the male's genitalia, but stands still on the site once insertion is achieved. During the early stage of copulation, the male supports his body with his hind-legs by arching the body with the free fore- and mid-legs. Positions are reversed as the copulation proceeds, the male standing still on the ground and the queen supporting her arched body with her hind-legs. The copulation continues for 26–42 seconds until the queen starts

Fig. 2.5. Mating behavior in some Japanese vespines. **A** *Vespa mandarinia*; **B** *Vespa dybowskii*; **C** *Vespula flaviceps*

to bite the side of the male's gaster violently. This violence, however, hardly seems to damage the male, according to most reports on the behavior of new queens of *V. simillima, V. analis* and *V. crabro* (Matsuura 1984) and *Vespula* (Richards 1937). New queens released from mating fly off from the site immediately, but on rare occasions are caught again by males and copulate further. Males, on the other hand, clean themselves and rejoin the group in front of the nest entrance. The rate of copulation in *V. mandarinia* has been estimated to be only 38–50% in spite of the high male density (Matsuura unpubl.). Since the rate of fecundated queens after hibernation is around the 30% level, it is thought that new queens losing the chance of copulation around the nest entrance just after their departure start hibernation without making any further attempt at mating (Matsuura 1969).

Copulation profiles similar to that of the above-described *V. mandarinia* have been reported for *Vespa simillima, V. analis* and *V. crabro* observed under artificial conditions (Matsuura unpubl.), *Vespula germanica* and *Vl. vulgaris* in Europe and *Vl. pensylvanica, Vl. atropilosa* and *Dolichovespula* in North America, and the copulation time seems to range from one to several minutes (Marchal 1896; Gaul 1942; Schulz-Langer 1954; Thomas 1960; Kemper and Döhring 1967; MacDonald et al. 1974).

4.3 Hibernation

The new queens who have left the nest and completed copulation in the field start searching activity for a hibernaculum on the same day. When 620 nest-leaving new queens of *V. simillima* were marked with paint, 14 hibernating ones were recovered and the distances from the nest to their hibernacula were within 20 m (3 queens), 100–150 m (7), 350 m (2), 520 m (1) and 1200 m (1) (Matsuura unpubl.). In *V. analis*, 7 of 120 marked queens were recovered in hibernacula and the linear distances from the nest were within 50 m (3 queens), 120 m (1), 350 m (1), 820 m (1) and 1100 m (1) (Matsuura unpubl.). The new queens become widely scattered after leaving the nest, and the distances to their hibernacula vary from several tens of meters to over 1 km. These migration distances seem to be affected partly by the sites where copulation takes place after they leave the nest. However, it is considered that, in order to conserve their physical strength for hibernation, the new queens do not make any long-distance migration, since there is no feeding activity after leaving the nest.

Hibernation sites for *Vespa* in Japan are in the soil and rotten wood of old stumps or logs lying on the ground. Hibernating queens of *V. simillima, V. analis* and *V. tropica* have been found in both of these sites, the two former species being especially frequent in rotten wood (Table 2.2). *V. crabro, V. mandarinia* and *V. dybowskii*, on the other hand, are found in soil, although the numbers observed are small. Some queens of *V. mandarinia* have also been found in a straw heap. The environments of these hibernation sites are always dim and rather moist places in forests, mostly facing north so that changes in temperature and moisture are minimal (Matsuura 1966b). The cell prepared for hibernation is dug by the queen herself in a spherical or oval form, and the tunnel (a few centimeters long)

Table 2.2 Hibernation sites of *Vespa* species (Matsuura 1984)

Species	Site in			Total
	Soil	Rotten Wood	Straw Heap	
V. simillima	17	59	0	76
V. crabro	6	0	0	6
V. analis	15	69	0	84
V. mandarinia	10	0	2	12
V. tropica	54	15	0	69
Total	102	143	2	247

leading to the cell is firmly plugged with earth or wood chips from within. The cells of *V. analis* and *V. simillima* are 4–6 cm in diameter, and the inside wall of the cell is finished very smoothly.

As for *Vespula*, queens of *Vl. flaviceps lewisii* have been found hibernating in soil several centimeters from the surface and have also been found sometimes in various spaces inside and outside wooden houses, as well as in spaces inside pieces of straw matting tied around the trunks of pine trees to prevent damage by pine caterpillars (Matsuura unpubl.). An observation has been reported of the hibernation of *Vl. shidai* within rotten wood (Sk. Yamane and Kanda 1979). In Europe, queens of *Vl. rufa* hibernating within the bark of dead fir trees and pre-existing insect-made tunnels in stumps have been recorded (Evans 1903). Many observations on the hibernation of *Vl. vulgaris* in various spaces inside and outside houses have also been reported (Free and Butler 1959). In addition, an observation has been reported of the use of a disintegrated nest as a hibernaculum by *Vl. germanica* (Edwards 1980). Judging from these observations, the queen of this genus seems to hibernate by creeping into a pre-existing space, unlike *Vespa*. In the Japanese *Dolichovespula*, on the other hand, some queens of *D. media* hibernate in rotten trees (Sk. Yamane and Kanda 1979) and those of *D. saxonica* have been found to use spaces in the wooden pillars of a garage (Matsuura unpubl.). The former species seems to build its own hibernation cell by gnawing a hole.

Queens of vespine wasps adopt hibernation-specific poses, the most remarkable of which is to conceal their folded wings under the ventral surface of the abdomen (Fig. 2.6). The antennae and all legs are closely drawn up against the body surface, the gaster arches and tergites are very stretched and swollen, and the tip of the gaster is turned towards the surface to which the queen clings. Such a pose is similar to that of a pupa and is considered to offer the least resistance during hibernation. A queen in a cell inside a piece of rotten wood clings to the ceiling or wall, but many of the queens of *V. mandarinia* and *V. tropica* hibernating within the soil lie on their backs on the floor. The latter situation can be explained by assuming that at some stage during hibernation a queen clinging to the ceiling drops to the floor for some reason, but continues in hibernation without changing her pose.

Fig. 2.6. Queen posture during hibernation. The wings are pressed close to body. *Top*: *Vespa tropica pulchra*; *bottom*: *Vespula shidai*

The *Vespa* queens generally hibernate one per cell, except for certain cases in *V. analis* and *V. simillima,* where two or three queens have been found in the same cell (Matsuura 1966b; Akio Kawazoe, personal communication). In Europe, it has been reported that ten queens of *V. crabro* were found hibernating together inside a piece of rotten wood (Sanderman 1936). In the case of plural or group hibernation, it can be assumed that the cell is not constructed by a joint effort, but that the cell containing one queen is invaded by other queens entering at different times. When new queens of *V. simillima* or *V. analis* just after copulation are transferred to a large cage containing rotten wood, they immediately start to gnaw holes in the wood. In this case, a few queens hibernate in one group by building a space which is several times larger than the usual cell. These queens in the same cell hibernate by keeping a certain distance from each other in order to avoid contact, thus they do not pile one on top of the other (Matsuura 1980b), which is common in the group hibernation of polistines such as *Polistes* and *Parapolybia.*

The poses of hibernating queens of *Vespula* and *Dolichovespula* are basically the same as those of *Vespa,* except in the case of *Vl. flaviceps,* whose queens are sometimes found firmly holding on to the wall with their mandibles. No such habit has been found in *Vespa*. Group hibernation is also common in these two genera. In *Vl. germanica*, group hibernations consisting of 38 (Billups 1882) and 68 (Sanderman 1936) queens have been reported, with the greatest number, 400 queens, being reported in New Zealand (Thomas 1960).

In temperate regions, including Japan, the duration of the dormant period for new queens is 6–8 months. These queens, therefore, possess unique physiological characteristics enabling them to prepare for the long dormancy. The most remarkable example is the accumulation of large amounts of fat body within the gaster. The fat body is enclosed in storage cells which are called trophocytes and are filled with a large amount of lipid in addition to glycogen and protein. Fat content as a percentage of the dry weight in queens of *Vespula* just before their hibernation is about 40%, which is remarkably high compared with the content (below 10%) in workers during the same season (Spradbery 1973a).

Ovaries of hibernating new queens remain in an immature condition, although the size of the egg cell at the end of the each ovariole is almost the same as the size of the nutritive cell. Copulated and uncopulated queens can be distinguished, since the sperm in the spermatheca of the former queens give the spermatheca a slightly opaque appearance while the spermatheca of the latter queens are transparent. No differences, however, can be found among them in terms of size and shape.

Chapter 3

Nesting Habits

1 Nesting Materials

Vespine wasps, like polistine wasps, construct nests by biting off plant fibers or other fibrous materials, mixing them with saliva to turn them into a pulpy form and then stretching the pulp to turn it into a construction material (Fig. 3.1). Such a process is similar to that used in commercial paper manufacture, but the nesting materials vary depending on the species. Plant fiber is the best material because it can be obtained near at hand, it facilitates construction of a light but solid nest, using a small amount of material, and modification of the nest is easy. Social bees belonging to the Apidae use bees wax secreted from their wax glands as the common nesting material, but the nesting materials used by vespine wasps have many characteristics which are not at all inferior to bees wax.

Vespa collects nesting materials from the woody parts of rotten trees and the dead parts of live trees, but sometimes bites the bark off live trees. *V. simillima* and *V. analis* visit both logs and living trees of Japanese cedar, pine, white cedar, bamboo, oak, chinquapin and peach (Matsuura 1984). As for *V. crabro*, individuals collecting nesting materials from the dead parts of live peach trees have been observed in an orchard in Japan, and some reports indicate that, in Europe and North America, a subspecies bites the bark off ash trees (Kemper and Döhring 1967) and lilac (Wray 1954). There is also a report of *V. tropica* collecting nesting materials from the dead parts of a chinquapin tree trunk and abandoned pine timber. Since the nest of this species is always reddish-brown, there is a possibility that this species, in contrast to other species, uses specific kinds of tree as the source of its nesting materials. *V. affinis* on Ishigaki Island has been observed to bite off nesting materials from the weathered part of the surface of a log of an unknown type of tree, which had been standing on the ground (Matsuura 1973c). In the Palestine area, on the other hand, *V. orientalis* is known as a pest which bites the bark off fruit trees, including orange, pear, peach and grapevine. Especially in the case of orange trees, only young ones 3–4 years old are damaged by this species (Bodenheimer 1951). As an unusual example of nesting material, *V. analis* has been observed to use a piece of newspaper left in the field (Matsuura 1984).

Among Japanese *Vespula* species, *Vl. flaviceps, Vl. shidai* and *Vl. vulgaris* collect nesting materials from the surfaces of rotten logs, pillars, wooden walls and fallen trees in addition to occasionally using corrugated cardboard and the paper from sliding screen doors (Matsuura unpubl.), indicating the adaptability of this genus to various materials. Only one example has been observed of *Vl. schrenckii* biting fibrous parts from the surface of an old Japanese larch log. Since the nest

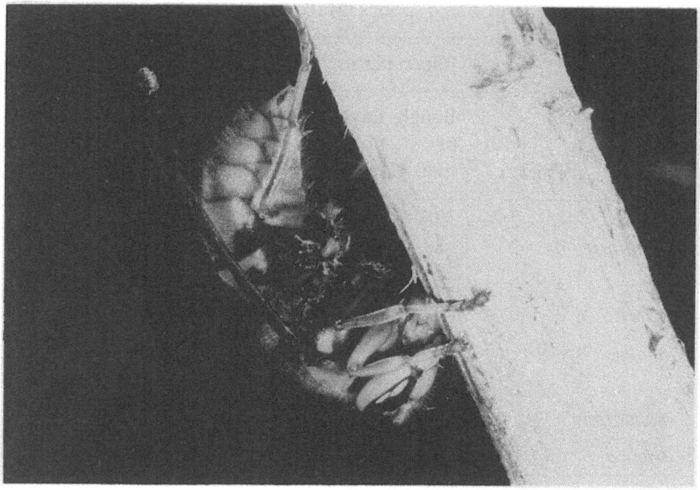

Fig. 3.1. *Dolichovespula media* queen gathering nest material

of this species has a fiber-rich structure in comparison with those of the three above-mentioned *Vespula* species, its preference with regard to nesting materials will probably differ slightly from other species.

The nests of *Dolichovespula* are made of fiber-rich "Japanese paper"-like materials which are different from those of the above two genera. Founding queens of *D. media* bite fibers from the surfaces of dead stems of mugwort and eulalia, and workers of the same species bite the bark off purple willow as their nesting material. A founding queen of *D. norvegicoides pacifica* has been observed to collect fibers from the surface of corrugated cardboard left in the field (Matsuura unpubl.). It has been reported that *D. norwegica* in Scotland collected tomenta growing thickly on the underside of rhododendron leaves and used them as a nesting material (Laidlaw 1930).

2 Nest Sites

2.1 Nest Sites of Each Genus and Species

Vespine wasps have nest site preferences which depend on species. In the case of *Vespula* and *Dolichovespula*, their preferences are genus-specific: the former nest mainly in underground situations and the latter in open situations above ground (Duncan 1939; Kemper and Döhring 1967; Akre et al. 1981). *Vespa*, on the other hand, is classified into two types of species, one having a preference for open situations above ground and the other for covered situations under or above ground (Table 3.1; Matsuura 1971a).

Table 3.1. Nest sites of *Vespa* species (Matsuura 1984)

Species		Open place				Covered place				
	Thicket	Bough of tree	Eaves of building	Rock wall	Other sites	Attic	Hollow tree	Other sites	Underground cavity	Total
V. simillima	A[a] 2	0	2	0	0	24	3	8	34	73
	B[b] 2	8	95	13	14	39	2	3	2	178
V. crabro	A[a] 0	0	0	0	0	13	4	7	15	39
	B[b] 0	0	0	0	0	25	4	5	7	41
V. analis	149	0	13	2	0	0	0	0	0	164
V. mandarinia	0	0	0	0	0	0	0	0	56	56
V. tropica	0	0	0	0	0	10	4	7	24	45

[a] A: Nest sites before or without relocation of nests.
[b] B: Nest sites after relocation of nests.

Among Japanese *Vespa* species, *V. analis* and *V. affinis* belong to the species preferring open, above-ground situations, while *V. crabro, V. dybowskii, V. tropica* and *V. mandarinia* have a preference for covered situations. *V. simillima,* on the other hand, nests both in covered and open situations, and one colony sometimes moves between the two types of location.

V. analis: This species nests only in free spaces above ground and hangs its nest on a twig or from the eaves of a building. In a study performed in southwestern Japan, 86% of such nests were found on twigs and the stems of plants, but no preference for specific plants was observed, although 23 different species of plant were recorded (Matsuura 1971a). The thickness of the nest-supporting twigs varied from 2 to 8 mm in diameter (3.6 mm on average), and 61.9% of the nests were located at a height of below 1 m from the ground. Since the nests are built at relatively low levels, most of them are hidden by bushes. Twigs and leaves near the nest envelope are transferred into the nest or attached to its surface by workers as the nest develops. Consequently, the whole nest is supported at various points, thus creating a reinforcement effect against wind. This species also builds nests in open spaces other than tree twigs, such as the eaves of houses (11.6%) and under overhangs on rock walls (2.5%). Subspecies in Taiwan (*V. analis nigrans;* Matsuura 1973c; S. Yamane 1977) and Sumatra (*V. a. analis;* Matsuura unpubl.) have been confirmed to be typical tree-nesting types.

V. affinis: Eight nests on Ishigaki Island were found attached to tree twigs and stems and the leaves of plants 0.5–3 m above ground (Matsuura unpubl.), but the same species in Taiwan nested on tree twigs 7–8 m above ground or under the eaves of houses (Matsuura 1973c; S. Yamane 1977). Nests of each subspecies in various districts in Southeast Asia have been commonly observed on trees and sometimes under the eaves of houses, but there are no known examples of nests in a completely covered situation (van der Vecht 1957; Matsuura 1984). This species and *V. analis* seem to have a competitive relationship with regard to their nesting sites in areas where their distributions overlap, since they are considered to be typical tree-nesting species and share many common features in their nest-site preferences.

V. simillima: In a study in Wakayama Prefecture, the nest sites of this species (subspecies *xanthoptera*) were predominantly located in open situations above ground, mainly under the eaves of buildings. The nests were also found in covered situations under and above ground, but the ratio of open to covered nest sites was about 8:2 (Matsuura 1971a). In Kyoto, not so distant from Wakayama, this ratio was reversed, and over 80% of the nest sites were observed in covered situations above ground, mainly in attics (Makino et al. 1981). According to an investigation on another subspecies, *V. simillima simillima,* in Otaru, Hokkaido, more than 80% of the nest sites were found in covered situations, the predominant nesting site (43%) being gaps in weatherboards (Ban 1980a). Within these fields of investigation, natural habitation environments such as mountain villages and mountainous regions were selected for the investigation in Wakayama. In Kyoto and Otaru, on the other hand, investigations performed in urban areas were only on the nests of vespine wasps which had been approved for extermination as a nuisance by administrative organizations. Even in Hokkaido, 50% of nest sites of this species were found in open situations when they were investigated in a near-natural environment, a botanical garden in Sapporo (Makino et al. 1981). In contrast to Wakayama, this species in Hokkaido usually builds its nest in open situations only 1–2 m above ground. These results indicate that this species not only has the most flexible nest site preference among Japanese vespine wasps, but is also a species with considerable adaptability, being able to change its nest site in response to the habitation environment.

All the above results are based on investigations of mature nests. In Wakayama, 97% of young nests of this species were found to be built in covered situations both above and under the ground, and when a nest became large after the emergence of workers, the whole colony moved to a more spacious site (Matsuura 1976b). Such a "relocation", however, does not occur when a nest is built in a space wide enough to accommodate maturation of the nest. Consequently, a high rate of mature nests in covered situations can be expected in an area where this species can use wide covered spaces above the ground from the beginning as the nest sites or, as in Hokkaido, when the colony size is comparatively small.

V. crabro: This is a typical covered-space nesting species, and the Japanese subspecies *V. c. flavofasciata* builds its nest in tree hollows, covered spaces inside houses (such as attics, spaces under floors and inside shutter-boxes), large empty bottles, straw piles, empty beehives or underground cavities (Matsuura 1971a). As a unique example, it was observed on 11 September 1974, that an exposed nest of this species was linked to a discarded nest of *V. analis* which had been constructed the previous year under the eaves of a house in Naga-chô, Wakayama Prefecture (Matsuura unpubl.). As shown in Fig. 3.2, the young nest of *V. crabro* was first founded inside the space of the discarded *V. analis* nest. As the nesting

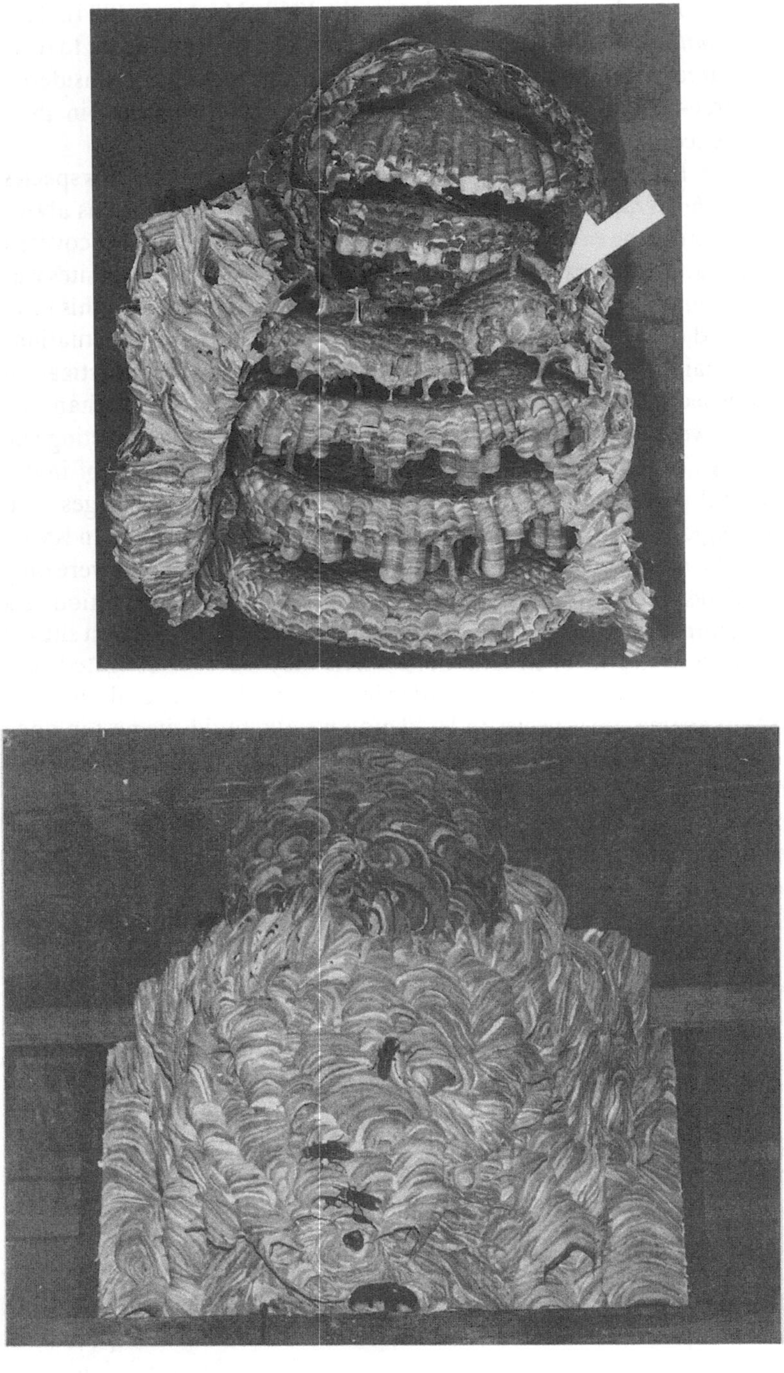

Fig. 3.2. *Vespa crabro* nest constructed under a dissolved nest of *V. analis*. *Arrow* shows the queen nest of *V. crabro*

scale increased after the emergence of workers, the lower side of the envelope of the discarded *V. analis* nest was broken and four *V. crabro* combs were constructed. The queen in the young nest used the pre-existing entrance to the discarded nest, but a new entrance was constructed by workers on the lowest side as the nest was extended outside the old nest.

Other subspecies in Europe and North America (*V. crabro crabro* and *V. c. germana*) also build their nests mainly in covered situations above ground but occasionally in underground cavities (Duncan 1939; Kemper and Döhring 1967; Akre et al. 1981).

As an exceptional example of open space nesting, a discarded queen nest built before the emergence of workers was obtained from under the eaves of a hut on the mountain side of Mt. Kariba, Hokkaido in August 1980. There are no reports of mature nests in open situations.

This species, like *V. simillima*, relocates its nest to a more spacious covered space when the space available for the first nest becomes cramped.

V. dybowskii: The nesting habits of this species are quite similar to those of *V. crabro*. Attics in houses, gaps in wooden walls and tree hollows have been known to be its nest sites (Sakagami and Fukushima 1957a; Matsuura unpubl.), but no observations of its nesting in open or underground situations have been reported. The queen of this species can found a nest by herself, but she sometimes invades a nest of *V. crabro* or *V. simillima* with just-emerged workers, which are nesting in a covered space, and she takes over the nest by killing its queen (Sakagami and Fukushima 1957a).

V. tropica: A subspecies in Japan (*V. tropica pulchra*) builds its nest in various covered situations above ground, preferring relatively spacious sites including attics, spaces under floors, the inside of wooden walls, baskets made of straw and vacant beehives, in addition to subterranean cavities. As a unique case, it has been observed that this species used the inside space of a large discarded nest (ca. 50 cm in diameter) of *V. simillima* under the eaves of a house in Yôrô-chô, Gifu Prefecture in October, 1978 (Matsuura unpubl.).

Various subspecies in Southeast Asia build their nests in covered situations inside buildings (*V. t. leefmansi*: Singapore, Chan 1972) and in underground situations (*V. t. leefmansi*: Sumatra, Matsuura 1983b; *V. t. pseudosoror*: Taiwan, S. Yamane 1977).

V. mandarinia: This is a typical underground-nesting species. It nests in subterranean cavities formed either around rotten tree roots or made by small rodents, snakes, etc. and occasionally in tree hollows close to the surface or 1-2 m above the ground. As will be described later, many species of *Vespula* are of the underground-nesting type, but they sometimes build their nests in open or covered situations above ground. This species, unlike the others, cannot support and stabilize the whole nest only by the pedicel, because of the weak development of the envelope and pillars.

Vespula: Except for a social parasite (*Vl. austriaca*) which does not build its own nests, the Japanese species of this genus are basically of the underground-nesting type, although their nests are found occasionally in open or covered situations above ground.

Vl. schrenckii usually prefers preexisting underground cavities which are relatively narrow and close to the surface of the ground (Takamizawa 1981a), but in Hokkaido its nests are sometimes observed in covered situations above ground such as attics and wooden wall gaps (S. Yamane 1970).

We have observed the nest of *Vl. vulgaris* only seven times in Japan. Its nest sites were underground cavities and tree hollows, and the subterranean nests were located within 20–30 cm below the ground surface.

Vl. vulgaris and *Vl. rufa* (a species closely related to *Vl. schrenckii*) are also distributed in Europe, and both of them usually build their nests in underground situations. In England, 17–18% of the nests of each species have been found in covered situations above the ground: *Vl. rufa*, around plant roots, below piles of timber and in a bird nesting box; and *Vl. vulgaris*, mostly in attics but sometimes under floors and in tree hollows (Edwards 1980).

Vl. flaviceps and *Vl. shidai* build their nests in underground situations ranging from just below the ground surface to depths of 40–50 cm, but sometimes in covered situations above the ground such as attics, under floors, inside wooden walls and piles of straw and tree hollows (Matsuura 1975b; Sk. Yamane et al. 1980b). Some nests of *Vl. flaviceps* have been observed in open situations including under the eaves of a house and on the underside of the floor of a fire escape on the side of a building (Sk. Yamane et al. 1980b; Matsuura unpubl.).

Dolichovespula: Of the four species of this genus in Japan, *D. adulterina* is a social parasite and does not build its own nests. *D. media* builds its nest only in open situations including on tree twigs, under the eaves of houses and on the window frames of buildings (Makino 1982; Ban and Matsuura unpubl.), but does not nest in covered situations, and this also applies to the same species in Europe (Janet 1903; Kemper and Döhring 1967). A closely related species in North America, *D. maculata*, is also an open space-nesting type, with only a few exceptions having been observed (Akre et al. 1981). Many nests of *D. saxonica nipponica* have been observed in both open (on trees, under eaves, etc.) and covered (attics, inside houses, etc.) situations (Makino 1982). Only a few nest sites of *D. norvegicoides* have been observed in covered situations above and near the ground surface, including in a corrugated cardboard box, a large empty can left on the ground and on the inside of a discarded vehicle, as well as in a subterranean nest just below the ground surface (Matsuura and Sk. Yamane unpubl.). Like *D. saxonica*, the nest site preference of *D. arenaria*, which is common in North America, is flexible (Greene et al. 1976). *D. norwegica albida* builds its nest in sunken sites within layers of moss close to the ground surface in the Arctic region of Canada (the tundra), which is close to the northern limit of the distribution of this genus (Sk. Yamane et al. 1980a). *D. sylvestris* in Europe frequently nests in underground situations (Brian and Brian 1952).

2.2 Intra- and Interspecific Competition for Nest Sites

The heights or depths of nest sites above or under the ground for five *Vespa* species are shown in Fig. 3.3 as vertical distributions. With regard to *V. simillima* and *V.*

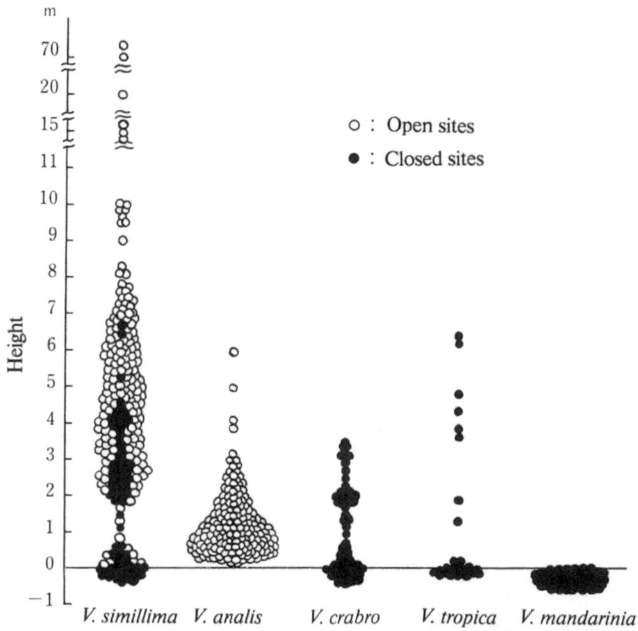

Fig. 3.3. Frequency distribution of distances of nests from ground surface in five *Vespa* species in southwestern Honshu (After Matsuura 1971a)

analis, which build their nests in open situations, the aerial nests of the former species are mainly built 2–7 m above ground, while those of the latter are concentrated within a height of 2 m, 61.9% of the nests being within 1 m from the ground (Matsuura 1971a). Consequently, competition between these two species for nest sites rarely occurs because, at least in southwestern regions of Honshu, the sites and heights of their nests differ from each other. In contrast, the three species (*V. simillima, V. crabro* and *V. tropica*) which build their nests in covered situations above and under the ground not only have similar nesting situations, but the heights and depths of their nest sites are also almost identical. Especially in subterranean situations, competition for nest sites can be expected among the four species, including *V. mandarinia*. Despite their similar preferences for nest sites, however, almost no observations have been made on competition for nest sites except for intraspecific competition. It seems probable that direct conflict among queens of these four species is prevented by the fact that the queens begin founding their nests at different periods (see Chap. 2) and by the presence of a remarkable interspecific dominance relationship.

As for *Vespula*, hardly any differences in nest sites can be found between *Vl. flaviceps* and *Vl. shidai*, which build their nests in the same situation. It is still unclear, however, that there is a competitive relationship between these two sympatric species.

Among *Dolichovespula* species, *D. media* is a typical open space-preferring species and *D. saxonica* in Hokkaido and *D. norvegicoides* in Honshu and southward also co-exist with *D. media.* Interspecific competition for nest sites, however, is considered to be rare because the latter two species build their nests not only in open situations but also in covered spaces.

There is a possibility that there is a high rate of competition between founding queens of the same species wishing to secure nest sites. In *Vespa* species, many observations have been reported of conspecific queens visiting a freshly founded nest one after another and fighting with the nest's queen. For example, eight queens of *V. simillima* visited a nest in one day and two dead *V. tropica* queens were found in a tunnel which led to the nest space (Matsuura unpubl.). A dead conspecific queen was found in a young nest of *Vespula flaviceps* (Shida 1959), and a similar observation has been reported for *Vespula* in North America (Akre et al. 1976). These observations can be considered to be examples of intraspecific competition for nest sites among queens during the nest-founding period.

3 Colony Foundation

In this section we deal with the behavior of foundress queens and colonial development in the preemergence period, i.e., from the initiation of the colony by an overwintered queen to the eclosion of the first workers. The structure and function of embryo (queen) nests will be discussed in the next section.

3.1 Haplometrotic Nest Foundation

The wasps of the subfamily Vespinae, with a few exceptions, found their nests haplometrotically (a nest is founded by a single queen). This method of nesting has both advantages and disadvantages. In haplometrosis, the single queen may maximize her reproduction, but when more than one queen shares a nest competition for the right to lay eggs may occur, which could impede colony growth or lead to a dominance system, where one queen is suppressed by the other, dominant one. Consequently, the mean productivity per foundress may decrease as compared with that in the haplometrotic system. However, this is the case only when predation on nests is not severe. For example, in single-queen nests the chance of her brood suffering from predation or parasitism will increase when the foundress is out of the nest. The loss of the foundress for any reason during the early stage of colony growth is another threat to reproductive production of the colony. In the tropics, where ant predation may be serious, pleometrotic nest foundation (nest is founded by more than one queen) or colony multiplication by swarming is a better way to decrease colony mortality. Indeed, tropical polistine wasps often adopt these strategies. In the case of pleometrotic nest foundation, co-foundresses are preferably sisters (or close relatives), which maximize their production of a similar genotype through kin selection, readily continued by the partner(s) if one of them dies.

Among the Vespinae, recent investigations by Matsuura (1983a) in Sumatra revealed that *Vespa affinis indosinensis* initiates nests by pleometrosis, and that colonies of two *Provespa* species are multiplied by swarming (see also Chap. 7). Almost all the other forms of this subfamily are thought to found nests haplometrotically. This may be the main factor that prevents Vespinae from extending their range into the lowland tropics (Jeanne 1975). Vespine foundresses generally behave intolerantly toward each other, showing a remarkable independence. Although this trait has sometimes been considered to indicate that the Vespinae are evolutionarily advanced (e.g., Yoshikawa 1963), it is instead a correlate of haplometrotic nest foundation. This intolerance becomes weak or "disappears" only among multiple sister queens that live in large overwintered *Vespula* nests (e.g., Thomas 1960; Spradbery 1973b; Ross and Matthews 1982), and among laying workers (which, in a sense, can be regarded as queens) in the queenless, orphaned nests.

3.2 Early Colony Development

The time of nest initiation by overwintered queens varies among genera, species and localities. In Hokkaido, *Vespa* starts nesting activity in late May, and *Dolichovespula* (probably also *Vespula*) in early and mid-May (Makino et al. 1981). In the Kanto district of Honshu, nest initiation by *Vl. flaviceps* queens is observed in late April (Shida 1959). In southwestern parts of Japan, *Vespa* initiates colonies in early May (*V. simillima, V. crabro, V. analis*), mid-May (*V. mandarinia*), or early June (*V. tropica*) (Matsuura 1971b).

Only a few case studies of Japanese species have followed colonial growth from the beginning. For example, one *V. analis* colony, studied in Sapporo, Hokkaido, was initiated about May 26, 1976, and foraging by the first workers began between July 2 and 7 (their eclosion may have been 2–3 days before the commencement of foraging). In this case, it took 35–40 days from colony initiation to the first worker eclosion. Another nest of *V. analis* observed in 1977, also in Sapporo, was initiated on June 7 and produced the first workers on July 15; an incubation period of 39 days was required.

In southwestern Japan, Matsuura made observations on the early growth of colonies of two species of *Vespa,* from their initiation. From nest initiation to first worker eclosion it took 38 days (May 23–June 30, 1983) for a *V. mandarinia* nest in Mie Prefecture and 35 days (June 8–July 13, 1974) for a *V. tropica* nest in Wakayama Prefecture. The number of cells constructed until the first worker eclosion was 64 and 42 respectively.

A *V. crabro* nest, observed by Janet (1895) in Europe, was started on May 15, 1894, and 56 days later produced the first workers (July 9). This may be the longest period so far recorded for *Vespa.*

There is no field data for *Vespula*. Matthews et al. (1982) kept embryo nests of four species of Nearctic *Vespula* in the laboratory. The mean duration of the preemergence period was 34.2 days (n = 7) for *Vl. maculifrons,* 27.4 days (n = 4) for *Vl. germanica*, 37 days (n = 1) for *Vl. vulgaris*, and 30 days (n = 1) for *Vl. vidua*.

According to Potter, the incubation period for the first workers was 23 days in a *Vl. vulgaris* colony in England (cited in Edwards 1980).

With the Japanese *Dolichovespula*, though no precise record is available, the length of the preemergence period is estimated at approximately one month (Makino et al. 1981). In the four *D. sylvestris* embryo nests observed in England, the eclosion of the first workers after nest initiation was on the 20th day in one nest and the 25th day in the remaining three (Brian and Brian 1952; Bunn 1982).

These data show that the length of the preemergence period varies among species and genera, and probably also varies according to the foundress ability and nest site condition.

Two types of envelope construction are recognized. In species constructing one large envelope sheet, such as *Vespa analis*, it takes 16–17 days to complete it. In species constructing several sheets, such as *Dolichovespula sylvestris*, only 3–4 days are required to complete the first sheet, and the subsequent ones are built at a similar speed (Brian and Brian 1952). *Dolichovespula* species, whose nests are built in open sites, probably must construct small sheets quickly, which serve as a shelter keeping the immature wasps warm (see next section). Some underground-nesting hornets build an envelope not wholly covering the comb.

The rate of increase in cell number, which may be an important indication of colonial growth, has been studied in some species. In southwestern Japan, in a nest of *V. mandarinia* and one of *V. tropica*, occupied only by the foundress, the number of cells constructed per day was 1.64 and 1.24 respectively (Fig. 3.4; Matsuura unpubl.). It was 0.70 in one such nest of *V. crabro* in France (Janet 1895),

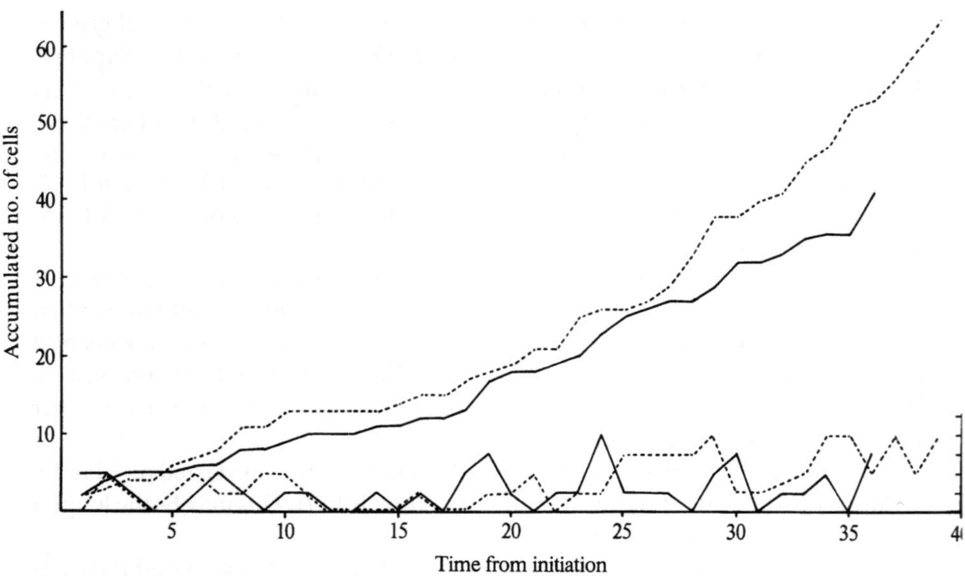

Fig. 3.4. Cell construction and oviposition rates during preemergence period in two Japanese *Vespa* species. *Dotted line*: *V. mandarinia* (Mie Pref., initiated on May 23, 1983); *solid line*: *V. tropica* (Wakayama Pref., initiated on June 8, 1974) (After Matsuura unpubl.)

1.43 and 1.75 in two such nests of *D. sylvestris* in England (Brian and Brian 1952); and 1.17–2.76 in four Nearctic *Vespula* species (Matthews et al. 1982). Thus, the cell construction rate by foundresses varies to some extent among species and nests.

The oviposition rate by foundresses was studied in two hornet species in Japan (Matsuura unpubl.). In a subterranean nest of *V. mandarinia,* the number of eggs laid per day was between 0 and 5 (mean: 1.53) during the 38 days until the first workers eclosed. There was a tendency for the number to increase during the second half of the period, when cell construction rate also became higher. It fluctuated between 1 and 4 (mean: 1.06) in a *V. tropica* nest during a period of 35 days (Fig. 3.4).

There is little reliable information on the total number of cells constructed during this period, mainly due to the difficulty of discovering embryo nests and observing them from outside the envelope. Examples are listed in Table 3.2.

Table 3.2. Number of cells constructed by the foundress queen[a]

Species (locality)	No. cells	Reference
Vespa s. simillima (Hokkaido)	36 (n = 1)	Sk. Yamane and Makino (unpubl.)
V. s. xanthoptera (SW Honshu)	43, 48 (n = 2)	Matsuura (unpubl.)
V. crabro flavofasciata (ditto)	29, 31 (n = 2)	Matsuura (1971b, unpubl.)
V. c. germana (France)	39, 60 (n = 2)	Janet (1895, 1903)
V. analis insularis (SW Honshu)	13–37 (m = 27.2, n = 5)	Matsuura (1971b, unpubl.)
V. mandarinia japonica (ditto)	37–60 (m = 44.3, n = 4)	Matsuura (1971b, unpubl.)
V. tropica pulchra (ditto)	17–43 (m = 34.3, n = 5)	Matsuura (1971b, unpubl.)
Vespula flaviceps lewisii (ditto)	39–60 (m = 45.5, n = 6)	Matsuura (unpubl.)
Vl. vulgaris (England)	40–56 (m = 46, n = 4)	Archer (1981a)
Vl. vulgaris (USA) in labo	36 (n = 1)	Matthews et al. (1982)
Vl. maculifrons (USA)	25–51 (m = 40.1)	Matthews et al. (1982)
Vl. maculifrons (USA) in labo	49–83 (m = 62.2)	Matthews et al. (1982)
Vl. germanica (New Zealand)	23, 30 (n = 2)	Sk. Yamane (unpubl.)
Vl. germanica (USA) in labo	65–79 (m = 71.3)	Matthews et al. (1982)
Dolichovespula arenaria (USA)	39–44 (m = 41.3, n = 3)	Greene et al. (1976)
D. norwegica norwegica (England)	35 (n = 1)	Ritchie (1915)
D. norvegicoides pacifica (Honshu)	36 (n = 1)	Matsuura (unpubl.)
D. saxonica nipponica (Hokkaido)	35, 49 (n = 2)	Makino (1980); Sk. Yamane (unpubl.)
D. media media (ditto)	37, 44 (n = 2)	Makino (1980); Sk. Yamane (unpubl.)
D. sylvestris (England)	42 (n = 1)	Brian and Brian (1952)
D. sylvestris (ditto)	31–48 (m = 40)	Archer (1981b)

[a] Among samples from Japan, nests with at least one cocooned cell are included in data, while nests with at least one emerged worker are excluded. This treatment may cause slight underestimations.

Authors might have underestimated the value if they counted cell number just at the time of first worker eclosion, because the foundress engages in foraging for some time afterwards. This table shows a considerable variation among species. It is evident that with colonies kept in the laboratory the nest size rapidly becomes bigger (Matthews et al. 1982).

3.3 Behavior of Foundress Queen

The behavioral repertoire of the Vespinae foundress does not differ fundamentally from that of the temperate *Polistes.* However, some traits appear to be peculiar to this subfamily. In this section we discuss some important components of foundress behavior.

Nest Construction. This comprises pedicel, cell and envelope construction. The process of envelope construction is difficult to observe in the subterranean species of *Vespula* and *Vespa*. In Japanese *Dolichovespula* and *Vespa analis*, the foundress coming back with a load of pulp directly enters the nest, malaxates it on the comb for a while, then appears on the envelope to begin construction. She puts the pellet onto a suitable part to be enlarged, then slowly walks backward and forward, while applying and thinning the material with her mandibles (Fig. 4.1A). After repeating these acts, *V. analis* foundress leaves the position, but before the applied pulp dries she usually returns to give it careful additional treatment (Sk. Yamane and Makino 1977). This additional thinning is not seen in *Dolichovespula* (Makino 1980).

A clear distinction in cell construction exists between *Dolichovespula* and *V. analis*. In the former, the comb cells are at first made with pulp collected from outside the nest, but once the first envelope sheet is near completion, subsequent cells are constructed only with the material scraped from the envelope (Makino 1980). On the other hand, in *V. analis*, the material for cell construction is carried in from outside the nest throughout the preemergence period. The foundresses of species constructing multi-layered envelopes (e.g., *Vl. flaviceps lewisii* and *V. simillima*) also reutilize the envelope for cell construction. This behavior is also observed in *V. mandarinia* and *V. tropica,* whose embryo nests have only one incomplete sheet (Matsuura unpubl.). In these two species the removal of the envelope seems not to adversely affect colony growth, because their nests are built in places sheltered from open air and protected from predators. Reutilization of envelope paper has not been positively demonstrated in the European *V. crabro* (Sandermann 1940). Matthews et al. (1982) reported that with *Vl. maculifrons* nests kept in the laboratory the removal of envelope was observed, but that this material was never reused.

Curling. This is a remarkable foundress behavior known only in the Vespinae. Since Janet (1895) described it in *Vespa crabro*, it has been confirmed to exist in nine species in three genera, but not observed in *Provespa*. The foundress queen, when resting within the nest, often curls herself around the pedicel between the comb and the innermost sheet (or disk) of the envelope (Fig. 3.5A). Time spent

Fig. 3.5. Embryo (queen) nests of Japanese *Vespa*. **A** *V. mandarinia*; **B, C** *V. analis* in Hokkaido; **D** *V. crabro*; **E** *V. tropica*; **F** *V. simillima xanthoptera*. Note that the foundress of *Vespa mandarinia* (**A**) is engaged in curling (incubating her eggs)

in this attitude occupies 40–80% of all intranidal activity. Nixon (1934), Brian and Brian (1952) and Ishay (1973) believed that this behavior has a function of warming (incubating) eggs and larvae in cells. In order to test this hypothesis, Makino and Sk. Yamane (1980) measured the temperature of cell walls, using a *Vespa simillima* embryo nest in Sapporo. The probe of an electric thermometer was inserted into the cell so as to touch the wall near the bottom of the cell, on which an egg had been deposited. While the foundress was engaged in curling, the temperature of the cell wall rose by 2.5–4°C, and during curling the temperature was kept constant (Fig. 3.6). As seen in the figure, however, it began to rise shortly

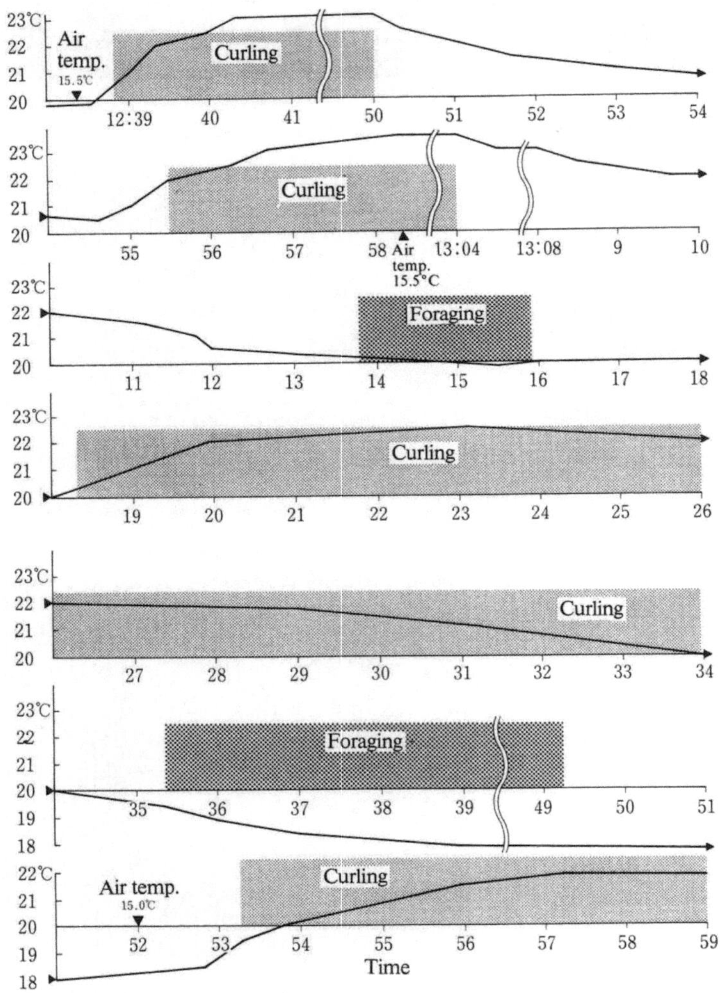

Fig. 3.6. A series of chronological changes in temperature on cell wall near ceiling in a *Vespa simillima* nest (Hokkaido, May 29, 1978). Corresponding activities (curing and foraging) are indicated by *shading* (After Makino and Sk. Yamane 1980)

before the commencement of curling; this might have been due to the foundress' activity in turning around the pedicel, which almost always occurred prior to the attainment of the curled, inactive posture.

Furthermore, Makino (1980) found that, in *Dolichovespula saxonica nipponica,* this characteristic behavior is seen only during the early half of the preemergence period, then disappears when some larvae have matured or pupated (Fig. 3.7). Similar observations were made for *V. analis, V. tropica* and *V. mandarinia* (Matsuura unpubl.). When the colony enters this latter phase, heat production in the nest may depend on energy metabolism by larvae and pupae, as suggested by Gibo et al. (1977).

Gastral Vibration. In vespine colonies, the worker is often observed to tap the gaster up and down on the comb. This behavior (gastral vibration) is usually associated with feeding the larva, and is considered to be a signal to hungry larvae. A behavior probably homologous to this has been reported also for the foundress

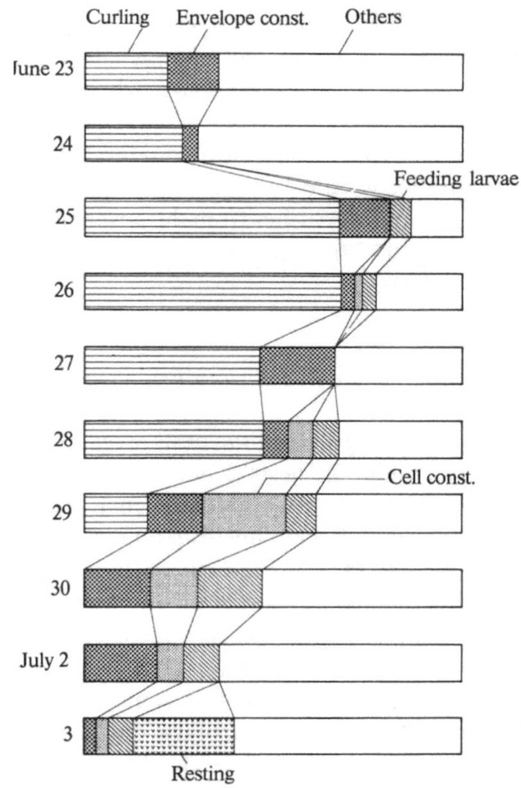

Fig. 3.7. Chronological changes in intracolonial activities in foundress of a *Dolichovespula media* nest (in Hokkaido, 1979). The curling disappeared after the appearance of six mature larvae on June 30 (After Makino 1980)

queen of *Dolichovespula* (*arenaria*: Greene et al. 1976; *saxonica nipponica*: Makino 1980). A recent study by Akre et al. (1982) also demonstrated the occurrence of gastral vibration in *Vespula consobrina*.

Tapping. The foundress of *Vespa tropica pulchra* taps mid- and hind legs on the dorsal face of the nest comb (Matsuura 1983b). This act, which lasts from 10 s to ca. 2 min, can be heard at a distance of about 1 m from the nest. It is observed when the foundress returns to the nest with a load of prey or rests on the comb. Among the Vespinae this peculiar behavior has so far been known only in *V. tropica*, and is thought to be a signal to the larvae.

Defense. The vespine foundress is generally not aggressive toward large objects. However, she assumes a posture of threat with her fore legs raised, or sometimes even rushes upon animals (such as insects and spiders) smaller than herself that are approaching the nest in flight or along a twig to which the nest is attached. She often captures the approaching animals and flies out of the nest to dump them. When she senses the approach of foraging ants, she inspects the envelope surface especially at the attachment point of the nest to the twig. There is a remarkable difference in attitude between *Vespa* and *Dolichovespula* when the nest is disturbed by man. In *V. analis insularis*, when a physical stimulus is given to the nest, the foundress rushes out from the nest, rustles, runs about on the envelope, but rarely flies out (Sk. Yamane and Makino 1977). This attitude is common to *Vespa* species studied to date. In *D. saxonica nipponica* and *D. media media*, on the other hand, the foundress rushes from the nest and flies around it, or may suddenly fall to the ground. The latter behavior has been found only in *Dolichovespula* (Makino 1980).

3.4 Conspecific Queen's Visit to Established Nests

Excavation of *Vespula* embryo nests reveals the presence of corpses of conspecific queens just under the nests or in the tunnels leading to them. These corpses may have been nest owners (resident queens), or individuals which had attempted to usurp these nests. Recent investigations show that the intraspecific usurpation is not rare, though there are few cases where it is known what circumstances drove spring queens to make usurpation attempts, which often lead to their deaths. It is possible that usurpation is a consequence of competition over nesting sites, rather than for the nests themselves, particularly among subterranean nesters. But the fact that such aerial nesters as some *Vespa* and *Dolichovespula* (for which potential nesting sites are virtually unlimited) do practice usurpation strongly suggests that the nest itself is an object of struggle as an important resource (cf. Sk. Yamane 1978). In view of yearly fluctuations, Archer (1980a) argued that interference between spring queens adversely affects colony survival. Here we give examples of nest-visiting by conspecific foreign queens observed in Japan. The adaptive significance of intraspecific usurpation will be discussed in Chapter 8.

Vespa analis insularis. On June 8, 1977, a very initial nest with only two cells was located on a young tree of *Abies sakhalinensis* at 50 cm above the ground, in a todo-fir plantation in Sapporo, Hokkaido. Owing to frequent visits of conspecific queens, as well as the delay in nest initiation (normally in late May), this nest only became stabilized with the first workers enclosed after July 20. The first visit of a foreign queen was seen on June 25, 2 days after the completion of the entrance tube. In total, up to July 22, three queens and three workers were found dead on the ground below the nest. During this period, the foundress was replaced by a foreign queen at least once. In the early interactions the foreign queens only alighted on the nest envelope, and as soon as they became aware of the presence of the resident they flew away. Later, they began to attempt intruding into the nest and wrestled with the resident within the nest, often resulting in both falling to the ground. Even after the first workers eclosed, invading foreigners continued to kill teneral workers. It is interesting to note that on June 27, when another nest of *V. analis* situated 50 m from the nest observed was destroyed, the queen abandoned it 2 days later. Although she had not been marked, it is very probable that this is one of the queens that had engaged in the usurpation attempts mentioned above.

The visit of foreign queens during the preemergence and cooperative period has been reported also for *V. tropica pulchra* (Matsuura unpubl.), *V. crabro flavofasciata* (Matsuura 1970b; Fig. 3.8), and *V. simillima* (Makino et al. 1981; Matsuura unpubl.

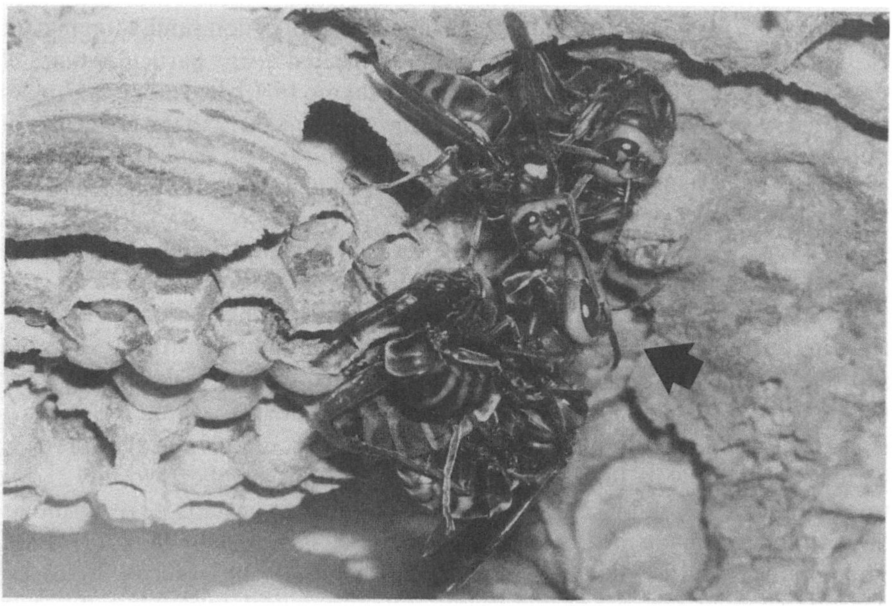

Fig. 3.8. Invasion of a *Vespa crabro* nest in the cooperative period by a conspecific foreign queen (*arrow*). The owner queen and workers are seen to counteract vigorously

Two Species of Dolichovespula. In the Japanese *D. media* and *D. saxonica*, the visits of conspecific queens are frequently observed during the preemergence and cooperative period (Makino 1980). They behaved nervously around the nest that they were visiting, as seen in *V. analis*. When they became aware of the nest owner, they often fell suddenly to the ground and flew off. This behavior is very similar to that practiced by the normally nesting foundress of *Dolichovespula* when her nest is disturbed. In young nests with several workers, the queen and workers cooperate to flight against intruders, which are occasionally killed.

3.5 *Daily Activity of Foundress*

In order to know how the foundress survives the fragile preemergence period, it is very important to study the details of the foundress' daily activity. Investigators have not yet paid much attention to this interesting aspect. Here we summarize the data, mainly of extranidal activity, obtained in Hokkaido, Japan, with *Vespa analis* and *Dolichovespula* embryo nests.

Extranidal activity includes pulp collection, hunting prey (or scavenging), feeding on sugars (flower nectar, tree sap, honeydew, etc.) and going out for other reasons not understood by observers. Attempts to usurp more developed nests by queens who have already constructed their own nests, if a common occurrence, should be included in extranidal activity, though at present positive evidence for it is not available.

The time spent in a single foraging trip does not greatly differ among the three categories of foraging, i.e., pulp collection, prey collection, and foraging for carbohydrates. Certain activities are not concentrated in any particular times of day (Fig. 3.9; Sk. Yamane and Makino 1977; Makino 1980). Because foundresses are often observed to feed on tree sap or flower nectar prior to the collection of pulp or prey, it is probable that sugar intake continues all day. Time spent in extranidal activity fluctuates from day to day, but generally amounts to around 50% or less of the total active time.

A *Vespa analis insularis* foundress, after having completed the envelope on June 12 (corresponding to the appearance of the first larvae), showed no conspicuous change in the number of foraging trips, but the total time spent in foraging decreased (Fig. 3.10). This means that the mean duration of trips became shorter. Figure 3.11 shows the number of trips plotted against the duration of each trip on May 30 (A), and from June 21 to July 2 combined (B). From this figure it is clear that on May 30, though a peak was reached between 300 and 500 s the trips varied considerably in duration. On the other hand, from June 21 to July 2 trips were confined to within 700 s with only one exception (1335 s). A similar observation was made for *Dolichovespula saxonica nipponica* by Makino (1980). Though this foundress activity has been confirmed only in the above two cases, it suggests that an embryo nest in its initial stage has a relatively low value, but gradually becomes more valuable. For predators such as ants, the net gain from invading an initial nest with eggs alone may be small. For conspecific queens intending to usurp the nest, it may not be worth fighting against the resident, which often leads to injury or death, because the nest is not expected to soon

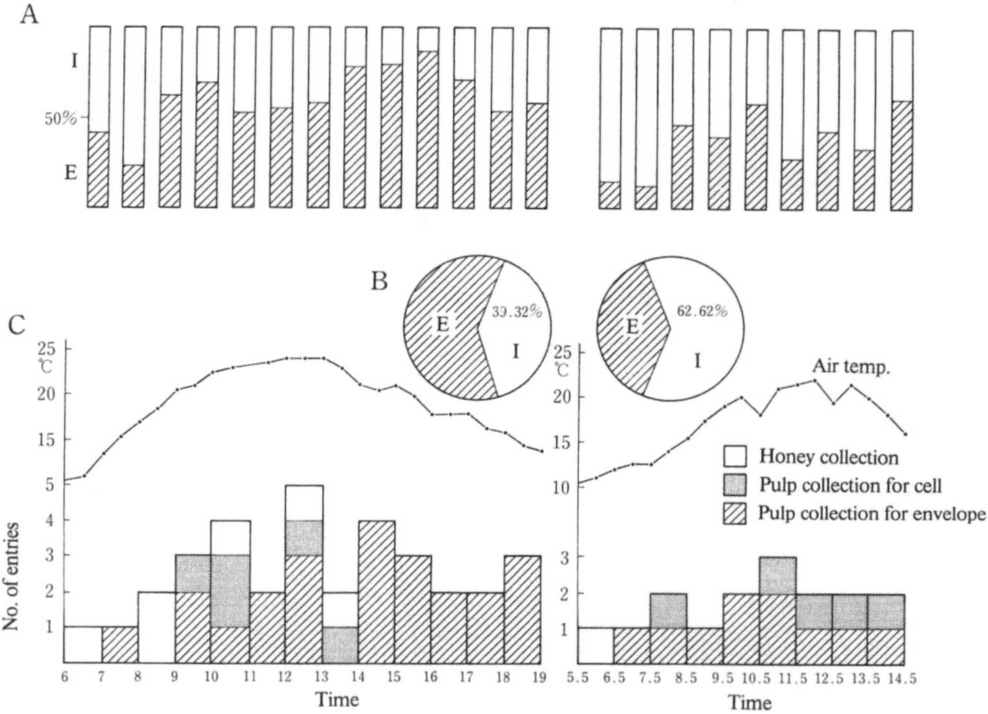

Fig. 3.9. Daily activity of foundress of a *Vespa analis* nest in preemergence period. **A** Intra- (*I, blank*) and extranidal (*E, shaded*) activities in each hour in percent. **B** Intra- and extranidal activities for the whole observation time. **C** Number of entries per hour. May 30 (*left*), and June 4 (*right*), 1976 (Hokkaido) (After Sk. Yamane and Makino 1977)

produce the first workers, which she will use as her labor force. However, as the larvae appear, the nest becomes increasingly more valuable to both the resident and predators (or usurpers). The resident queen will thus benefit by reducing not only the total amount of extranidal activity (for example by refraining from lounging in the field) but also the duration of each trip.

The threshold temperature necessary for a foundress queen to engage in foraging may vary among species and localities. For Japanese species, only a few data are available in this respect. In Hokkaido the foundress of *V. analis* does not forage at temperatures below 10°C (Sk. Yamane and Makino 1977). In *D. saxonica* and *D. media* in the same locality, the threshold was 7°C, and the foundress was quite active at 10°C (Makino 1980). L.R. Taylor (1963) reported that there was a flight threshold of about 15°C in *Vl. germanica* queens. However, since his result was not obtained by all-day observations but by capturing spring queens with suction traps, it remains doubtful if the result precisely reflects foundress flight activity. In the worker of *Vl. germanica*, on the other hand, the threshold was reported to be 2°C (Potter 1964). Perhaps workers, which have the advantage of living in a warm nest, have body temperatures above the minimum limit for flight when outside temperatures are relatively low (Spradbery 1973a).

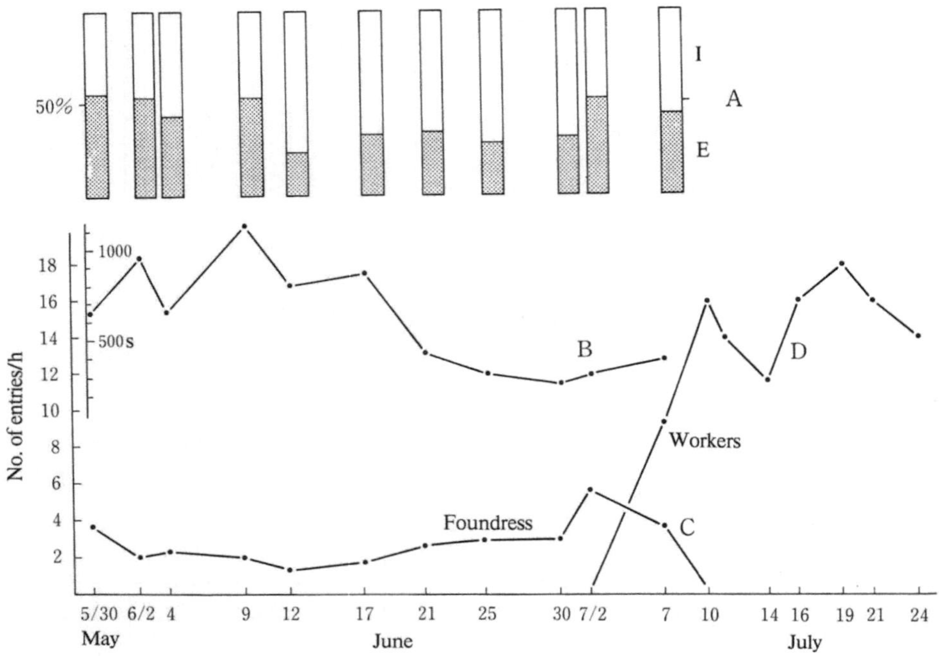

Fig. 3.10. Extranidal activities of *Vespa analis* foundress during first 40 days from nest initiation (the same nest as in Fig. 3.9). Observations were largely made within the period from 10:00 to 14:00. **A** Change in intra- (*I*) and extranidal (*E*) activities in percentage of time spent in each. **B** Change in mean duration of foraging trips. Change in number of entries into nest is shown by **C** (foundress) and by **D** (workers) (After Sk. Yamane and Makino 1977)

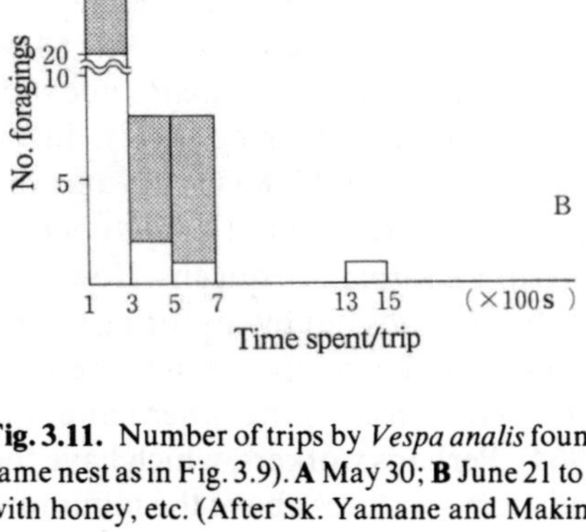

Fig. 3.11. Number of trips by *Vespa analis* foundress against time (× 100 s) spent in a single trip (the same nest as in Fig. 3.9). **A** May 30; **B** June 21 to July 2 combined. *Dotted*: Returned with pulp; *blank*: with honey, etc. (After Sk. Yamane and Makino 1977)

3.6 Abnormal Nestings

As mentioned above, the preemergence (or solitary) period is the most vulnerable stage in the life cycle of a colony. The rate and factors of colony mortality at this stage will be discussed in detail in Section 3 of Chapter 6. Here we deal with two situations that obviously deviate from the normal nesting pattern.

Simultaneous Maintenance of Two Nests. A very interesting case where one foundress simultaneously cared for two nests built close to each other was reported for *Dolichovespula saxonica* in Hokkaido, Japan (Makino and Sk. Yamane 1979). This type of behavior was also observed in Japanese *Polistes* (Kasuya 1980). In both cases, the foundress' action seems not to have been adaptive but rather caused by her misconception, owing to similar topographic features around the nests.

In the case of *D. saxonica*, the two nests were built under eaves which were latticed, as seen from below and were situated 40 cm apart in similar positions on adjacent squares. Observations were made from May 29 to June 10, 1976. The foundress cared for both the nests nearly impartially (Fig. 3.12), and by the time she disappeared, for unknown reasons, 2 envelope sheets and 19 cells, and 1.5 envelope sheets and 12 cells had been constructed respectively. The nest construction rate and the time spent in extranidal activity for the two nests combined did not surpass that of normal nesting.

Construction of Atypical Nests in the Laboratory. It is difficult, but not impossible, to induce overwintered queens to initiate nests or adopt nests already established by other queens, under controlled conditions (Ross et al. 1981). Nests initiated by such queens sometimes lacked the pedicel (1.5% in *Vl. maculifrons*, 9.1% in *Vl. vulgaris*; Ross 1982). A queen of *Vl. maculifrons* constructed two cells along a

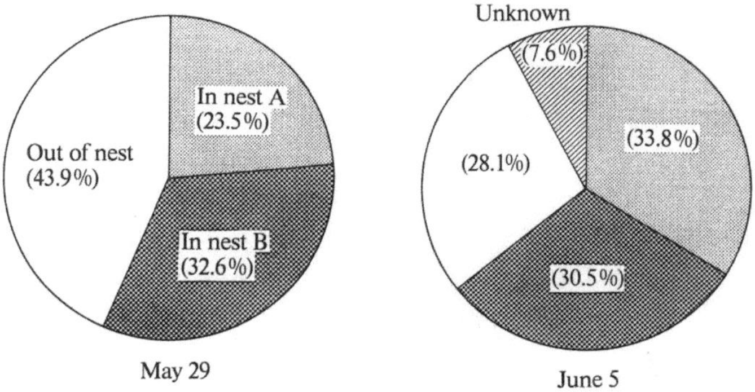

Fig. 3.12. Simultaneous maintenance of two nests by a single foundress of *Dolichovespula saxonica* (in Hokkaido, 1976). Percentage ratio of time spent by foundress in each extra- (*blank*) and intranidal (**A, B**) activities to total observation time (764 min for May 29, and 600 min for June 5). (After Makino and Sk. Yamane 1979)

depression in a wood chip in the nest box. A *Vl. consobrina* queen and two *Vl. vulgaris* queens built a typical hanging sheet, but rather than suspending a pedicel from this, they built cells directly on the side of the sheet. All these nests were abandoned within one or two months. It is not known whether these atypical nests are mere results of stress imposed on the queens under artificial conditions or occasional manifestations of behavioral potency, which was a normal trait in the ancestors of Vespinae. If the latter is the case, atypical nests may occur in natural conditions and contribute to the mortality of colonies.

4 Architecture and Function of Embryo Nests

The morphology and function of embryo or queen nests of Vespinae have not been thoroughly studied, since locating some species is quite difficult. In this section we try to compare these aspects among species and genera, chiefly with Japanese species.

4.1 Embryo Nest Architecture

As mentioned in the previous section, two principal methods of founding nests occur among the Vespinae: founding by one or more queens, and swarming. Embryo nest, as used here, means the nest founded by a queen(s) (therefore *Provespa* is excluded) and not yet producing worker adults. The vespine embryo nests are usually made of plant fibers mixed with salivary secretion. The typical embryo nest consists of the following parts (Fig. 3.13): (1) Pedicel (petiole or spindle), which is club-shaped or is a narrow strip of paper with a spiral twist,

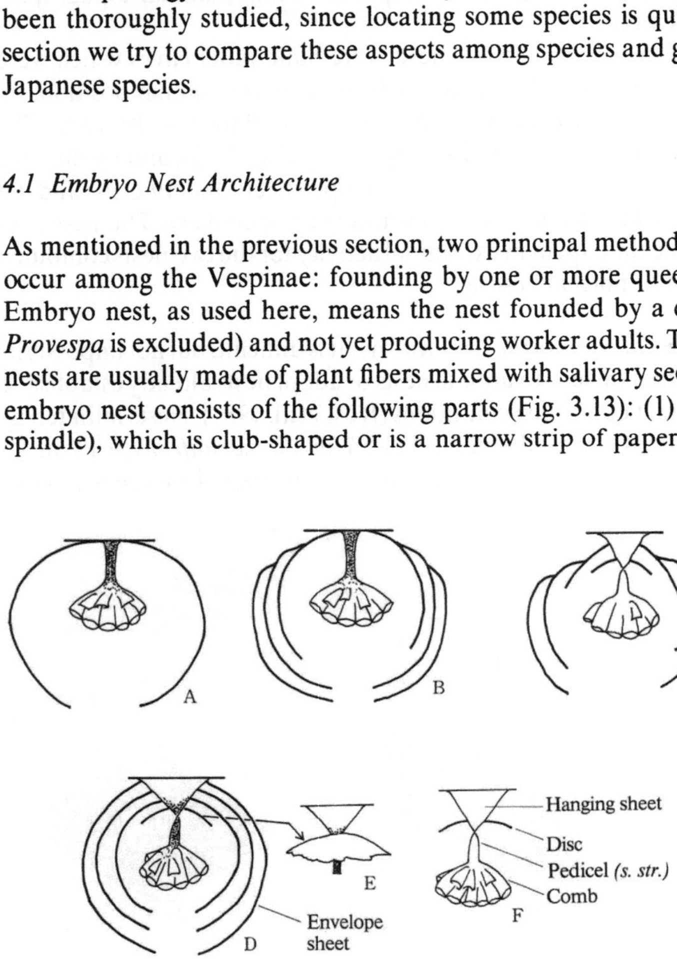

Fig. 3.13. Schematic drawings of embryo nests of Japanese Vespinae. **A** *Vespa crabro*; **B** *V. simillima*; **C** *Vespula flaviceps*; **D** *Dolichovespula media*; **E** first envelope sheet scraped off by the foundress (*D. media*); **F** details of pedicel and comb

supporting the comb. There is always only one pedicel, which is attached to the center of the comb and often coated with an oral secretion. In *Vespula* and *Dolichovespula* nests, the upper triangle part of the pedicel (*s. lat.*) is often called "hanging (or suspension) sheet" and the lower one "pedicel" (*s. str.*); (2) Comb. A queen constructs only one comb, consisting of 30–40 cells to rear workers. No marked difference exists in the architecture of the comb among species and genera; (3) Envelope. One or more paper sheets cover the comb and pedicel. In some species the envelope is more or less reduced. The texture of the envelope sheet is usually much looser than in the other parts of the nest. In *Dolichovespula* and *Vespula rufa* group, the sheets have many small holes in them, resembling lace (Makino 1980; Robin Edwards pers. comm.; Matsuura unpubl.). The pattern of sheet multiplication and the number of sheets in the completed embryo nest vary among species and genera.

Vespa. All the species construct a brownish, fragile nest, with a club-shaped pedicel that lacks a hanging sheet. A queen of *V. analis insularis* applied pulp to the twig from which the pedicel was started, and carefully coated this part after the completion of the pedicel. A similar behavior was also observed in *V. simillima simillima* (Fig. 3.14A). It is not clear whether this lump of pulp represents the ancestral condition of the hanging sheet or not. Obviously the most carefully masticated pulp hardened with the aid of oral secretion is used to construct the

Fig. 3.14. Initial nests and their queens of *Vespa simillima* (**A**), and *V. analis* (**B**) in Hokkaido

pedicel. There is no spiral twist, as is seen in the other two genera. The foundress repeatedly applies oral secretions to the surface of the pedicel, which gradually becomes shining black. The length and thickness of the pedicel in five Japanese *Vespa* are shown in Table 3.3. When the first three cells or so are built, the construction of an envelope starts. The first sheet is always begun from the substrate, never directly from the pedicel (Sk. Yamane and Makino 1977). Three types of envelope are found (Matsuura 1971b; Figs. 3.5A-F): (1) A bowl-shaped type. The envelope comprises a single, inverted bowl-shaped sheet. The greater part of the comb is exposed, as seen in *V. tropica pulchra* and *V. mandarinia japonica*; (2) A ball-shaped type. The envelope, as seen from the outside, is nearly spherical, and completely covers the comb. The number of sheets constructed varies among species, being always one in *V. crabro flavofasciata* and three to four in *V. simillima,* in which the simultaneous construction of more than one sheet is common (Makino and Sk. Yamane 1980). The embryo nest of the Philippine *V. luctuosa luzonensis* possibly falls within this type, but the number of sheets constructed is unknown (Kojima and Sk. Yamane 1980); (3) A flask-shaped type. The main body of the envelope is spherical and single-layered, with a long entrance tube (vestibule) produced below. The length of the vestibule varies among nests. Although data available are not sufficient for conclusions, in *V. analis* generally it is short in Hokkaido and longer in the southwestern parts of Japan (Sk. Yamane and Makino 1977). The embryo nests of *V. affinis affinis* in the Yayeyama-shotô (southern Ryukyus) and Taiwan also belong to this type (Matsuura unpubl.; S. Yamane pers. comm.).

Some measurements of envelope sheets and pedicels in some Japanese *Vespa* are given in Table 3.3.

Table 3.3. Sizes of envelope and pedicel in embryo nests of some *Vespa* species (in mm)

	Envelope		Pedicel (No. examined)	
Species	Diam. of 1st sheet[c]	Thickness	Length	Thickness
Vespa simillima simillima[a]	43–45 E_2: 46–54, E_3: 58–65, E_4:63–74		10.5	1.8 (3)
V. s. xanthoptera	44–46	0.29–0.44	7.6	2.1 (5)
V. crabro flavofasciata	68–85	0.43–0.62	9.0	2.0 (5)
V. analis insularis	61–74 (m = 66)	0.39–0.61	7.1	2.5 (21)
V. tropica pulchra	50–55	0.63–0.70	17.8	2.3 (5)
V. mandarinia japonica	70–75	0.71–0.93	15.5	3.4 (2)
V. luctuosa luzonensis[b]	50–57	0.42	13.0	2.6 (2)

[a] Hokkaido (Makino and Sk. Yamane 1980).
[b] Philippines (Kojima and Sk. Yamane 1980). Others are from SW Japan (Matsuura 1971b).
[c] E_2–E_4: 2nd to 4th envelope sheets.

Vespula. The nest material falls into two types, which, however, do not correspond to a phylogenetic grouping. The species of the *Vl. rufa* group (subgenus *Vespula*), except for *Vl. austriaca* and the *Vl. squamosa* group studied to date, construct an embryo nest made of gray, relatively pliable material. Among the species of the subgenus *Paravespula, Vl. germanica* and *Vl. pensylvanica* construct a similar nest, though in the *Vl. vulgaris* and the *flaviceps* group the embryo nest is brownish and fragile, as in *Vespa*. No information is available as to the embryo nest architecture of the subgenus *Rugovespula.*

The shape of pedicels agrees among species, though only some of the known species have been studied for this character (Sk. Yamane et al. 1981). The hanging sheet is a reversed triangle, and its lowermost part usually changes to a spiral twist and then to the pedicel, without having a short "neck" just above the twist and not constituting the hanging sheet itself, as observed by Brian and Brian (1948) in *Dolichovespula sylvestris.* In *Vl. germanica*, the hanging sheet slightly varies in shape among nests, and the twist may be either clockwise or counterclockwise and varies somewhat in the degree of twisting. Janet (1903) and Potter (1964) illustrated embryo nests of *Vl. vulgaris* that had no twist, though Edwards (personal communication 1981), who examined many embryo nests of this species, could confirm the twist. In this genus, carefully coating the pedicel with oral secretion does not seem to be a general trait in at least the initial part of the preemergence period (Greene et al. 1976; Sk. Yamane et al. 1981; but see Reed and Akre 1983 for *Vl. acadica*); during the second half of the period, some embryo nests have a slightly darkened pedicel, indicating a slight application of secretion.

The embryo nests of *Vespula* always have a ball-shaped, multi-layered envelope. The first sheet is started from the hanging sheet at the mid-length from its base, or at positions slightly above it. The subsequent sheets each start from the preceding one at around its basal part so that all of these become dependent on the first sheet (Sk. Yamane et al. 1981). More than one sheet may sometimes be constructed simultaneously (Duncan 1939). Sheets made earlier are in turn scraped off, and they may be reprocessed and converted into cells. One feature peculiar to this genus is the construction of a small disc just above the twist and below the first sheet (Figs. 3.13F, 3.15; Matthews et al. 1982, named this "paper button"). The lower margin of the disc usually has a smooth periphery, implying that the disc has not been shaven after once having covered the comb. Shida (1959) was the first to notice the occurrence of the disc. He concluded, based upon his close examinations of *Vl. flaviceps lewisii* embryo nests, that the disc was constructed after the completion of the first sheet. Sk. Yamane et al. (1981) supported him, because in the New Zealand *Vl. germanica,* most of the nests with more than one sheet had the disc (Fig. 3.16). In this species, the distance between the disc and the comb was always 5–6 mm. Matthews et al. (1982), on the other hand, mentioned that the disc is first constructed. Until now this structure has been confirmed only in five species, though they represent all the groups (subgenera and species groups) other than *Rugovespula* and the *Vl. squamosa* group, for which this character has not yet been studied. This suggests that the disc occurs nearly throughout the genus.

Fig. 3.15. Incomplete embryo nest of *Vespula schrenckii.* Note that the disk is exceptionally constructed before the completion of the first envelope sheet

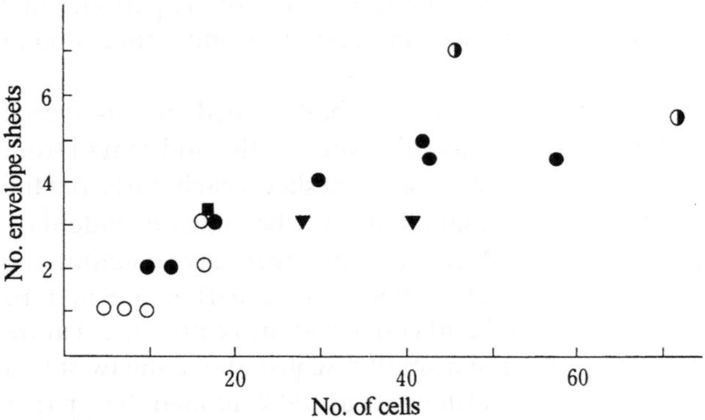

Fig. 3.16. Relation between number of envelope sheets and cells, and disk formation in *Vespula* embryo and young nests. ○ *Vespula germanica* nests without a disk; ● those with a disk with smooth margin; ◑ those with a disk with a rugged margin; ▼ *Vespula* sp. nests with a disk with a smooth margin; ■ *Vl. schrenckii* nest with a disk with a smooth margin. *Vl. germanica* nests were obtained in New Zealand, the others in Hokkaido, Japan

Dolichovespula. The embryo nest of this genus is always gray and more or less pliable. In this respect, it is often difficult to distinguish it from those of the *Vespula rufa,* and the *Vl. squamosa* and the *Vl. germanica* groups. The *Dolichovespula* embryo nest is generally less fragile, lacks the true paper disc, and has differently arranged envelope sheets.

The pedicel (s. lat.) is similar in shape to that of *Vespula*, always with a spiral twist. Brian and Brian (1948) observed in *D. sylvestris* a short "neck" that could not be regarded as a part of the hanging sheet, between the triangle and twist, and considered it to be a part of the true pedicel. This short "neck" is occasionally seen also in Japanese species, though in many cases the condition is not clearly distinguished from that in *Vespula* embryo nests. One important feature peculiar to *Dolichovespula* is the coating of the lower part of the twisting area and of the true pedicel with a blackish brown oral secretion of a rubbery consistency (Greene et al. 1976; Jeanne 1977a; Makino 1980). As mentioned above, the embryo nests of *Vespa* are also coated with secretions, but these are not viscous.

The envelope is ball-shaped or flask-shaped, and always composed of several paper sheets (Fig. 3.17). The first sheet is started at the lower portion of the

Fig. 3.17. Embryo nests of Japanese *Dolichovespula*. **A** *D. saxonica* (not completed); **B** same species (completed nest, seen from below); **C** *D. media media*; **D** *D. norvegicoides*

hanging sheet near the spiral twist. The subsequent sheets are started from the hanging sheet, each above the preceding one; the last one or two sheets are sometimes attached directly to the substrate such as tree twigs (Sk. Yamane and Makino 1977). The envelope in this genus clearly differs from those in *Vespa* and *Vespula*, because the sheets are completely independent of each other. Brian and Brian (1948) figured two *D. sylvestris* embryo nests in which the third sheet starts from the second sheet, and the subsequent sheets each start from the preceding one. If their figures are correct, *D. sylvestris* forms the only exception to this rule in the genus. In the figures presented by Janet (1895, 1903), however, mature nests of this species have the first several sheets independent of each other. In the embryo nest of this genus, the first sheet is largely scraped off to leave the basal portion after the completion of the second and third sheet so that the remaining part assumes a "disc". This "disc", however, differs in appearance from the true disc observed in *Vespula*, due to its rugged periphery. The simultaneous construction of two sheets is often seen in two Japanese forms (*D. media media* and *D. saxonica nipponica*; Makino 1980).

The European *D. media media* and Nearctic *D. maculata* (both in the subgenus *Dolichovespula*) construct nests with the outermost sheet produced at the lower margin to form a long vestibule (Rau 1929; Bohart and Bechtel 1957; Kemper and Döhring 1967). In *D. maculata*, this vestibule sometimes attains a length of nearly twice the diameter of the main body of the envelope (Akre et al. 1981; Fig. 40a,b). In Japan, *D. media* rarely constructs a short vestibule (Makino 1980), and in *D. saxonica* the entrance tends to be produced only slightly (Sk. Yamane and Makino 1977).

4.2 Function of Embryo Nests

The embryo nests of the vespine wasps are, as a matter of course, built in order to rear the first workers. But we do not precisely understand what are the functions of the details of architecture, and whether differences observed among species and groups have adaptive significance or not. We are discussing here the supposed functions of some important character conditions.

Spiral Twist of Pedicel and Application of Oral Secretion. Shida (1959) examined the function of the spiral twist, and thought that *Vespula* embryo nests could respond with flexibility to physical forces by having the pedicel made of very carefully processed pulp and provided with a spiral twist. He argued that this peculiar device makes it possible for the pedicel to support the comb without applying oral secretion, as is seen in *Vespa*. It must be noted here that most of the *Vespula* species nest in covered places, and their embryo nests may well be protected from physical disturbance from outside, compared with those of aerial nesters. The pedicel need bear up only against the queen's movements on the comb and the weight of the comb filled with larvae and pupae. The nests of *Dolichovespula*, on the other hand, are usually built in open sites such as tree twigs, and therefore are more severely exposed to external forces such as wind. *Doli-*

chovespula foundresses amply apply oral secretion to the pedicel in addition to constructing a spiral twist. Jeanne (1977a) argued that the rubbery secretion absorbs the shock produced by the queen moving around on the comb. As suggested above, this may only partly explain the variation in the shape and nature of the pedicel among vespine genera, because the *Vespula* embryo nest has the pedicel free of secreted coating in spite of a similar queen behavior in the nest. It is reasonable to consider that *Dolichovespula* species are primarily twig-nesters, so that their exposed nests frequently swing during windy days in spring, and that the rubbery secretion may act to protect against shocks caused mainly by winds (Sk. Yamane et al. 1981). The *Vespa* foundress, often nesting in open sites, constructs a thick, strong, pillar-like pedicel which is coated with an oral secretion without a rubbery consistency. In this respect, *Vespa* species adopt a strategy similar to that in *Polistes*, but clearly different from that in *Dolichovespula*.

Envelope. The envelope is generally thought to have functions of retaining warmth and of defending the nest from parasitoids and predators. The shape of the envelope is fairly stable in *Vespula* and *Dolichovespula*, but a considerable variation exists in *Vespa*. Among the species of *Vespa*, those always nesting in covered spaces such as underground cavities and tree hollows (*tropica, mandarinia*) construct an embryo nest with a very incomplete, single-layered envelope. *V. crabro*, mainly nesting in tree hollows, constructs a single-layered envelope wholly covering the nest comb. On the other hand, *V. simillima*, frequently nesting also in open spaces such as on tree twigs, constructs a multi-layered envelope wholly covering the comb. These facts suggest that there is a correlation between the development of the envelope and the nesting site. Covered places are thermally regulated and not easily accessible for parasitoids and predators, while for open places the reverse is the case. It is doubtful that the multi-layered envelope better serves to defend broods against predators than does the single-layered one. In this respect, it is interesting to note that *V. analis* and *V. affinis*, in spite of their tree-nesting habits, construct only one envelope sheet. The center of distribution range of these species lies in the tropics, where predators such as ants are more abundant than in the temperate regions, while thermoregulation is not so urgent. Instead of building multi-layered envelopes, these species make a long entrance tube. Though not confirmed experimentally, this vestibule seems to contribute to decreasing the chance for predators (e.g., ants) to invade the nest (Sk. Yamane 1976; Hermann and Blum 1981). The fact that in Japan *V. analis* has a relatively short vestibule in Hokkaido and a longer one in southwest Honshu may support the above hypothesis (Sk. Yamane and Makino 1977). However, the fact that in the tropical *V. affinis* and *V. analis* in Sumatra the vestibule is rather short (Matsuura unpubl.) contradicts it. Elaborate experiments should be done to test the hypothesis. At any rate, it may be concluded that increasing the number of sheets serves for thermoregulation, and that constructing an envelope wholly covering the comb serves for both thermoregulation and defense against parasitoids and predators.

The multi-layered envelope built by *Vespula*, most species of which build subterranean nests, seems to be inconsistent with the idea mentioned above, because underground cavities themselves are more or less thermally regulated. The only explanation for this may be that the main range of this genus lies in relatively higher altitudes and latitudes, where multiple protection from the cold air is advantageous. Multi-layered envelopes built by *Dolichovespula* are more easily interpreted from the view of thermoregulation, since the species of this genus inhabit the cold temperate and subarctic regions, and commonly nest in open places. Among them, *D. media* and *D. maculata* often construct an extraordinarily long vestibule, of which the adaptive significance is also unknown. Their distributions (more southerly than in the other congeners), and nesting habits (almost exclusively preferring open sites) of these species may be consistent with the defense hypothesis. Greene (1979), on the other hand, considered this structure to have a thermoregulatory function. If he is correct, conspicuous vestibules must be observed in other species inhabiting the range of these two species, or in colder regions. Such a report has not been available to date.

Finally, a short comment will be made on the disc observed in *Vespula* embryo nests. The attaching point of the disc to the hanging sheet is roughly equivalent to that of the first sheet in *Dolichovespula*. In the latter the foundress queen has been often known to curl around the pedicel between the comb and the first envelope sheet; Edwards (1980) and Makino (1980) think that the queen warms the comb with eggs and larvae while curling (this behavior has been more intensively studied for *Vespa*; see Sect. 2 of this chapter). Recently we have noticed that Nixon (1934) had already discovered this behavior in *Vl. germanica*, and had regarded it as brood incubation. (Nixon's paper had long been missed by students of embryo nests.) The function of the disc can be thought to be a partition keeping the heat produced by the queen during curling from escaping directly upward. This reasoning is consistent with the behavior of *Dolichovespula* foundresses that break down the first envelope sheet after the completion of the subsequent sheets but leave a disc-like structure.

5 Development and Structure of Nest

The completed nest of Vespinae consists of circular combs (from several to ten-odd) which are first expanded horizontally and then piled up vertically with the aid of pillars. The whole combs are then covered and protected by an envelope which usually has only one open part as the entrance. Each comb is separated from the envelope in principle, but in the later period of nesting, combs of the upper layer are connected to the envelope, thus effectively reinforcing the whole nest. A nest structure similar to that of the Vespinae is found in *Angiopolybia* and *Pseudopolybia* of the Polistinae (Richards 1971), but the vespine nests can be distinguished from other envelope-possessing nests of the Ropalidiini and many Polybiini by the direction of development of the combs, the existence of pillars, and the structure of the envelope (de Saussure 1853–58; Richards and Richards 1951). Genus- and species-specific differences in the structure of nests after emergence of workers can also be found in the four Vespinae genera (*Vespa*,

Vespula, Dolichovespula and *Provespa*). The general characteristics of the nests of the former three genera (mainly in Japan), are as follows (those of *Provespa* are described in Chap. 7):

5.1 Changes in Shape of Nest

After the emergence of workers, the shape of the young nest founded by the queen is changed by the new architects. In *Vespa,* the sheet-like envelope of the queen's nest is bitten off gradually by workers and transformed into cell walls. In *V. crabro,* about four-fifths of the queen nest envelope is removed within 2 months of the emergence of the workers, and the whole combs are exposed, leaving the rest of the envelope, which covers only the upper layer of combs in the shape of an umbrella (Matsuura 1973a). During and after August, when the rearing of new queens starts, reconstruction of the envelope becomes active, and the new envelope develops into the shape of a hanging bell and covers most of the combs, leaving only part of the lowest layer of combs. Also, in *V. tropica* and *V. mandarinia,* most of the nest-constructing activities are concentrated on increasing the number of cells during a period of about one month after the first emergence of workers, and the envelope in this period covers only the back of the first comb in an umbrella shape. In these two species, one sheet of the envelope is extended almost to the lower combs when a few cocoons appear in the second comb, and then shell-like aerial chambers are added to the surface. The shape of the whole nest of *V. tropica* becomes oval, white that of *V. mandarinia* becomes almost spherical if there are no obstructions. In *V. analis* and *V. affinis,* the first workers remove the tubular vestibule from the queen nest and construct a new entrance on the lower side of the nest envelope (Sk. Yamane and Makino 1977). After this, large shell-like aerial chambers are attached in order from the top of the nest toward the lower side and the whole nest becomes spherical. From September to October when the rearing of new queens and males proceeds, the lower part of the nest swells slightly and the shape of the nest becomes oval. In *V. analis,* during the period of emergence of new queens, the envelope around the entrance is densely coated as the nest becomes large and the shape of the nest becomes irregular, having a swelling at the side. No such change, however, occurs in middle- and small-sized nests of this species or in the nests of other *Vespa* species. The nest of *V. simillima* after the emergence of workers becomes irregular in shape when the nest site is a narrow covered space, because the nest is expanded along the inner wall. The nest, however, becomes an oval shape with a swollen lower side when the nest site is spacious.

In the case of *Vespula,* the nest of *Vl. schrenckii* during the period of emergence of new queens is spherical but slightly flat, the widest area being around the top layer of the combs. The shape of the nest before this period is almost spherical with no such irregularities. The nests of *Vl. flaviceps* and *Vl. shidai,* on the other hand, are developed into a spherical shape even after emergence of the workers, but their lower parts swell during the period of emergence of new queens. In nests built in attics or under eaves, however, there is a tendency to extend horizontally along the comb layer as the nests become large.

Among *Dolichovespula* species, the nest of *D. media* and *D. saxonica* are constructed in a nearly spherical form after the emergence of workers, but the lower side becomes slender during the period of emergence of new queens and the entrance turns sideways. The surface in this region has thickly and spirally-arranged labra attached.

5.2 Modification of Nest Site

In species preferring covered situations, after the emergence of workers, a nest site requires a certain degree of enlargement to adapt it to the needs of growing colony. Among social wasps, the nesting habit in underground situations has shown limited development in three genera of Vespinae (*Vespa, Vespula* and *Dolichovespula*), indicating not only their advanced behavioral specialization but also their various responses to nest site modification.

Among *Vespa* species, queens of *V. mandarinia,* which is a typical underground-nesting species, prefer rather narrow cavities for nest foundation. After the emergence of workers, soil adjacent to the nest is constantly scraped off by workers with their mandibles and placed outside the nest, the soil particles being formed into balls of 5–7 mm in diameter. Pebbles 1.0–1.5 cm in size are also removed occasionally. With the ball between the mandibles, the workers leave the nest through the entrance of the ground surface on foot and drop the ball a few to several tens of centimeters away from the entrance, thus forming a tongue-like platform of soil around the entrance. This platform reaches 2 m in length, 50 cm in width and 15 cm in maximum depth in the later stages of nesting. The expansion of the nest site and removal of soil are performed by the same individuals, showing the persistence of specific workers (Matsuura unpubl.). Large stones more than several centimeters in size are dropped to the bottom of the nest cavity following removal of the earth beneath them. In *V. tropica*, nest site expansion after the emergence of workers is quite rare, since queens prefer a rather spacious site for nest foundation and the colony size is small. In the case of underground nesting, soil adjacent to the nest is occasionally scraped with the mandibles and placed outside the nest, but the soil that has been removed does not become a platform like that made by *V. mandarinia*. When *V. crabro* and *V. simillima* nest in covered situations above or under the ground and the sites become narrow, they relocate colonies in more spacious situations, seldom undertaking the expansion of their original nest sites. Even in the case of open-situation nesting, such as under eaves, *V. simillima* relocates its colonies when there are obstacles, such as a pillar, in the direction of construction of the original nest (Matsuura 1984).

All Japanese *Vespula* species usually build their nests in underground cavities. *Vl. schrenckii* seldom undertakes expansion of its nest site because it prefers spacious sites for nest foundation. Species of the subgenus *Paravespula*, in contrast, found their nests in relatively narrow situations and expend a considerable amount of labor in excavating the soil after the emergence of workers. In *Vl. flaviceps*, *Vl. shidai* and *Vl. vulgaris*, workers in charge of extranidal activities usually scrape away the soil adjacent to the nest and make balls 3-4 mm in diameter prior to their departure. They fly off with the ball between their mandibles and drop it from the air several tens of meters away from the nest. Apart from these extranidal workers, some other workers repeat the process of removal of soil, as in the case of *V. mandarinia*, by holding soil between the mandibles, leaving the nest on foot and dropping it outside the nest. Unlike *V. mandarinia*, however, the excavation labor essential for the nest site expansion is thought to be performed by the extranidal workers.

Dolichovespula usually builds its nest in open situations but, when a nest is built inside a house or in underground cavities, this species prefers a spacious place, so that expansion of the nest site is not required.

5.3 Envelope

There are remarkable differences in the shape of the envelope between open space- and covered space-preferring species of *Vespa* (Matsuura 1971a). In *V. analis* and *V. affinis*, which belong to the former type, the envelope completely covers the combs, leaving one entrance at the side of the nest (Fig. 3.18). The envelope consists of shell-like aerial chambers which are small, compact and thin at the top of the nest, but become large, rough and thick at the middle and bottom. The envelope of *V. analis* at the peak of colony activity is 2.0–2.5 cm thick with 4–6 layers of aerial chambers at the top and middle of the nest and 3.0–3.5 cm and 5–8 layers at the bottom. *V. simillima* is considered to be an open space-preferring species by nature, since the envelope of its nest generally also completely covers the combs (Fig. 3.20), even when the nest is built in a narrow covered situation (Fig. 3.21) (Matsuura 1971a). *V. simillima* constructs the biggest nest of all Japanese vespine wasps, and the thickness of the envelope reaches 3.5–5.5 cm in the middle and ten-odd centimeters at the bottom, with 8–12 layers of aerial chambers. In *V. crabro*, *V. dybowskii*, *V. tropica* and *V. mandarinia*, which belong to the covered space-preferring group, the envelope is simple with imperfect aerial chambers (Fig. 3.18) and, in the middle of the nesting period, it covers the combs like a cylinder made of 1–2 sheets of paper (Fig. 3.19). In the later nesting period, large shell-like aerial chambers are added, mainly to the upper and middle parts, but the whole or central part of the lowest comb is exposed because the bottom of the nest is always open. The bottom opening is usually used as the entrance, but another entrance is sometimes established as the nest develops.

In *Vespula*, the envelope of all known species usually covers the combs completely (Fig. 3.22) but, in the nests between wooden walls, part of the envelope adjacent to the walls is removed during the later nesting period. The bottom of the nest of *Vl. schrenckii* is opened widely until the emergence of males and new queens, but is then narrowed to about 1 cm in diameter after the middle of August. The top of the envelope of this species consists of 6–8 layers of compact shell-like aerial chambers 0.6–0.9 cm in thickness, but the chambers become rough and

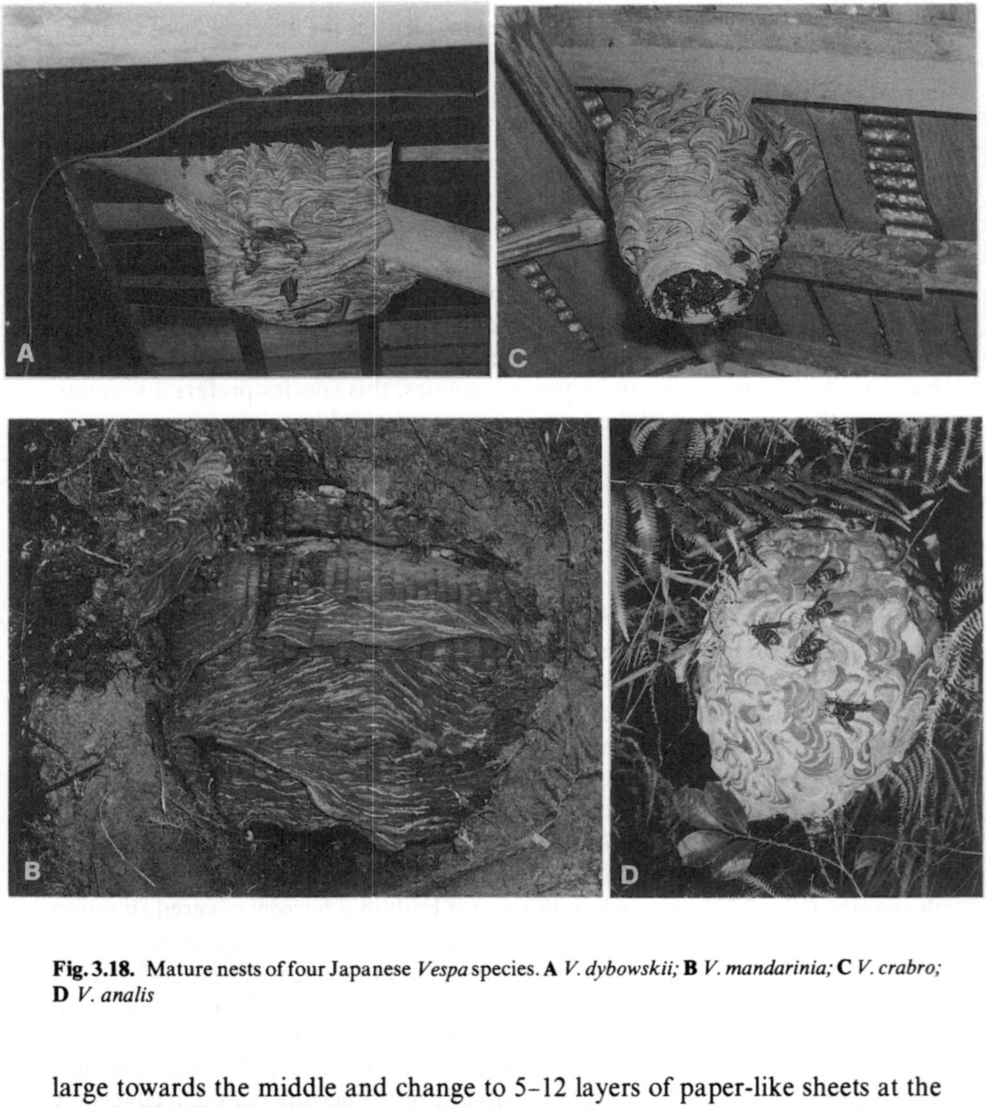

Fig. 3.18. Mature nests of four Japanese *Vespa* species. **A** *V. dybowskii;* **B** *V. mandarinia;* **C** *V. crabro;* **D** *V. analis*

large towards the middle and change to 5–12 layers of paper-like sheets at the lower side, like those made by *Dolichovespula*. Such a structure is considered to be characteristic of this subgenus, since similar structures have been observed in *Vl. consobrina, Vl. vidua, Vl. atropilosa* and *Vl. acadica* in North America, which belong to the *Vl. rufa* group (Akre et al. 1981). In the Japanese *Vl. shidai, Vl. flaviceps* and *Vl. vulgaris*, which belong to the subgenus *Paravespula*, the whole of the envelope consists of piles of compact shell-like aerial chambers 0.5–1.2 cm in thickness, and the chambers increase in size towards the lower part of nest. The entrance is made on the bottom or the side of the envelope, and the size (0.8–2.5 cm in diameter) enlarges as the nest develops.

The envelope of *Dolichovespula* covers the whole of the combs, leaving one entrance at the bottom (Figs. 3.23 and 3.24). Unlike the cases of *Vespa* and the subgenus *Paravespula*, the envelope does not form shell-like aerial chambers but

Fig. 3.19. Nest of *Vespa tropica* in the polyethic period

consists of 5–10 layers of bag-like sheets which are separated from each other. The structures of the nests of *D. media* and *D. saxonica* are basically the same, and the thickness of the envelope of both species is 2–5 cm at the top, 2–4 cm in the middle and 4–6 cm at the bottom at the stage of maximum development (Makino 1982). Envelopes of *D. sylvestris* and *D. norwegica* in Europe are also sheet-like (Janet 1903; Spradbery 1973a; Edwards 1980), and only the envelope of *D. maculata* in North America has shell-like aerial chambers at the top, which develop remarkably in the later period of nesting (Greene 1979; Akre et al. 1981).

5.4 Combs

The shape of each comb in all genera becomes almost circular through centripetal extension around the central mainstay, but it changes in various ways, depending on the nest site and the presence of obstacles.

Vespa species can be divided into two groups on the basis of the shape of the combs: one forming an umbrella-like comb with a dorsal surface swollen at the center around the mainstay, and the other showing a flat dorsal surface. *V. mandarinia, V. analis* and *V. tropica* belong to the former group, and each comb of the former two species becomes conical (Matsuura and Sakagami 1973). The latter group includes *V. simillima*, *V. crabro* and *V. dybowskii* in Japan, and *V.*

Fig. 3.20. Nests of *Vespa simillima*. *Top*: Ssp. *simillima* in Hokkaido (not matured). *Bottom*: Ssp. *xanthoptera* in Honshu (dissolved)

affinis, *V. velutina* and *V. basalis* in Taiwan also belong to the same group (Matsuura 1973c). The number of combs in the Japanese *Vespa* species changes in the following order: *simillima* >> *crabro* > *mandarinia* > *dybowskii* > *affinis* > *analis* > *tropica*. Ten combs are not rare for *V. simillima* but five combs have never been recorded for *V. tropica* (Table 6.4). The size of combs generally increases in species having larger numbers of combs, and the largest comb observed so far in Japan is one of *V. mandarinia* measuring 61 × 48 cm in diameter with 1191 cells (Matsuura unpubl.).

In *Vespula*, each comb of all species swells slightly at the center around its mainstay but is, on the whole, flat. The number of combs for *Vl. schrenckii* is two

Fig. 3.21. *Vespa simillima* nest constructed in a wall void in Hokkaido. Envelope is constructed outside the wall

to four per nest and is usually three (Takamizawa 1981a), but a nest having five combs has been reported in the closely related species, *Vl. rufa,* in England (Spradbery 1973a). *Vl. consobrina, Vl. vidua* and *Vl. acadia* in North America, which belong to the *Vl. rufa* group (subgenus *Vespula*), each possess a maximum of four combs, but this number is less than half that of the subgenus *Paravespula*, as will be described below, and their combs are also small in size. On the other hand, nests of *Vl. shidai, Vl. flaviceps* and *Vl. vulgaris* contain 6–12 combs. In England, the average number of combs in *Vl. vulgaris* has been reported to be nine, which was slightly higher than that in *Vl. germanica* in the same district (Spradbery 1973a), but the number of combs of the same species in North America is limited to three to five (Akre et al. 1981). Huge nests about 2 years old have often been found in species of the subgenus *Paravespula* in countries other than Japan. In the case of *Vl. vulgaris*, 21 combs were recorded in a nest found in California (Duncan 1939). Nests of *Vl. germanica*, a naturalized species in New Zealand, have been found to contain 30 or more combs (Spradbery 1973a). In addition, two nests of *Vl. squamosa* containing 17 and 39 combs have been dug out in a tropical zone, although this species, which inhabits the eastern coast of North America, does not belong to the above two subgenera (Tissot and Robinson 1954).

Each comb of all known species of *Dolichovespula* is typically circular and surrounds the central mainstay, where a slight swelling is apparent. The comb, however, does not become flat but bends upwards rather steeply toward its edge, which is a characteristic feature of this genus. The number of combs is three to six in both *D. media* and *D. saxonica* and the largest comb (25 cm in *D. media*, for

Fig. 3.22. *Vespula flaviceps* nest made under the eaves (*top*), and *Vl. shidai* mature nest, envelope removed (*bottom*), in Hokkaido

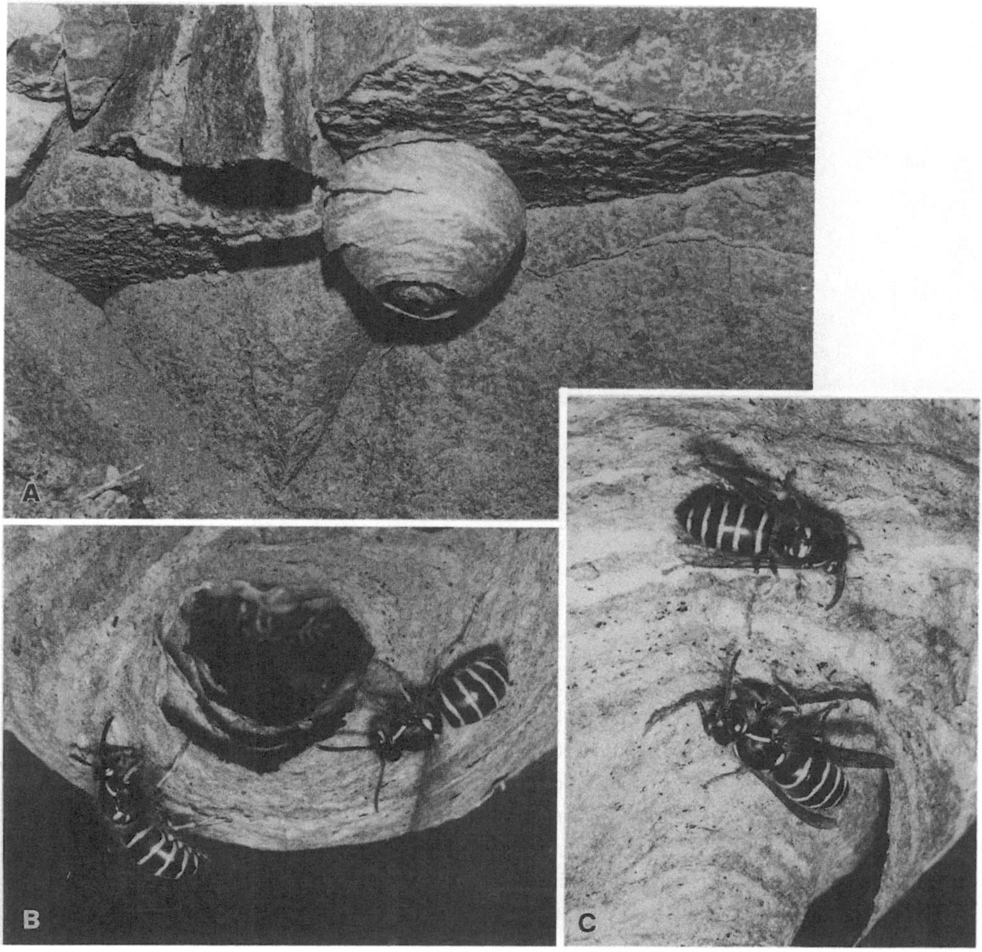

Fig. 3.23. Nests of *Dolichovespula saxonica* in Hokkaido. **A** Young nest made on a cliff; **B** fanning by workers; **C** envelope construction by workers

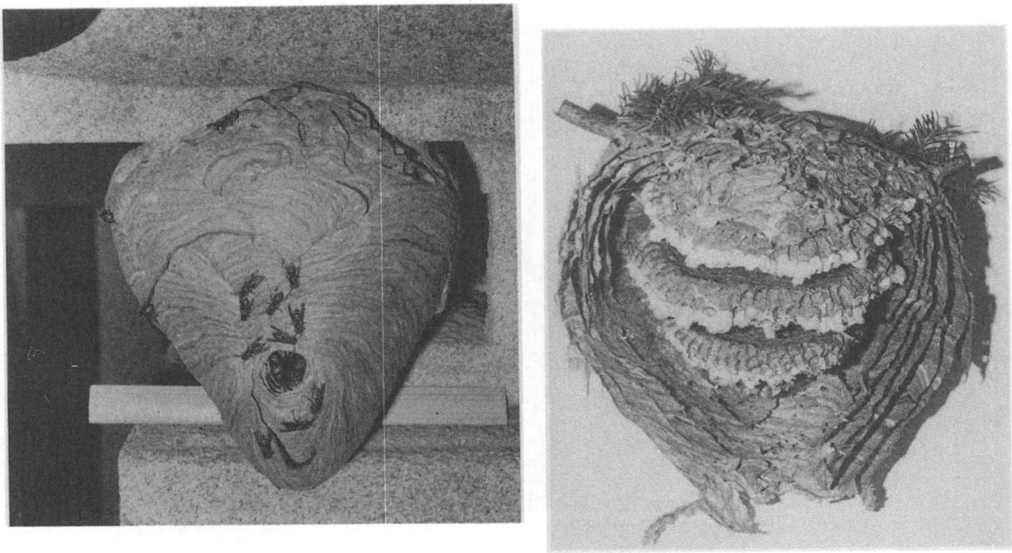

Fig. 3.24. Mature nests of *Dolichovespula media* in Hokkaido (Left photo T. Ban)

example) in any nest is usually located in the center. However, no differences can be found between the two Hokkaido species in terms of the number, shape or size of combs (Makino 1982). It has been reported for this genus that the number of combs in each nest in the later period of nesting is also limited to three to six for *D. norwegica* and *D. sylvestris* in Europe and for *D. norwegica albida, D. arenaria, D. norvegicoides* and *D. maculata* in North America (Spradbery 1973a; Greene et al. 1976; Edwards 1980; Akre et al. 1981).

5.5 Pillars

In all known Japanese and foreign species of *Vespa*, except for *V. analis*, each comb is connected to the others by one thick main pillar (mainstay) in the center and by a few to several tens of auxiliary pillars (pillars) sporadically constructed on the dorsal surface. Each comb of *V. analis* is generally supported by only one mainstay; the first comb sometimes has a few pillars in the later period of nesting. All mainstays and pillars in this genus are in the shape of a club, but as nests become enlarged, the pillars on the upper few combs change gradually to zonal shapes by the addition of plaster to their surface, and are connected to each other, thus strengthening the combs. Such a change is conspicuous in the species with large colony sizes, especially in *V. simillima*. In *V. mandarinia*, which nests in underground situations, zonal development of pillars usually does not occur in spite of its large-sized nest. This can be explained by assuming that, in the case of an underground nest, the load of the nest weight is not concentrated on the upper combs but rather is dispersed through the whole nest, since the nest is supported from the outside at several points by plant roots and stones. In the latter period of nesting, all *Vespa* species exhibit the habit of strengthening the pillars by gnawing off the cocoon caps that they left after emergence, and plastering them on the pillar surface.

In Japanese *Vespula* species, a few differences in the shapes of pillars can be observed between *Vl. schrenckii* and the three species of the subgenus *Paravespula*. The combs of *Vl. schrenskii* have a hard cord-like mainstay at the center and, in the later period of nesting, connections between the combs are strengthened by the addition of a few to ten-odd pillars which are shaped like a rod or a cord (or a ribbon in some nests). Plastering of cocoon caps on the pillar surface is rare in this species. In *Vl. rufa* in England, the mainstay in the first comb has the shape of a rod or a cord but the pillars resemble thin ribbons similar to those in *Dolichovespula* (Spradbery 1973a). The pillars in some species of the *Vl. rufa* group in North America also show a ribbon-like shape (Akre et al. 1981). In contrast, in three Japanese species of the subgenus *Paravespula* (*Vl. flaviceps, Vl. shidai* and *Vl. vulgaris*), the mainstay at the center of the comb and several tens of pillars sporadically located on the comb, are rod- or cord-shaped, and their surfaces are plastered with a mass of paste made of gnawed-off cocoon caps left by the workers after emergence. The mainstay constructed in the nest-founding period is left on the first comb, and the few to ten-odd supporting pillars on the comb are rod-shaped but rarely ribbon-shaped (Fig. 3.25).

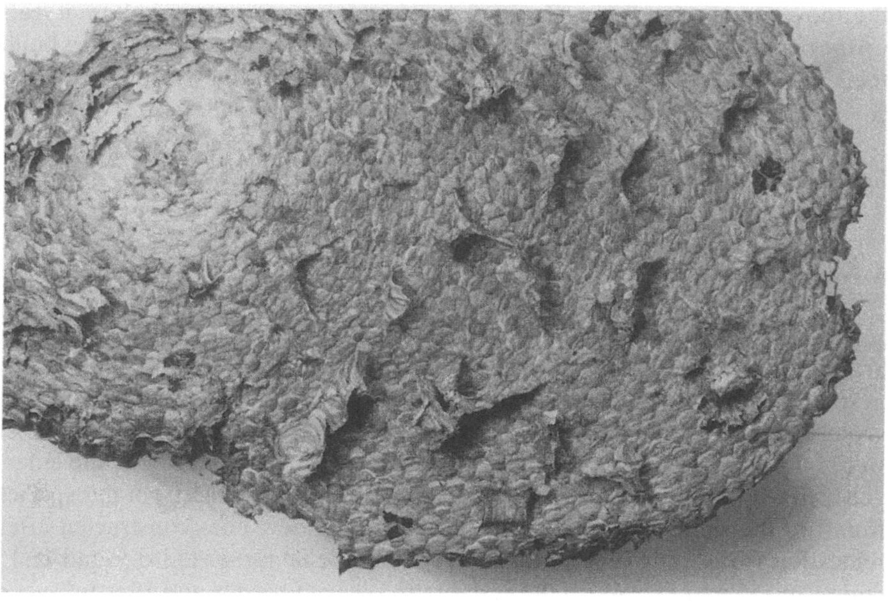

Fig. 3.25. Dorsal view of the first (*top*) comb of a *Vespula flaviceps* nest in Hokkaido. Note that in this nest the suspensoria connecting the ceiling and first comb are exceptionally ribbonlike

In *Dolichovespula*, only one rod-like and hard mainstay is located at the center of combs in nests which are at the stage for the pre-rearing of males and new queens, but as nesting progresses, thin ribbon-like pillars, each a few centimeters long, are constructed surrounding the mainstay. Three species of this genus in Japan (*D. media, D. saxonica* and *D. norvegicoides*) have the same type of pillar, and no interspecific differences are found among them. This type of pillar is considered to be characteristic of this genus, since the same pillar structure has also been found in five species of this genus in North America, including *D. maculata*, and in *D. sylvestris* and *D. norwegica* in Europe.

5.6 Weather Stripping

This habit can be seen in species of *Vespa* nesting in covered situations. Spaces around the nest site through which light enters are sealed up with sheet-like or shell-like materials independently of the envelope. This habit is not observed in queen nests or nests just after the emergence of workers, but appears mainly when the rearing of new queens starts.

When *V. crabro* or *V. dybowskii* builds its nest in a spacious, covered situation like an attic, it sometimes seals up most of the light-admitting spaces within 2–3 m around the nest. Completely sealed spaces are also sometimes observed, such as the gap between a wall and roof board, a ventilation window or a knothole (Fig. 8.1). In *V. tropica* nest, this sealing activity is performed on a considerably smaller scale (several tens of centimeters around the nest) compared with the activity of the above two species. When *V. mandarinia* builds its nest in a tree hollow, it seals up the opening of the hollow, leaving an entrance. When these *Vespa* species build their nests in subterranean cavities, the opening part on the ground surface is covered with an envelope-like sheet if the opening is too wide.

Nothing has been recorded on whether *Vespula* and *Dolichovespula* seal up light-admitting spaces or cover the opening parts of subterranean cavities.

6 Nest Relocation

As stated in the previous section, nests of *V. simillima* in the later stages of nesting are observed frequently in open situations such as under eaves. Most of the queen nests, however, are built in narrow covered situations above and under the ground. In an investigation in Wakayama Prefecture, 30 of 50 queen nests (60%) were relocated, following a search for a more spacious nest site after the emergence of workers. In addition, 88.6% of the nests (132 of 149 nests) of this species found in the same district in and after July were new ones constructed after relocation (Matsuura 1984). Similar relocation behavior was also found in *V. crabro* and 55.6% (25 of 45) of the nests of this species found in and after July were new nests which had been relocated. Since this species always nests in covered situations, its relocation is not of the same type as that of *V. simillima* (covered place → open place). It is a movement to a more spacious covered place (Matsuura

1984). These habits are considered to be a specialized nesting strategy in the above two Japanese species, since no similar behavior has been observed, either in the other species of the same genus or in *Vespula* or *Dolichovespula*.

6.1 Process of Relocation

Relocation of *V. simillima* and *V. crabro* colonies can be seen from late June to August when the number of workers increases, but when the rearing of males and new queens has not yet started. A common factor in the relocation of colonies is that the site of the queen nest has become cramped, hindering expansion of the nest, and this is seen in nesting situations such as in an underground cavity, a hollow in a stone wall, a wooden box or an empty can. Workers in these nests begin to search for a new site, flying about in all directions within scores of meters up to 200 m from the nest, suspending their nesting activities, including the job of increasing the number of cells and the extension of the envelope. The process of relocation after this is shown in Fig. 3.26.

When a suitable site is found, the scout workers settle at the location and spend most of the time at rest there, except for an occasional return to the original nest. Different scouts settle at different sites simultaneously, and other workers visit the sites for several days. The number of visiting workers gradually increases at the most suitable site, and several to ten-odd workers repeatedly settle at the site and revisit the original nest. After 3 to 4 days of this searching work, the queen starts to search for a new site herself, leaving her nest. Workers in the nest, however, do not follow the queen, which is quite different from the swarming behavior seen in honeybees. The queen's searching behavior is a sort of trial-and-error process. She flies around her nest site in the same manner as the

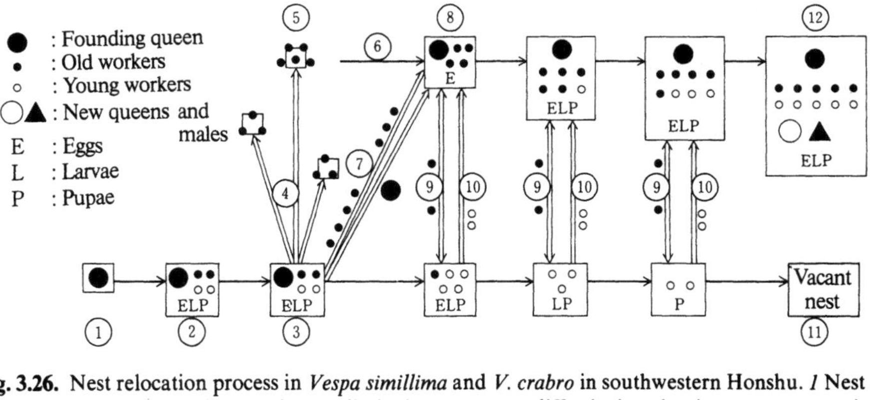

Fig. 3.26. Nest relocation process in *Vespa simillima* and *V. crabro* in southwestern Honshu. *1* Nest initiation in a small space by foundress; *2* limited space creates difficulty in enlarging nest; *3-5* search for new nest site by scouting workers; *6* selection of new site; *7* moving of workers and foundress to it; *8* construction of new nest; *9* intercourse of workers between the two nests; *10* moving of newly emerged workers in primary nest to new one; *11* decline of primary nest; *12* production of males and new queens in new nest (After Matsuura 1984)

scouts, and checks candidate environments one by one. When she finds the workers gathered at the site which has been located by a scout worker, she settles at the site and does not return to the original nest. The details of the process of the queen's discovery of the site, however, are still unclear, because workers on the candidate site show no special behavior upon the arrival of the queen, and workers have not been observed to take any special action in guiding the queen to the site. It has been reported that a nest of *V. crabro*, after the emergence of workers, was visited frequently by the queens and workers of other nests (Matsuura 1970b). It seems likely that they were searching for their new nest sites.

With the arrival of the queen, the behavior of workers at the new site becomes active. Workers construct the first comb, consisting of 5–12 cells, within the first day, and the queen lays an egg in each cell on the following day. In *V. simillima,* the number of cells reaches 30–40 about 2–3 days after arrival of the queen, and construction of the envelope starts at the same time. The first envelope, built in the shape of a sheet, covers the comb cylindrically. In this species, the second comb is constructed after the first comb has been completely covered with the envelope. After this, construction of combs and extension of the envelope are performed simultaneously. In *V. crabro*, however, three to four combs are first constructed, and then extension of a sheet-like envelope is generally started (Fig. 3.27, top).

Workers in the original nest become nervous when the queen migrates to the new site, and spend a long time searching for it. However, once they have found the queen's new site, they frequently come and go between the two nest sites.

Thus, new and old nests now exist. The distance between the two nests is usually small (within 10 m), but the maximum recorded is 180 m for *V. simillima.* Workers in the old nest migrate to the new nest one by one, leaving the young workers and the brood. Workers in the new nest, on the other hand, visit the old nest to feed larvae and newly emerged workers and, at the same time, take the secretions of fifth instar larvae and use them as a food source. The workers emerging from the old nest become capable of flying when they are 2–3 days old and leave the nest independently in search of the new nest. In this way, relocation usually lasts for about a month until all eggs, larvae and pupae mature and migrate to the new nest. The queen, however, never visits the old nest again. If the new nest is taken away, together with all its population, at night during the relocation period, another new nest is constructed the following day on the same site by workers from the old nest. Such re-building is repeated again and again unless workers in the old nest are captured simultaneously. When all the brood in the old nest mature and leave the nest, emergence of workers starts in the new nest, the original nest is abandoned and there are no further visits by workers.

6.2 Function of Relocation

This habit of relocating the whole colony in the middle of nesting is known, among the Vespinae, only in *V. simillima* and *V. crabro.* The functions of this behavior are discussed here. A common feature of these two *Vespa* species is that their queens usually found their nests in narrow covered situations, causing cramping

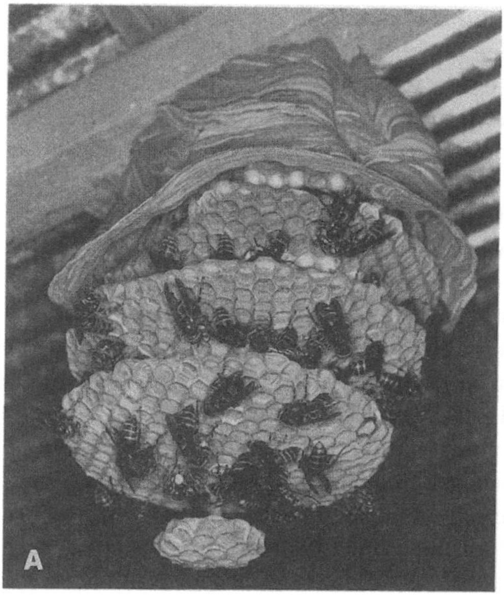

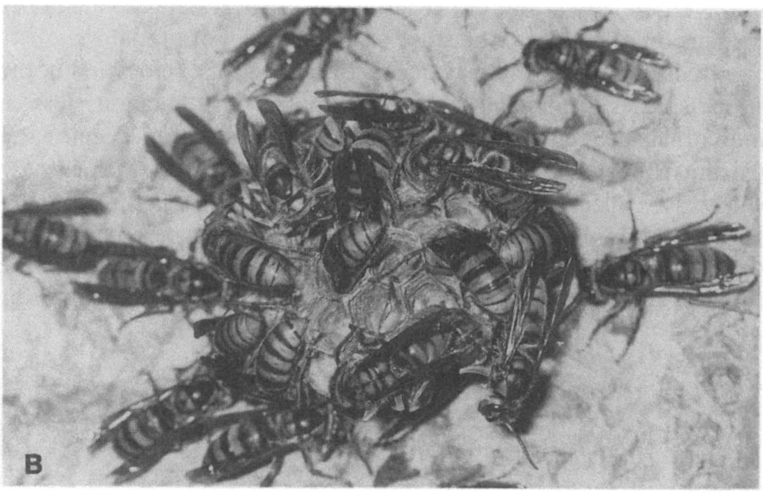

Fig. 3.27. Relocated nests of *Vespa crabro* (*A*) and *V. simillima* (*B*). In the former species three to four combs are constructed first, and then envelope is added

of space at the nest site as the colony size increases after the emergence of workers. Since neither species relocates its nests when the queen nests are founded in spacious open situations, a relationship can be noted between relocation behavior and the size of the nest space. Among *Vespa* species, *V mandarinia* also builds its queen nest in a narrow space, but workers of this species have a highly developed habit of gnawing off soil adjacent to the nest with their mandibles and removing it to a location outside the nest site. In contrast, behavioral adaptation to underground nesting, as seen in *V. mandarinia*, is not developed in *V. simillima* and *V. crabro* because these species rarely undertake excavation activity, even in underground nests. Consequently, the two species presumably shift their labor force from nest site expansion to new nest construction by moving to a more spacious place.

The preference of queens of *V. simillima* and *V. crabro* for a narrow space as their nest-founding site is considered to be related to the fact that their nest-founding time is slightly earlier than that of the other *Vespa* species (Fig. 2.1). Under the unstable weather conditions prevalent in spring, the microclimate in a narrow space above or under the ground functions to protect the eggs and larvae in the nest from drastic changes in temperature. It is also possible that a narrow and covered nest site makes it difficult to discover, and that this is a strategy for protection against natural enemies during the queen's solitary period.

Social bees are frequently observed to abandon their nests and construct new ones in different situations when the nesting environment deteriorates. This is called "absconding". Such behavior is especially well known in *Apis cerana* in Japan (Sakagami 1970), and the case of nest reconstruction by polistine wasps attacked by *V. tropica* also belongs in this category. In these cases of absconding, however, the larvae and pupae left in the old nest starve to death because of the suspension of rearing or because of the activities of predators.

In contrast with these examples of absconding, relocation of the above two *Vespa* species is efficient in the sense that very little colony loss occurs, as no individuals are left behind in the old nest. In addition, continuation of brood rearing in the old nest by workers which have migrated to the new nest enables the workers to collect larval secretion, which is an important food source for them. As a result, thermoregulation activities begin in the new nest for brood rearing, since there are enough workers during the hot season, and after rapid development of the nest at the spacious new site the emergence of males and new queens starts.

7 Repair

7.1 Factor Dependency

Vespine wasps exhibit the habit of repairing damaged nests immediately in order to prevent complete breakdown. The repair process, however, varies depending on the development stage of the nest and on the season.

During *Vespa*'s solitary period, a queen will restore her nest immediately if part of the envelope is damaged by a storm. *Vespa* queens, however, abandon their nests when more than half of the combs are removed artificially (*V. tropica*) or when the envelope is entirely removed, exposing all combs (*V. analis*). In nests after the emergence of workers, damage to the envelope and combs is repaired immediately by workers. In the case of *V. analis, V. simillima* and *V. crabro*, when combs in a nest were damaged by a stone which was thrown and which passed through the envelope, the hole in the envelope was first mended, and then the damage inside the nest was restored. When a nest of *V. analis* was attacked by a honey buzzard and about a quarter of its envelope lost, temporary repairs were carried out on the damaged parts, using a sheet-like envelope, after which a shell-like envelope was added to the sheet (Matsuura unpubl.).

7.2 Reconstruction

In *Vespa*, when a nest is removed artificially or lost in a storm during the solitary period, the surviving queen does not usually reconstruct another nest at the same site. After the emergence of workers, however, a queen and her workers who have lost their nest sometimes construct a new nest at the same site. According to an observation at Kibi-chô in Wakayama Prefecture in June 1968, a nest of *V. analis* attached to a tree twig was blown off in a rainstorm, but the surviving queen and twelve workers constructed a new nest on a twig about one meter distant from the original site and brought up new workers afterwards. When a nest of the same species, which was built under the eaves of a house, was beaten down with a stick at night and removed from the site, leaving its inhabitants behind (at Misugi-mura, Mie Prefecture; August 1964), the queen and about 30 workers assembled at the original site the following morning, built a sheet-like envelope first, then constructed cells and produced workers and males 1.5 months later. In *V. simillima*, when the entrance of a nest in a tree hollow was blocked with a clod of earth in the daytime and its inhabitants were locked in the hollow (Yuasa-chô, Wakayama Prefecture; July 1974), about 60 workers which were left outside the hollow tried to enter it for several days, but later constructed a new nest under the eaves of a temple about 3 m distant from the tree, 4 days after the blocking up of the hollow. The new nest consisted of three combs (with 12, 18 and 21 cells respectively) arranged in a line, with a few centimeters of space between them, and the circumference of the combs was covered with a zonal envelope (2–3 cm in width). This nest, however, was abandoned afterwards without oviposition. In an observation on *V. mandarinia* (Anô-chô, Mie Prefecture; July 1983), the fine roots of the pedicel of a nest snapped, due to increased nest weight, and the combs fell to the ground, but the queen and eight workers constructed new combs, making a new pedicel under the remaining envelope (Matsuura unpubl.).

Also, when all the combs of an underground *Vespula* nest were dug out for food, leaving the queen and workers behind, new combs were constructed at the same site or in an adjoining cavity, and the combs were covered with an envelope afterwards. In this case, the exposed nest was completely protected by the construction of a mud wall. Such a nest is called a "second nest" by *Vl. flaviceps* and *Vl. shidai* nest collectors in Nagano and Gifu Prefectures. The second nest, however, is rarely collected again, because larvae in the nest are poor in quality compared with those in the normal nest.

For *Dolichovespula*, no observations have been reported on the reconstruction of nests by Japanese species. In Europe, on the other hand, nest reconstruction has been reported for *D. media* and *D. sylvestris* (Janet 1903), *D. norwegica* (Latter 1915) and *D. saxonica* (van Lith 1956). Development of the envelope and combs in each case was quite inferior to that of the normal nest.

Chapter 4

Social Structure and Behavior

1 Division of Labor Between Queen and Workers

1.1 Caste Differentiation

In the four genera belonging to the Vespinae (*Vespa, Vespula, Dolichovespula* and *Provespa*), differentiation of caste between queens and workers is very clear in terms of morphology, behavior and physiology.

In the external morphology of *Vespa*, the castes of *V. analis* and *V. tropica* are sometimes indistinguishable on the basis of their body sizes, and this tendency is especially strong in the queen and large-sized workers which emerge in the later stage of nesting (van der Vecht 1957). However, caste distinctions can be drawn from external appearance in *V. crabro, V. dybowskii, V. mandarinia* and *V. simillima*.

Morphological differences between queens and workers are unmistakable in the Japanese species of *Vespula* and *Dolichovespula*, and are especially remarkable in *D. media*. In *Vl. germanica* in Europe, the length of the queens is 1.4–1.5 times greater than that of workers and 1.3–1.4 times greater in *D. sylvestris* and *D. norwegica* (Weyrauch 1935).

In internal morphology, the number of ovarian tubules is seven to ten pairs in both queens and workers, but individuals whose ovarian tubules are one to two pairs shorter are sometimes found in small-sized workers (Iwata 1971a; Kugler et al. 1976; Sk. Yamane 1974; Matsuura unpubl.).

The queen and worker of 12 *Vespula* species, including *Vl. flaviceps* in Japan, has six pairs of ovarian tubules, and four species of *Dolichovespula* also show the same number (Kugler et al. 1976), thus exhibiting no caste-specific differences in the two genera.

The size of the corpus allatum in the queen of *Vl. flaviceps*, however, is about five times larger than that of workers (Takamatsu 1952), and the size of the van der Vecht's organ in the gaster of the *Vl. germanica* queen is also larger than that of workers (Spradbery 1973a).

Hibernating queens accumulate a large amount of fat body in the gaster, thus showing physiological adaptation to cold. The amount of fat body, however, is quite small in workers and "worker-like" new queens taking part in extranidal activities, but the external morphology and body size of the latter are otherwise similar to those of the queens. In *V. simillima*, the fresh weights of new queens collected during their hibernation were 540–860 mg, showing the accumulation of fat body, but the amount of fat body in workers during the fall season was quite small, even in large-sized individuals, when their fresh weights were below 460 mg. In *V. tropica*, variation in fresh weight among hibernating queens was great, the weight ranging from 709.5 to 1273 mg (999.2 mg on average, n = 24), though their gasters were all replete with fat body (Matsuura 1966b). Although individuals with a fresh weight of 800–900 mg are not rare among workers of this species, they are distinguishable from queens on the basis of accumulation of fat body.

1.2 Behavioral Changes in Queen in Relation to Worker Emergence

Behavioral differentiation between queens and workers has developed into a typical form of division of labor: oviposition by the former and all other work by the latter. However, in *Vespa, Vespula* and *Dolichovespula* in temperate regions, the nest is founded by a single queen and, therefore, the queens are engaged in every type of work during the first 35–45 days, like subsocial wasps, until the emergence of the first worker. As a result of worker emergence, the labor of the queen is reduced, and her activity changes both in quality and quantity along with the development of her colony until her tasks finally become restricted to oviposition.

According to Matsuura (1974b), behavioral change in the queen after the emergence of workers in *Vespa*, mainly in *V. crabro*, may be explained as follows:

Collection of Food. Food collection activity for larvae is the most important form of labor for the queen during her solitary period. For the queen of *V. crabro*, the percentage of time taken for food collection among the total activities performed in one day was 60 and 65.3%, 7 and 5 days before the emergence of workers. Food collection activity by the queen continued for 16 days after the emergence of workers, but the time so spent was on average reduced to 16.7%, and most of the source of the collected food became restricted to carbohydrates, such as tree sap, only 3% of the total sources being prey.

Also in *V. analis, V. mandarinia* and *V. simillima*, food collection activities by the queens continued for 10–25 days after emergence of the first workers, the main food source being tree sap (Matsuura unpubl.). The queen of *V. tropica*, on the other hand, continues her food collection activity for 20–30 days after the emergence of workers and preys on larvae and pupae in the nests of polistine wasps (cf. Chap. 5).

Collection of Nest Materials. Collection of nest materials by the queen of *V. crabro* was observed at a frequency of about twice per hour before the emergence of workers, but became rare rapidly after emergence. However, the last collection of nest materials was observed 24 days after the first worker emergence as the queen's final extranidal activity, occurring three times during 6.5 h, and the materials were used for the expansion of the envelope. In *V. analis, V. mandarinia* and *V. simillima*, collection of nest materials by the queens is completed 1–3 days after the emergence of workers. In *V. tropica*, retirement of the queen from extranidal activities is significantly slow compared with the above four species;

that is, collection of nest materials continues for 10–30 days after the emergence of workers (Matsuura unpubl.).

Feeding Larvae. Even after the emergence of workers, most of the food collected by the queen from the field is distributed to the larvae herself. Also, after retirement from food collection activities in the field, the queen continues to feed the larvae by receiving a proportion of various types of food collected by workers, including flesh pellets formed from prey and tree sap. In the case of *V. crabro*, distribution of liquid food and pellets to larvae by the queen was observed to continue for about 1 month and over 50 days respectively, after the emergence of workers (Fig. 4.1B), but the frequency of feeding by the queen decreased gradually. One month after the emergence of workers the number of occasions of feeding was 0.1–0.9 per hour, even for the combined total of liquid and pellets. The feeding larvae by queens after the emergence of workers continues for 40–70 days in *V. analis* but for 20–40 days in *V. mandarinia* and *V. simillima*. In *V. tropica*, on the other hand, feeding of liquid food and fluid from the prey by the queen generally continues throughout her life (Matsuura unpubl.).

Nest Construction. Nest-building activities in vespine wasps include construction and reinforcement of the mainstay and pillars, scraping the inside walls and expanding the outer wall of the envelope, initiation of cells and enlargement of cell walls. Of these activities, construction and reinforcement of the mainstay and pillars become the sole responsibility of the workers after their emergence, but the queen continues to take part in other activities. In *V. crabro*, the queen continues her nest-building activities for about 40 days after the emergence of workers. These activities are terminated in the following sequence: (1) initiation of cells, (2) enlargement of cells and envelope with nest materials which she has herself collected from the field, (3) enlargement of cell walls with nest materials collected by gnawing off the envelope, and (4) reprocessing moist cell walls (rethinning without use of nest materials) freshly enlarged by workers. The final form of activity (reprocessing of cell walls) can be observed even during the later stage of the polyethic period in nests containing 100–150 workers.

The nest construction activities by the queens after the emergence of workers continue for 30–50 days in *V. analis*, and 20–30 days in *V. mandarinia* and *V. simillima*. Initiation of cells by the queens of the above four species terminates completely 1–5 days after the emergence of workers, and the labor becomes restricted to workers. Among building activities performed by the queen of *V. tropica*, on the other hand, initiation of cells terminates 20–40 days after the emergence of workers but reprocessing of cell walls continues throughout her lifetime. However, the frequency of both activities tends to decrease after the emergence of workers as they take over the greater part of the labors (Matsuura unpubl.).

Points in which building activities of *Vespa* differ from the closely related group, the Polistinae, are: (1) the nest material collected inside or outside the nest by a queen or worker is always processed by the collector, and (2) cells originally initiated by the queen is generally undertaken by workers after their emergence.

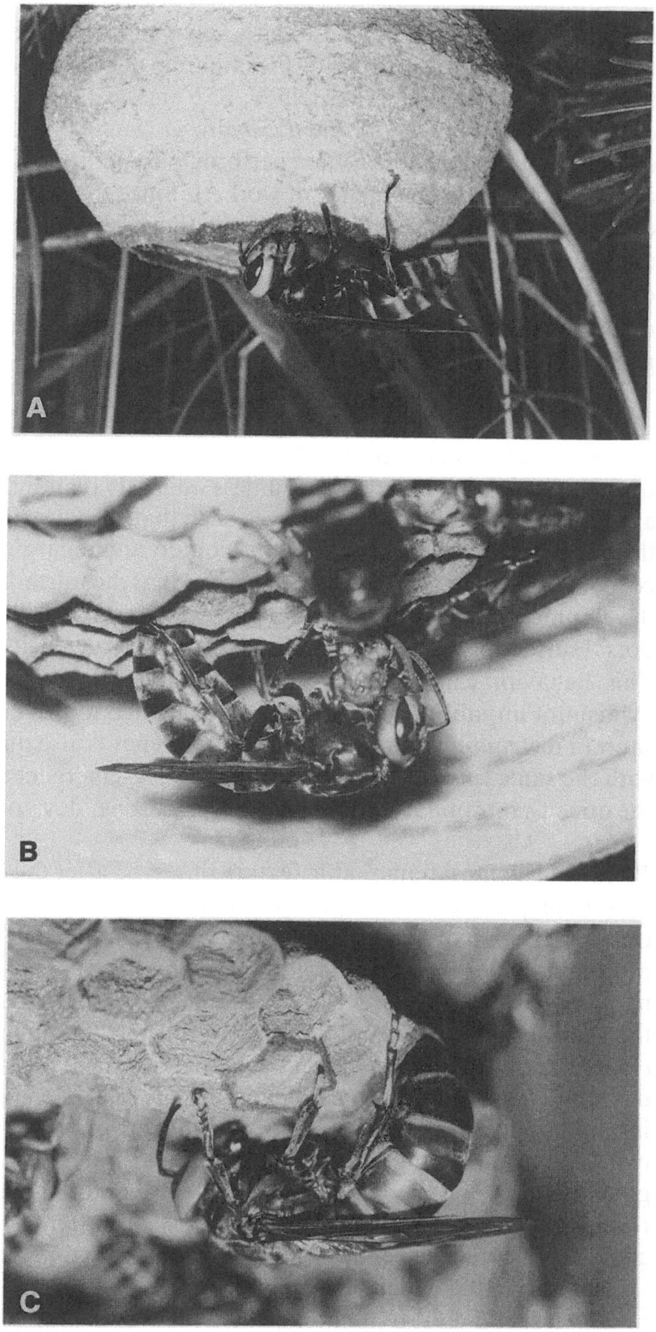

Fig. 4.1. Foundress behavior in some Japanese *Vespa* species. **A** Envelope construction by a foundress of *V. analis* in the solitary period; **B** *V. crabro* foundress malaxates food even in the cooperative period; **C** oviposition by a *V. mandarinia* queen

In some *Polistes* species, it is well known that the queen takes part in nest construction activities, including the initiation of cells, by receiving nest materials from workers (West-Eberhard 1969; S. Yamane 1976; Matsuura 1977a), *Mischocyttarus* (Jeanne 1972) and *Parapolybia* (Sugiura et al. 1983). In addition, queens of the above three genera perform most of the activity for the initiation of cells even after the emergence of workers. Since the activity of cell initiation by queens of polistine wasps is induced by ovipositional impulse (Deleurance 1957), the above behavior is considered to indicate immature differentiation of division of labor between the castes; that is, the behavioral chain relation between the ovipostional impulse and cell initiation cannot be discontinued after the emergence of workers in *Polistes* (S. Yamane 1976).

Nest Preservation. Nest cleaning, removal of cocoon caps left by emerged workers, nest defence and regulation of nest temperature are conducted by the queen after worker emergence, but less intensely than by workers. In *V. crabro,* a form of behavior has been observed such that, 43 days after worker emergence, the queen receives water collected by workers and regurgitates it over the caps of cocoons. The queen of *V. mandarinia* maintains the warmth inside the nest for the development of eggs and young larvae by curling herself around the pillar on the dorsal surface of the second comb, which is constructed after worker emergence. This behavior can be observed for about 20 days after worker emergence (Matsuura unpubl.). However, the time devoted to this curling activity is quite short (a few minutes per time) and the frequency is also quite low in comparison with the same form of behavior during the pre-emergence period when most of the queen's intranidal activity amounts to curling, day and night.

Resting. The most remarkable feature observed after the emergence of workers is that the queen comes to spend most of her time at rest on the comb, with only occasional cell patrols and trophallaxis with workers and larvae. This behavior can properly be regarded as resting, since the queen remains still for a long time, mainly on the dorsal surface of the comb, by concealing her folded wings under the ventral surface of her abdomen and turning her antennae downward. In *V. crabro*, nearly 70% of the queen's activities are spent on extranidal duties in the daytime before the emergence of workers, the total percentage including various intranidal activities being 95%, with only the remaining 5% being used equally for oviposition and resting. In contrast, extranidal and intranidal activities decrease greatly on the first day of worker emergence, and the amount of time spent in resting on the comb increases rapidly. Especially after termination of extranidal activities, the queen spends around 98% of the daytime resting.

Similar to the queen of *V. crabro*, queens of *V. analis, V. mandarinia* and *V. simillima* spend most of their daily activities resting for 30–60 days after worker emergence. As the foundation of cells by workers increases rapidly, the ratio of time spent in oviposition increases. The *V. tropica* queen, on the other hand, rests only one-half to one-third as much as the above four species, the queen continuing her various tasks even after the emergence of workers (Matsuura unpubl.).

Oviposition. In all *Vespa* species, the oviposition process consists of a sequence of: (1) discovery of empty cell by the queen herself, (2) inspection of the inside of the cell using her antennae, (3) repeated insertion of gaster into the cell, (reinspection of the inside of the cell), (4) fixing of the gaster in the cell, (5) oviposition, (6) cleaning of antennae with her mandibles and, (7) immediately upon oviposition, withdrawal of the gaster from the cell and taking in of secretion from mature larvae (Fig. 4.1C). The duration of oviposition from fixation to withdrawal of the gaster has been measured to be 82–182 s (m = 136.7) in *V. crabro*, 144–184 (m = 153.5) in *V. analis*, 120–181 (m = 159) in *V. mandarinia* and 82–157 (m = 120.5) in *V. simillima*. The number of eggs laid per day by queens of the above four species and *V. tropica* is 0–5 eggs (m≦1.6) in the preemergence period and 3–20 eggs (m≦4) during the 50 days after worker emergence. This shows a tendency to increase with the development of colonies. Oviposition is limited to the daytime in the preemergence period, but can be observed frequently even at night after the emergence of workers. Especially in the case of large nest-constructing species like *V. mandarinia* and *V. simillima*, construction of new cells is continued even into the night by workers, and oviposition by the queen is performed in new cells immediately (Matsuura unpubl.).

1.3 Relationship Between Size of Colony and Division of Labor

In the above five species of *Vespa*, the degree of development of division of labor between queens and workers after worker emergence varies depending on the species, and shows a close relationship to colony size in the process of colony development (Matsuura unpubl.). That is, division of labor between the castes as a result of worker emergence tends to progress more quickly in species with a larger colony size. In *V. simillima*, which has the largest colonies, the queen retires from extranidal activities around 10 days after the emergence of the first worker, and the same event occurs at around 15 days after worker emergence in *V. mandarinia*, the colony size of which is second largest. On the other hand, in *V. tropica*, for which the colony size is quite small, the extranidal activities of the queen terminate 20–30 days after worker emergence. Such a difference in the retirement time of queens is also related to the degree of nest development and the number of workers present. In *V. analis*, termination of extranidal activities by the queen varies greatly from one nest to another, ranging from 15 to 40 days after worker emergence, but termination has been also observed to occur when the number of workers increases to no more than between five and eight. Nests showing prolonged extranidal activities by queens are restricted to those in which the number of workers does not increase, or in which many workers are stylopized by *Xenos* spp.

It has been reported that a queen of *Vl. flaviceps* terminated her extranidal activities when the number of workers reached 11 (Shida 1959). In England, queens of *Vl. germanica* and *Vl. vulgaris* were reported to terminate extranidal activities when the number of workers reached four or more (Spradbery 1965). These three species of *Vespula* construct large nests. In *Dolichovespula*, which constructs small nests, there are two examples in European species. In one, a queen of *D. sylvestris* was engaged in extranidal activities for 33 and 50 days after worker emergence, but in the latter case, only three workers were produced (Brian and Brian 1952). In *D. media*, a queen continued her extranidal activities for 2 months after the first emergence of workers, but the number of workers in the nest was only 12 (Verlaine 1932).

2 Relation Between Labor and Age in Workers

2.1 Behavior of Workers Soon After Emergence

Workers on the day of emergence have no flying ability because their bodies are not hard enough, and they cannot yet fold their wings. They spend most of the first day thrusting their head into empty or egg cells without moving about, except for occasional visits to mature larvae to ingest secretion from the larval salivary gland. Although occasionally fed by the queen and workers, new workers rely mostly on the larvae as their food source.

Intranidal activities are very rare on the first day, but the following types of behavior have been observed in *V. crabro* workers (Matsuura 1974b): (1) reenlargement of the cell wall without use of nest materials, (2) enlargement of the cell wall with nest materials collected by scraping the inner part of the envelope, (3) malaxation of prey that is fed to larvae, and (4) cleaning of the bodies of foraging workers soaked with rain in the field. These activities are very rare here, and common to the other species of *Vespa*. On the second day, they begin to take part in a variety of intranidal activities, such as cell initiation, thermoregulation and guarding of the nest.

2.2 Orientation Flight

The worker of various species of *Vespa* leaves the nest and performs a flight in order to memorize the location of the nest, as soon as its body has hardened and flying has become possible. This is called the orientation flight, and in honey bees it is performed in a group at a certain time, though in vespine wasps flights are undertaken independently and at irregular times.

The orientation flight is performed by 1-4-day-old individuals (Table 4.1), but most frequently by those 2–3 days old. When a worker first leaves the nest, it appears from the entrance hesitantly, flies straight out of the nest for about 20–30 cm, and then turns to face the entrance. Still facing the entrance and flying slowly in every direction, the worker gradually lengthens the distance from the nest to some 3–5 m and then abruptly returns to the nest. A second flight is usually carried out on the same day in a similar manner to the first, but the distance from the nest is extended to 15–20 m and, at the farthest point, the workers circle in the sky more

Table 4.1. Age when first orientation flights were made by *Vespa* workers (Matsuura 1984)

Species	Days after emergence						No. of individuals observed
	0	1	2	3	4	5	
V. simillima	0	3	29	18	7	0	57
V. crabro	0	5	38	31	6	0	80
V. analis	0	3	19	10	3	0	35
V. mandarinia	0	2	17	18	2	0	39
V. tropica	0	1	18	14	3	0	36
Total	0	14	121	91	21	0	247

than ten times at levels up to ca. 10 m from the ground. Several individuals carry out a third flight on the same day. The first orientation flight takes 70–100 s and the second flight 30–150 s, hardly varying among species.

The first extranidal activities are usually performed after the second or third orientation flight, without the hornets returning to the nest. During this trip, the young worker collects various types of load, such as nest materials, water, tree sap or prey, showing no individual-specific tendency (Matsuura 1973b).

2.3 Day Age and Labor

The existence of age-dependent division of labor has been reported for social insects including ants (Sudd 1967) and *Apis mellifera* (Rösch 1925; Lindauer 1952; Sakagami 1953). The activities of marked workers in five species of *Vespa* (*simillima, analis, crabro, mandarinia* and *tropica*) during their life from just after emergence to final disappearance from the nest are summarized in Tables 4.2 and 4.3 and in Figs. 4.2, 4.3 and 4.4 (Matsuura 1984). These results indicate that, although the length of life of workers varies depending on the species and individual, workers of the five species are all endowed with intra- and extranidal skills 2–3 days after emergence, and performed a variety of activities within any one day. Subsequently, no relationship between age and labor can be seen as they continue to collect nest materials, water and food until the last day of their lives. A similar observation has also been reported for *V. orientalis* (Ishay et al. 1968).

In *Vespula*, workers of *Vl. flaviceps* start various intranidal activities 2 or 3 days after emergence, in a similar way to *Vespa*, and undertake all extranidal activities 1–2 days later. Such a tendency can be seen throughout the 2–3 weeks of their life (Matsuura unpubl.). In *Vl. vulgaris* in England, however, it has been observed that workers undertook nursing activities in the nest for about 10 days after emergence, then started to collect nest materials and food and then, when they were over 30 days old, they took up guard duty at the entrance (Potter 1964, cited in Edwards 1980).

In the Japanese *Dolichovespula media*, the workers have been shown to perform larval feeding, enlargement of cell walls and development of the

Table 4.2. Percent of time spent in each activity of single *Vespa simillima* worker (No. 9) during its lifetime (Matsuura 1984)

	Age in days															
Activity	0	1	2	3	4	5	6	7	8	9	10	11	12	13	14	15
Extranidal act.																Disappearance from the nest
Orientation	–	–	0.8	0.1	–	–	–	–	–	–	–	–	–	–	–	
Pulp collection	–	–	–	–	1.3	2.9	5.7	–	13.2	10.9	6.2	10.2	5.3	3.1	10.8	
Food collection	–	–	–	1.6	2.1	4.3	18.4	44.4	19.2	15.1	6.8	21.2	28.8	7.5	15.1	
Water collection	–	–	–	–	0.8	2.1	8.3	–	11.4	3.2	5.1	9.5	3.9	1.5	4.5	
No load collection	–	–	–	19.7	2.9	0.7	7.9	1.1	1.5	1.1	0.1	0.3	2.8	0.2	0.3	
Intranidal act.																
Pulp malaxation	–	–	0.6	1.5	1.3	2.2	1.3	–	6.4	1.5	2.3	2.8	1.1	1.3	2.8	
Scraping envelope	–	–	3.4	5.2	9.1	3.3	1.6	–	1.7	1.3	2.1	1.1	2.3	3.5	3.5	
Enlarging envelope	–	–	0.4	4.6	6.9	–	6.2	–	0.8	2.9	4.2	10.1	7.4	3.1	8.1	
Making cell	–	3.1	3.3	5.1	11.1	23.2	6.3	–	1.5	3.2	3.5	5.9	3.8	1.1	3.4	
Making pedicel	–	1.7	–	–	–	–	–	–	0.1	0.2	1.0	0.3	0.5	0.2	0.2	
Feeding larvae	–	4.1	1.5	2.8	0.1	–	1.5	1.2	1.2	2.2	2.2	1.8	2.1	3.1	3.1	
Prey malaxation	–	0.6	–	0.7	0.8	–	–	–	3.2	0.9	1.5	2.1	1.6	0.9	0.5	
Food exchange between adults	–	0.1	0.2	0.5	1.6	0.5	0.3	0.4	1.0	0.5	1.8	1.5	0.8	0.8	0.7	
Food intake from larvae	2.4	5.3	0.3	0.4	0.6	0.6	1.7	–	0.3	0.8	0.9	1.9	0.8	1.1	0.9	
Body cleaning	3.0	–	3.9	5.0	7.4	–	3.5	9.4	0.5	1.8	2.3	5.2	3.9	3.1	6.2	
Nest cleaning	–	–	–	1.3	–	–	4.9	2.2	4.1	1.2	4.1	0.3	1.1	3.1	1.1	
Spitting water	–	–	11.7	0.8	–	3.4	2.5	–	4.0	1.5	1.8	3.2	1.1	0.8	1.3	
Fanning	–	0.0	0.3	0.1	0.8	–	4.6	–	1.5	0.4	5.5	0.8	0.5	3.5	5.1	
Resting with occasional cell patrol	94.6	85.1	73.6	50.6	52.4	56.8	25.3	41.4	28.4	51.3	48.6	21.8	32.2	62.1	32.4	
Duration of observations (h)	4	4	5	9	5	3	9	3	2.5	5	8.5	3	5	8	4	

Table 4.3. Percent of time spent in each activity of single *Vespa analis* worker (No. 4) during its lifetime (Matsuura 1984)

Activity	Age in days 0	1	2	3	4	5	6	7	8	9	10	11	12	13	14	15	16
Extranidal act.																	
Orientation	–	–	0.2	–	–	–	–	–	–	–	–	–	–	–	–	–	
Pulp collection	–	–	–	1.2	7.5	6.6	10.6	9.5	3.4	4.5	6.8	5.2	6.0	3.5	7.2	5.5	
Food collection	–	–	–	3.1	9.3	5.9	5.8	16.2	11.2	5.3	14.3	7.3	11.0	4.9	7.8	5.8	
Water collection	–	–	–	1.2	1.1	1.2	0.2	0.5	1.0	–	–	1.1	0.2	–	–	1.0	
No load collection	–	–	–	3.5	1.4	3.4	5.4	3.5	3.5	2.1	5.5	2.4	5.7	3.1	2.5	3.0	
Intranidal act.																	
Pulp malaxation	–	–	1.2	2.4	3.2	4.5	8.3	7.8	2.5	6.0	7.5	6.2	7.2	4.5	9.5	6.2	Disappearance from the nest
Scraping envelope	–	1.2	4.1	3.1	2.1	0.5	1.5	2.5	1.5	0.2	2.1	1.5	1.1	0.3	3.4	2.5	
Enlarging envelope	–	–	2.2	6.2	6.5	7.2	12.1	11.1	4.0	7.5	10.2	7.4	7.8	5.3	10.2	8.0	
Making cell	–	4.2	2.4	3.8	2.1	1.5	1.1	2.8	2.2	0.1	0.5	1.8	1.2	0.5	2.9	1.5	
Making pedicel	–	–	0.1	0.2	0.1	–	0.3	0.2	0.1	–	–	0.1	–	–	0.2	0.1	
Feeding larvae	–	5.5	3.1	5.9	3.7	6.1	4.1	3.3	2.1	4.4	5.8	3.3	2.5	3.1	6.0	5.2	
Prey malaxation	–	2.1	2.8	5.1	4.5	5.1	7.5	10.1	13.3	5.9	10.2	4.9	5.9	6.2	4.9	4.8	
Food exchange between adults	0.2	1.2	2.2	2.2	3.2	1.5	2.2	2.9	1.6	0.5	0.6	0.8	1.2	2.4	2.6	0.9	
Food intake from larvae	3.1	6.2	3.1	1.8	2.5	2.0	3.5	2.1	1.5	0.5	1.1	1.3	2.1	1.5	3.2	1.1	
Body cleaning	5.2	1.1	0.5	0.3	0.2	1.0	0.5	0.6	0.3	1.0	0.5	0.3	0.5	0.3	0.3	0.6	
Nest cleaning	–	1.1	0.3	1.3	2.2	0.5	0.6	0.4	0.4	1.1	0.2	0.4	0.1	1.1	1.2	0.5	
Spitting water	–	–	0.2	0.3	0.5	0.6	0.5	0.3	0.2	0.2	0.1	0.1	0.3	0.3	0.5	0.5	
Fanning	–	0.2	0.3	0.3	0.2	0.2	0.3	0.2	0.3	0.4	0.1	0.3	0.5	0.4	0.6	0.4	
Resting with occasional cell patrol	91.5	77.2	77.3	64.2	49.7	52.2	35.5	26.0	50.9	60.3	34.5	55.7	46.7	62.6	37.0	52.4	
Duration of observations (h)	3	3	3.5	4	2.5	8.5	9	4	3	4	3	8	9	4	4	5	

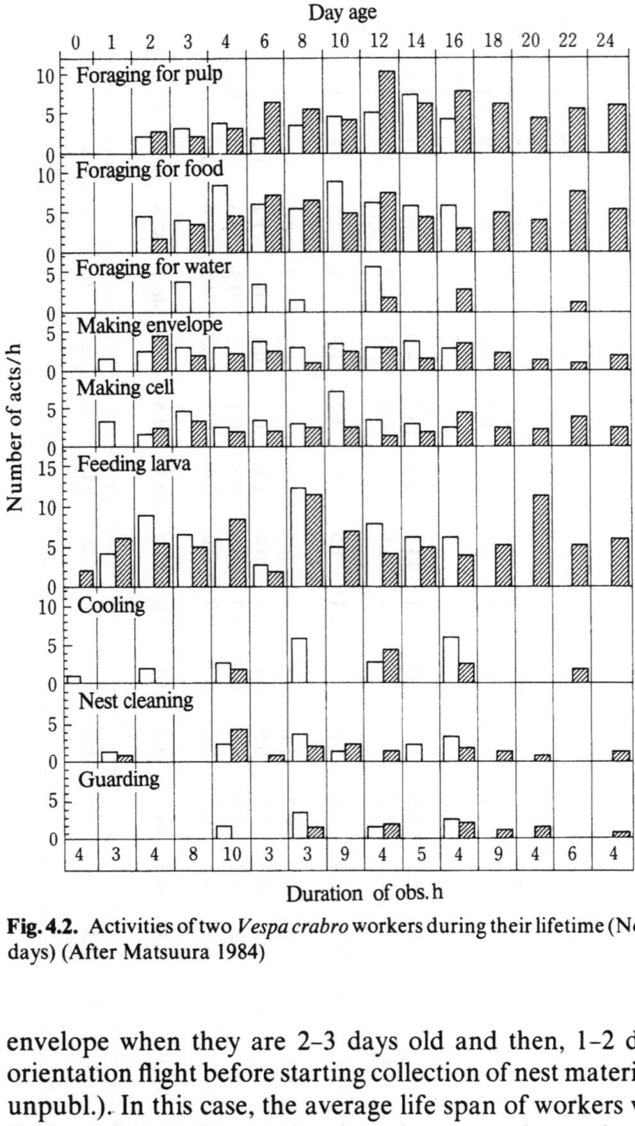

Fig. 4.2. Activities of two *Vespa crabro* workers during their lifetime (No. 38 □ : 17 days; No. 41 ▨ : 25 days) (After Matsuura 1984)

envelope when they are 2–3 days old and then, 1–2 days later, to take their orientation flight before starting collection of nest materials and food (Matsuura unpubl.). In this case, the average life span of workers was 21.2 days (n = 12). Task preference by restricted workers was observed regarding nest construction and collection of prey, but no relationships between age and labor were found.

Changes in labor according to age are well known in *Apis mellifera* and are closely related to secretion glands in worker bees. The secretory abilities of the hypopharyngeal glands which are related to nursing, and of the wax glands related to nest construction, vary depending on the age of workers, and such a dependence is the cause of activity related to changes in labor corresponding to age (Rösch 1925). In the case of vespine wasps nothing has been reported so far on the relationship between internal metabolic functions (including glandular secretion), and the activities in nursing and nest construction. It is considered that

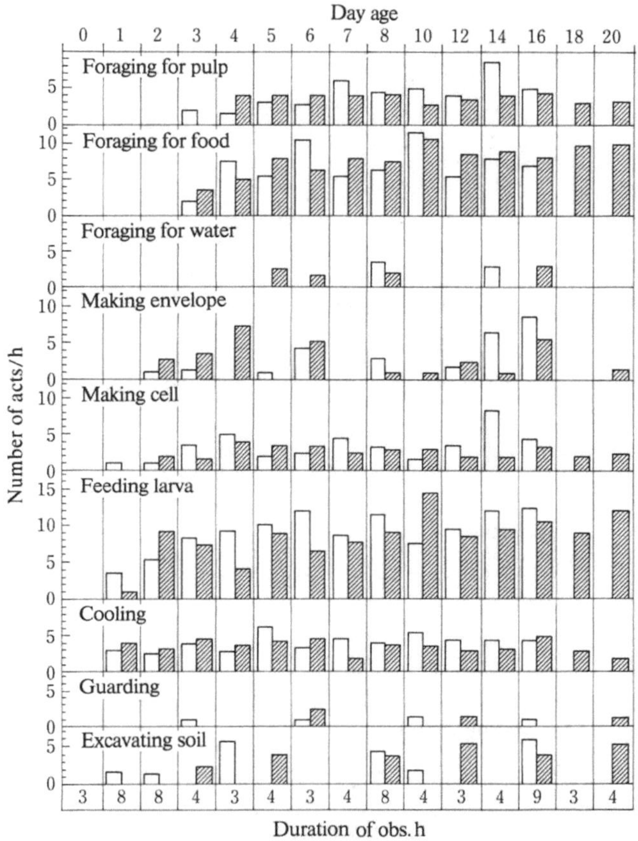

Fig. 4.3. Activities of two *Vespa mandarinia* workers during their lifetime (No. 5 □ : 18 days; No. 6 ▨ : 22 days) (After Matsuura 1984)

each vespine worker has the ability to respond to every labor task, except in the period just after emergence, since any food for larvae or nest building material collected in the field can be used directly as it is. Consequently, the regulatory functions essential for a colony are performed and maintained by workers of all age classes.

2.4 Proportion of Time Spent in Each Activity

Vespine wasps working in the field give any observer the impression that they have no interest in resting. Their activities, together with those of honeybees and ants, have often been used as a synonym for diligent industry. However, contrary to expectation, their activities inside the nest are quite different from those seen in the field.

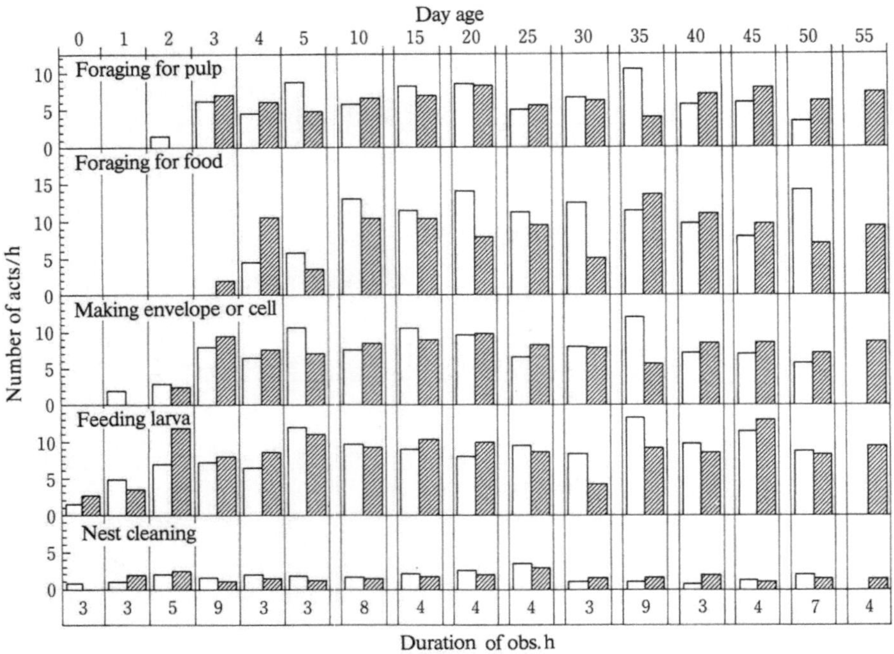

Fig. 4.4. Activities of two *Vespa tropica* workers during their lifetime (No. 2 □ : 53 days; No. 3 ▨ : 61 days) (After Matsuura 1984)

Tables 4.2 and 4.3 show the proportion of time spent in each activity in the daytime during the lifetime of a single *V. crabro* worker and a *V. mandarinia* worker, both of which were marked with paint just after emergence. They spent a little over 10% of the time in extranidal activity and spent the rest of the time inside their nests. Among intranidal activities, time spent in walking, with no specific form of labor, and resting was greatest, and the sum of these two forms of behavior amounts to about half the total daytime activity, including extranidal activities. The proportion of nest-related activities, including nursing, nest construction and thermoregulation, varies slightly between the two individuals (Matsuura 1973b), and also among other workers. Some intranidal activities which have no direct relationship to labor are in fact most costly in terms of time. Such a tendency is common to workers of other *Vespa* species.

The walking and standing of the worker inside the nest will include the individual's own resting time. In other words, individuals engaged in these apparently unproductive activities are regarded as providing a stand-by work force which can be recruited whenever conditions change and the colony needs to meet a new situation. The complement of workers in any nest is generally higher than the essential number required for colony development, but this can be explained as a pool of potential labor.

Similar results have been obtained in the honeybee *Apis mellifera*. According to Lindauer (1952), the workers spent about two-thirds of the entire observation time, including the night, either resting or wandering through the interior of the nest. He explained that these outwardly unproductive activities enhanced the capacity of the colony as a whole to respond to capricious changes in the environment. According to Sekiguchi and Sakagami (1966) and Sakagami (1970), the daily life of *Apis mellifera* workers is, in general, a repetition of a cycle consisting of a long resting time (R), patrolling (P) and performance of tasks (T), and necessary tasks are quickly found and performed through the basic behavioral system of RPTnR without close communications among workers. For this purpose, the possession of extra workers at all times is required in colonies of social wasps.

3 Oviposition by Workers

Vespine colonies enter a relatively stable phase after completing the solitary stage and producing worker adults. However, it is not a rare occurrence for the colony to lose the foundress, even in the polyethic period. With the approach of the end of the reproductive season, the foundress may die a natural death or may be evicted by the workers. These colonies are called "orphan(ed) nests". In the colonies of *Vespa*, worker oviposition usually does not occur if the foundress is lost during the cooperative period just after the eclosion of the first workers (Matsuura 1968a). In more advanced or mature colonies, if the queen is lost, some workers begin in a few days to lay eggs (Jeanne 1980).

3.1 Worker Behavior After Loss of Foundress

If the foundress dies or is removed, the colony becomes noisy, and intense worker-worker aggressions (biting, etc.) frequently occur (Fig. 4.5). Foraging considerably decreases in frequency, and the construction of the envelope and cells and care for larvae are also retarded (Ishay 1964; Landolt et al. 1977). Between the third and seventh day after the loss of the queen, some of the defeated workers leave the nest, which then becomes calm, and the intra- and extranidal activity returns to its normal level. Between the fifth and tenth day, some dominant workers begin to lay eggs (polyethism established), but aggressions among them continue. The laying individuals tend to cease foraging. In *V. orientalis* and *Vl. atropilosa*, laying workers divided the combs and defended their own territories to monopolize oviposition there. They often ate eggs deposited in cells of neighboring territories and replaced them with their own eggs, or pulled out larvae derived from queen-born eggs in order to lay their own eggs (Ishay 1964; Landolt et al. 1977). In *Vespa*, generally a cell receives one egg (but see below). In *Vl. flaviceps* several eggs are often laid in a single cell by the laying workers (Matsuura unpubl.). The number of laying workers varies among species

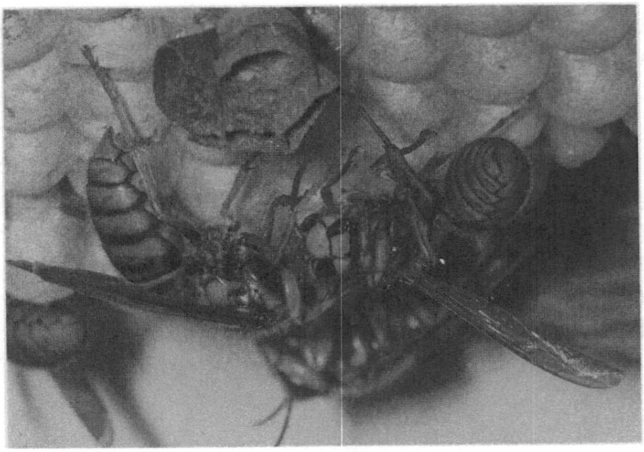

Fig. 4.5. Aggression among workers in orphaned colonies of *Vespa simillima xanthoptera* (*top*), and *V. crabro* (*bottom*)

and colonies. Usually only one laying worker appears in orphan nests of *V. analis* and *V. tropica*, which construct a small colony with a small number of workers. In larger orphan colonies of *V. simillima, V. crabro, Vl. flaviceps* and *D. arenaria*, there may be several to many laying workers (Matsuura 1968a; Sk. Yamane 1974; Greene et al. 1978). In a queenless colony of *Vl. atropilosa*, mainly young workers developed ovaries (Landolt et al. 1977). On the other hand, in colonies of *Vespa* orphaned during the early to mid-polyethic period, individuals that emerged earlier tended to become laying workers. Colonies at the late polyethic stage had larger and younger individuals that took part in ovipositing (Matsuura unpubl.).

3.2 Mechanism of Occurrence of Laying Workers

The occurrence of laying workers means that the factor(s) suppressing the development of worker ovaries has been eliminated. As mentioned in Section 3 of Chapter 2, the mechanism of queen control over worker ovaries varies among the vespine genera. This variation may have an effect upon the patterns of organization of polyethism among the workers of orphan nests. An explanation is needed for the fact that only part of the workers can develop their ovaries after the queen control has been lost; in other words, why do not all the workers become ovarially developed?

Two important observations are available in this regard. One concerns an orphan nest of *V. simillima,* collected in Sapporo on August 29, 1973, and analyzed in the laboratory (Sk. Yamane 1974). This nest comprised five combs and housed 238 workers and 55 males. Examining anesthetized workers under a dissecting binocular microscope revealed that 34 individuals (14.3%) had a polished body, with hairs, especially on the head, lost or damaged (Table 4.4). Dissections of 14 of them proved that all had well developed ovaries, with a mean of 16.4 mature eggs. The dissections of 29 workers, apparently normal in external appearance, showed that one individual (3.4%) possessed developed ovaries. Judging from these results, the total number of ovarially developed workers was estimated at 41 (17.2%). In external appearance, most of the ovarially developed workers resembled foundress queens from normal colonies in the polyethic period. Those individuals in which most of the body hairs were lost and gasters were polished and blotched, were in particular barely distinguishable from the foundress queen, except in size. This suggests that these "false queens" suppressed the ovarian development of the other workers and monopolized oviposition.

The other observation was made by Ishay (1964) and concerns an orphaned nest of *V. orientalis.* In this nest, laying workers appeared after the loss of the queen. As the colony calmed down, one to three normal workers gathered around

Table 4.4. Degree of loss of hairs on head in 16 ovarially developed workers in orphan nest of *Vespa simillima simillima* (After Sk. Yamane 1974)

	Observed parts on head[a]								Observed parts on head[a]						
Individual	a	b	c	d	e	f	T	Ind.	a	b	c	d	e	f	T
1	2	3	3	3	3	3	17	10	1	3	3	2	2	2	13
2	2	3	3	3	3	3	17	11	1	3	3	1	2	2	12
3	1	3	3	2	2	3	14	12	1	3	3	2	2	2	13
4	1	3	3	1	2	2	12	13	0	1	1	0	0	0	2
5	1	3	3	3	3	3	16	14	1	1	3	1	1	1	8
6	1	3	3	3	2	2	14	15	0	1	1	0	0	0	2
7	1	2	3	1	1	2	10	16	0	0	0	0	0	0	0
8	0	1	2	1	1	1	6	Total	14	36	40	25	25	28	168
9	1	3	3	2	1	2	12								

[a] a: clypeus, b: outer side of mandibular base, c: gena, d: vertex, e: frons, f: sinus of eye, 0: apparently normal, 1: less than 50% of hairs lost, 2: 50–80% lost, 3: more than 80% lost.

a laying worker as if attracted to the foundress queen, although fewer workers were attracted than in the normal foundress-worker "royal court". (According to Matsuura, unpublished, the attraction of workers by laying workers as observed by Ishay has not been confirmed in the Japanese species, *V. crabro, V. simillima* and *V. analis*.) On the basis of the two observations mentioned above, it is quite possible that, although less effectively than a queen, the laying workers control other workers with the aid of "queen substance".

What kind of workers, then, develop their ovaries first? Sk. Yamane (1974) showed that the ovarially developed workers were, surprisingly, the medium to smaller individuals (Fig. 4.6). The reason for this is not obvious, though a possible explanation is that smaller workers born in a mature colony tend to engage in more intranidal tasks, have more opportunities to receive regurgitation from larvae, and lose less energy than do the larger ones, so that they have by far the best chance to develop their ovaries.

Regarding *Dolichovespula*, fewer reports are available that suggest the mechanism by which the development of worker ovaries is controlled by the laying worker(s). Greene et al. (1978) mentioned that the laying worker in a *D. arenaria* orphaned colony in the beginning showed stereotyped, agonistic interactions with other workers. It remained with the nest and did not forage, and spent much of its time either inspecting cells or resting appressed to the comb. When the colony was alarmed, it stayed on the comb and never went out onto the envelope. These behavior patterns resemble those of the foundress queen in mature colonies. An interesting case was reported by Greene et al. (1978), where one worker in the process of laying an egg was pulled out of the cell by another worker, who bit her head and thorax.

Sk. Yamane (1982c) found one of the 30 workers in a *D. saxonica nipponica* orphan nest to have a hairless thorax and mandibular bases and an extensively stained gaster. This worker, marked with yellow paint on July 21, was quite active

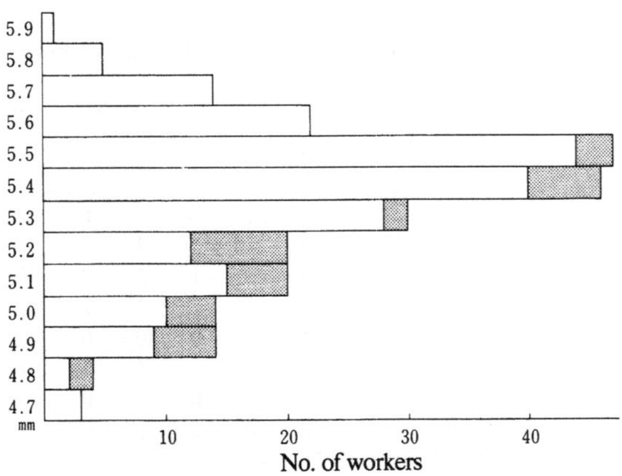

Fig. 4.6. Head width of workers from orphan nest of *Vespa simillima* in Hokkaido. *Blank*: Normal workers (n = 203, mean = 5.37 mm); *dotted*: ovarially developed workers or individuals supposed to be so by their external appearance (n = 35, mean = 5.16 mm) (After Sk. Yamane 1974)

on the comb but never left the nest on the next day. Her oviposition was not reliably observed, though she inserted her gaster into a cell. More interestingly, she was "attacked" and gnawed by other workers. The attraction of normal workers to her was observed three times on July 22–23. This suggests a possibility that also in this genus the laying worker controls other workers' ovaries with the aid of a pheromone.

Although a similar phenomenon has been reported to occur in *Vespula* orphan nests (Landolt et al. 1977; Akre and Reed 1983), the direct mechanism is unknown. In this genus, since workers are not attracted to the foundress queen in a normal colony, worker-worker interaction in orphan nests also may not be achieved through direct contact (see also Sect. 2 of Chap. 2).

4 Behavior of Larvae

4.1 Orientation Within Cell

As will be explained later, the larval Vespinae have five instars. The hatched larva hangs down from the upper portion of cell wall to which its caudal end is attached by means of the egg shell. In later instars, the larva is prevented from dropping by not wholly casting off its exuviae, which connect it with the egg shell. This way of hanging continues until it grows into the fifth instar. When the body becomes larger and fills up the cell during the later stage of the fifth instar, the larva supports itself by tensioning its dorsal and pleural lobes (protuberances) and by ventral swelling (tropholopade). At this point it becomes free of its attachment to the wall and is able to move freely within the cell. In many textbooks (e.g., Spradbery 1973a; Edwards 1980), the larva is said to be free of its attachment to the wall from the fourth instar onward, but according to our observations this is doubtful. The fourth instar larva is still too small to fill up the cell, and its pleural lobes are not yet fully developed. When anesthetizing a whole colony for collection, many fifth instar larvae drop out of cells, but no fourth instar larva falls. Thus, fourth instar larvae, and probably also the early-phase fifth instar larvae, are still attached to the cell wall at their caudal end.

4.2 Behavioral Repertoire

Vespine larvae lack legs, and are thus often likened to a simple digestive tube. Certainly they are under the care of nurses and need not search for food for themselves, so their behavioral repertoire is quite limited. Social life, which generally causes behavioral degeneration in larvae, has also evolved some specialized larval habits. As will be explained later, most important among them are feeding behavior and its associated communication acts, performed in response to stimulation by adults. The larvae, especially of the last instar, regurgitate sugary saliva when requested by the adults. After completing feeding and prior to pupation, the mature larva applies thin silken lines around the walls of the cell, and spins a thicker cocoon cap at the opening of the cell. The shape and development of the cocoon varies among castes, sexes, and genera.

4.3 Hunger Signal and Pattern of Feeding Habits

When rearing vespine colonies, we can hear a rasping noise which is emitted by the larvae scraping their mandibles against the cell wall. This act was first described in detail by Weyrauch (1935). Ishay and Landau (1972) and Ishay (1975) more closely analysed the noise and compared its pattern among several vespine species. The scraping by the larvae is no doubt a signal to communicate to the adults that they are hungry, since scraping larvae attract nurses with food (Ishay and Schwartz 1973) and nonscraping larvae often reject food (Sk. Yamane 1976). Sk. Yamane (1976) distinguished the following six elements in larval feeding behavior, and explained the behavior pattern from the emission of the hunger signal to the receiving and taking food in the mouth: (1) Backward bending. The larva bends back the head and thorax; in the final (fifth) instar the head is pressed to the cell wall. (2) Forward bending. The larva bends the head toward the ventral surface of the thorax and the following abdominal segments. In the fourth and fifth instars the labrum and mandibles are completely hidden beneath the trophalopade by this action. (3) Dropping of saliva from the salivary opening. This action is always associated with the preceding action. (4) Opening and closing of the mandibles. This action is associated with the above three actions. The mandibles do not touch or cross medially in the first and second instar, and are never used for biting in any instar. (5) Raising of the mandibles. This action to open and erect the mandibles at an angle of more than 90 degrees is always associated with the head shaking. (6) Shaking of the head. This action is easily distinguished from the backward-forward bending complex, for it is practised quickly and repeatedly with the mandibles opened.

Among these elements, (1)–(3) are observed throughout the larval stage, and (4) is seen from the second to final instar. The raising of the mandibles and shaking of the head (5, 6) are, on the other hand, exclusively characteristic of the final instar when it is emitting the hunger signal.

The typical behavior pattern shown by the hungry final instar larva is depicted below (nest combs with larvae isolated from the nurse wasps were used in the laboratory).

When the comb is scraped with the tweezers (substituting for the stimuli by the adult), the larva bends its head back toward the cell wall and opens and closes its mandibles violently. If it is fed, it receives the food pellet, bends forward, and opens and closes its mandibles in front of the trophalopade while secreting saliva, to take the pellet to its mouth. Thus, the opening and closing of the mandibles has two functions (e.g., food-reception and food-intake). On the other hand, the combination of two behavior elements, the raising of the mandibles and shaking of the head, has one function, the emission of a hunger signal.

4.4 Origin of Hunger Signal

The mouth parts of larval Vespinae are peculiar among social vespids in that they open at a right angle to the body axis (just facing cell wall), and have stout mandibles. Vespine larvae are morphologically rather similar to eumenid larvae (Reid 1942; Sk. Yamane 1976). In the eumenid larvae, the stout mandibles serve for masticating the food (Iwata 1953). Sk. Yamane (1976) has explained why the vespine larva is ancestral in morphology, yet quite specialized in behavior.

In the larva of a social species which receives food pellets masticated by the adult wasps, a mouth opening directed toward the cell opening (downward) may be advantageous for receiving food. Furthermore, the mandibles should have degenerated, because of the larvae having no need to masticate its food. To be sure, this tendency has been wide-spread in the Polistinae, but it is unlikely that this condition was immediately achieved at the early phase of social evolution. The common ancestor of the Polistinae and Vespinae probably reserved an eumenid-like appearance in the larva. In the line leading to Vespinae, the larva might have undergone the following behavioral changes; first, the larva of the vespine ancestor may have evolved a behavior pattern – bending backward to receive food, bending forward to hold it in front of the mouth, then taking it into the mouth –, as observed also in extant species. This was necessary because it was difficult for the larva with the mouth opening opposite the cell wall to receive food without bending backward. This behavior pattern incidentally increased the chance for the mandibles to touch the cell wall when hungry larvae repeated the act in demanding food. Once the worker used the sound produced by this act as a key to identify hungry larvae, the larva that emitted the sound more frequently – the one that raised its mandibles and shook its head – should have been selected for. The stouter mandibles were also no doubt advantageous. Thus, the mandibles, prior to any possible reduction, acquired a new function; that is, scraping to demand food. Thus, the seemingly disadvantageous trait (mouth opening opposite the cell wall) played a role in the evolution of wall-scraping behavior (Fig. 4.7; Table 4.5). When an organ with one function acquires in the course of evolution a new function, especially under changed selection pressures, it is called "functional change" (Sewerzoff 1931) or "pre-adaptation" (Bock 1959).

This explanation for the evolution of the hunger signal among the Vespinae has not been explicitly refuted or supported.

4.5 Comparison Between Species and Genera

Requesting food by scraping the cell wall has so far been found only in the Vespinae. In the paper wasps (*Polistes*), the larva-adult communication is much simpler (Evans and West-Eberhard 1970; Sk. Yamane 1976). Most workers agree that among the Vespinae this behavior is especially conspicuous in *Vespa* (Ishay and Landau 1972; Sk. Yamane 1976; Greene 1979; Jeanne 1980; Akre 1982). Though observed also in some *Dolichovespula* (Weyrauch 1935; Jeanne 1980), it is not so conspicuous as in *Vespa* (Sk. Yamane 1976). It is not obvious that it is

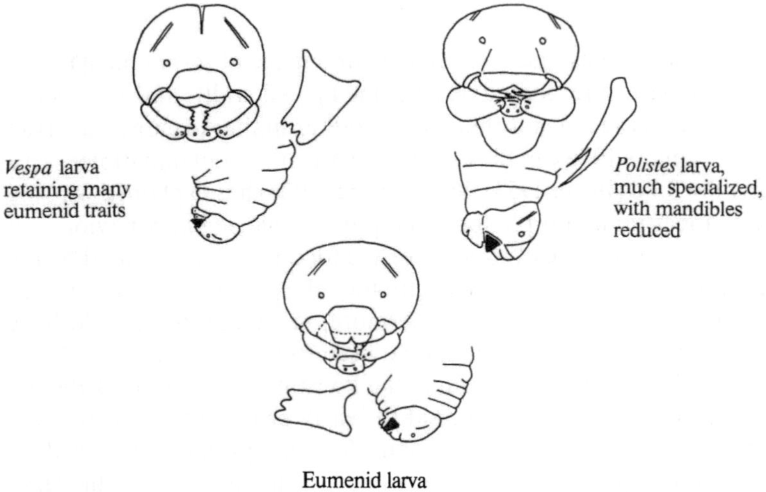

Fig. 4.7. Morphological comparison of larvae of *Vespa, Polistes* and Eumenidae (After Sk. Yamane 1976)

Table 4.5. Comparison of mandibular shape, behavior elements and their functions among mature larvae of Eumenidae, *Vespa* and *Polistes* (After Sk. Yamane 1976)[a]

Mandibular shape	Behavior elements	Functions	Taxonomic groups
Wide and stout	Crossing	Food mastication	Eumenidae
	Opening and closing	Food intake	
		Food reception	*Vespa*
	Raising	Scraping for food requirement	
Elongate and feeble	Opening and closing	Food intake	*Polistes*

[a] In Yamane (1976), following Bock and von Wahlert (1965), behavior elements are called "functions", and functions "biological roles".

commonly and constantly observed also in *Vespula*. Ishay and Brown (1975) did not confirm cell wall scraping in a *Vl. germanica* colony kept in the laboratory; this was erroneously cited by Akre (1982) thus: "the larvae scrape the sides of their cell walls when hungry, as in *Vespa* colonies". Akre et al. (1976) reported that larvae of *Vl. pensylvanica* and *Vl. atropilosa* twitch in unison to stimulate the adults, but have not been seen to actually scrape the cell walls, although under certain conditions they do so and produce an audible sound. More detailed observations are needed, but at any rate this behavior is less conspicuous in *Vespula* than in *Vespa* and *Dolichovespula*. Only one vespidologist, Weyrauch (1935, 1936), claimed that the cell wall scraping is more violent in *Vespula* than in *Dolichovespula*.

Ishay (1975) compared the frequency of scraping sounds produced by the larvae of three vespine species, *V. crabro, V. orientalis* and *D. sylvestris*, and showed that the sounds produced by *V. crabro* were bimodal, instead of the unimodal pattern exhibited by the others. The exact reason for this difference is at present unknown (Edwards 1980).

Finally, the claim that the scraping behavior is observed also in the fourth instar larva (Ishay and Landau 1972; Spradbery 1973a; Edwards 1980) will be examined. This claim is associated with the view that the fourth instar larva can move freely within its cell (the latter view has already been criticized at the beginning of this section). First, the identification of instar by body size alone is often difficult; head width and mandibular shape are the only reliable indicators to distinguish a large fourth instar larva from a small final instar larva. We have never observed fourth instar larvae constantly engaging in cell wall scraping. Of course, as they frequently engage in "backward and forward bending", their mandibles may happen to touch the cell wall and produce audible sounds. But the behavioral elements used solely in scraping, i.e., raising the mandibles and shaking the head, are observed only in the final instar. The correct identification of instar and close observations, with properly analyzed behavior elements in mind, are prerequisites to confirm the above allegations.

The relationship between the excellent acoustic nature of the material and structure of vespine nests and the sound-communication between the nest mates are discussed at length by Ishay et al. (1968).

5 Trophallaxis Between Larvae and Adults

Larvae of vespine wasps are not only fed by adults but also give adults their salivary gland secretion. Larvae at the final instar stage, when the amount of secretion is abundant, are frequently visited by adults because the secretion is a very attractive food (Fig. 4.8), not only for the founding queen and workers but also for males and new queens. In *V. tropica,* the frequency of secretion intake by adults increased as nesting proceeded, reaching 32–104 times per hour (m = 67.6, n = 16) per one final instar larva in the reproductive period, thus greatly exceeding the number of feeding times (7–14/h) by workers (Matsuura unpubl.). As a result, growth of these larvae stops, starveling larvae increase in number, and most of them die before pupation (cf. Chap. 6).

Exchange of food between larvae and adults was first recorded by Réaumur in 1742, and Janet (1903) subsequently made more detailed observations. Roubaud (1916) observed a similar phenomenon through his studies on polistine wasps, including *Ropalidia, Polistes* and *Belonogaster,* and called it "oecotrophobiosis", although the phenomenon was re-named "trophallaxis" by Wheeler (1928), which is now more prevalent. As a function of trophallaxis, Roubaud and Wheeler suggested that it would have formed the basic driving force in the evolution of sociality in wasps as a result of the effect of the close relationship between larvae and adults. Maschwitz (1966) has considered that the larval secretion might not only be used as a food by adults but might also function

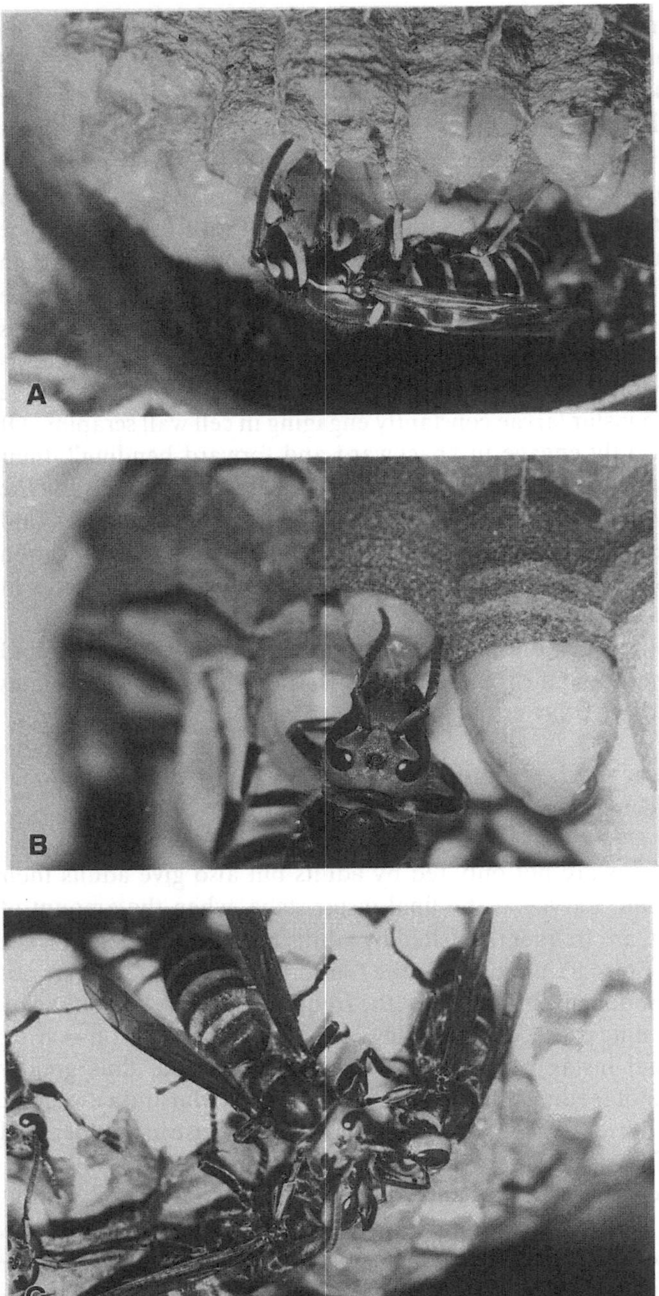

Fig. 4.8. Trophallaxis among colony members. **A** *Vespula flaviceps* foundress taking salivary secretion from a larva; **B** *Vespa tropica* worker taking salivary secretion from a larva; **C** *V. tropica* new queens soliciting a worker for salivary secretion

as a means of food preservation for the whole colony, since the larval secretion contains 9% sugars, including trehalose and glucose, the concentration of which is four times higher than that in the body fluid. Maschwitz has calculated that 1 ml of *Vespula* larval secretion could provide energy for 1.8 h of activity by a worker and that, therefore, a single worker would be able to survive for about half a day with only one intake from one final instar larva. According to studies on the analysis of larval secretion in *V. orientalis* by Ikan et al. (1968) and Ishay and Ikan (1968), the secretion contains 7% sugars and 1.3% protein and free amino acids, and they have concluded that, although these concentrations are low, the contents are essential for adults as nutrients, especially for the development of queen ovaries. At the same time, they have suggested that sources of nitrogen metabolism in adults are dependent on this larval secretion, since larvae of *V. orientalis* contain proteinases while these enzymes are undetectable in adults. However, Grogan and Hunt (1977) have disproved the theory of Ishay et al., since they have clearly demonstrated the existence of proteinases in adults of *Vl. germanica* and *Vl. maculifrons.* Further investigations, therefore, will be needed to test this theory.

6 Daily Activity of Workers

The vespine wasps are diurnal except for the completely nocturnal *Provespa,* and the hornet *Vespa crabro,* some of whose activity is seen after dark. In this section, we provide examples of worker daily activity, observed in two *Dolichovespula saxonica* colonies in Sapporo, Hokkaido (Sk. Yamane unpublished). Figure 4.9 shows the change in worker activity level as indicated by the total number of wasps leaving and entering the nest per hour. In one colony observed on July 31 – August 1, 1974, the number of wasps reached a peak during early morning between 4:10–5:10, then decreased until midday, with a sudden increase (second peak) between 12:10–13:10, then decreased to zero at 20:20. In the other colony, observed on July 12, 1975, the peaks were seen in the early morning and just after midday, though these were not as clear as in the previous example. In both cases we could not find a clear correlation between temperature and the activity level of workers. Envelope construction was observed all through the day, but was more frequent in the morning.

The presence of the early morning peak has been reported in various species, including *Vespula vulgaris* (Potter 1964), *Vl. rufa* (Edwards 1980), *Vl. maculifrons* (Gaul 1952) and *V. simillima xanthoptera* (Iwata 1971b). One of the reasons for this peak may be that the increase in the larval food requirements after a night without food increases the labor demands of the colony until the demand is filled. Since different activity patterns have been reported for *V. orientalis* and *Vl. vulgaris,* several factors possibly contribute to determining the pattern.

In the case of the *D. saxonica* colonies studied in Sapporo, foraging started 30 min before sunrise. On July 12, 1975 when sunrise was at 4:06, a worker appeared on the envelope at 3:19, and walked about with her wings buzzing, then

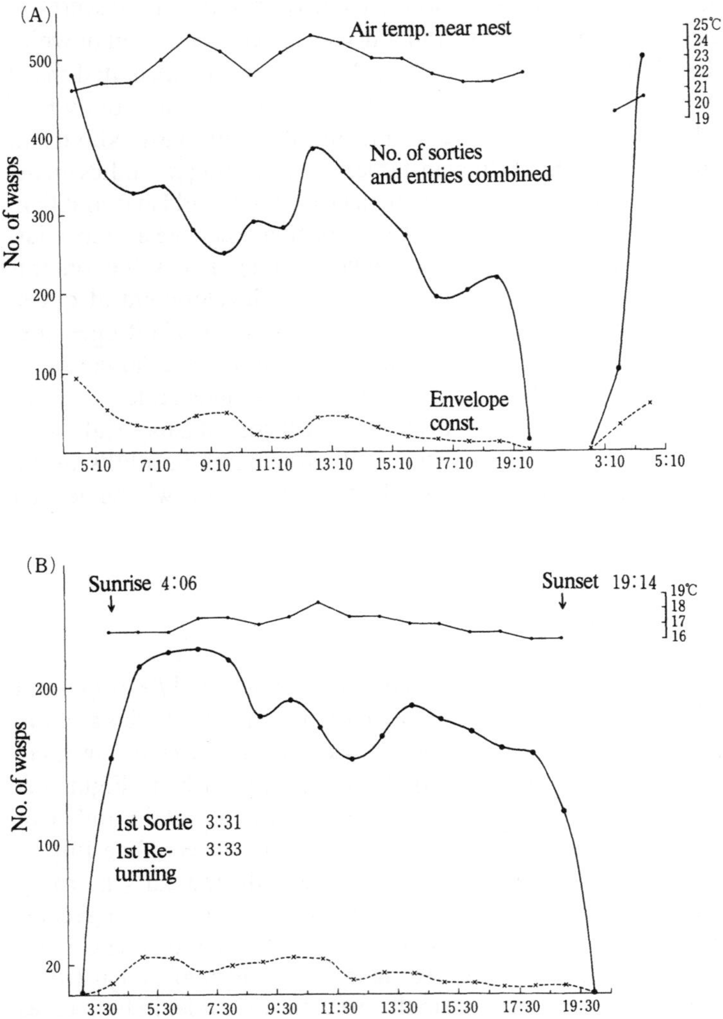

Fig. 4.9. Daily activities of *Dolichovespula saxonica* workers in Hokkaido. **A** July 31 to August 1, 1974, cloudy, sometimes with light rains; between 10:10–12:10 with some amount of rainfall. **B** July 12, 1975. Cloudy, later drizzling rain (Sk. Yamane unpubl.)

entered the nest. At 3:21 four workers appeared on the envelope. Among them was one individual that walked about for some time with her wings buzzing, entered the nest, then reappeared, and left the nest at 3:31 (first sortie). Similar phenomena were observed in the other colony in 1974. The count of workers leaving and entering the nest was made mainly based on the hum of wasps' wings and the noise of landing, because it was still too dark to see. When the sky was turning slightly gray, the returning wasps landed on lighter parts of the envelope.

Pulp collection started just after sunrise. It was not clear what were the purposes of the flights prior to sunrise. The extranidal activity nearly ceased approximately 30 min after sunset, and as in case of the sudden rising in early morning, the cessation of activity came suddenly during the last hour. The deciding factor that affects the beginning and cessation of flight activity seems to be light intensity, although moonlight as well as sunlight is possibly responsible for flight behavior (Edwards 1980).

Chapter 5

Foraging Behavior

The food of vespine wasps collected in the field consists of two defined classes; liquids and solids. The former is carbohydrate food consumed mainly by adults, while the latter consists of so-called "flesh pellets" eaten entirely by larvae. Vespine wasps, which occupy the highest rank in the food web among insects, utilize many different food sources and have evolved various strategies to collect them, thus each species can be classified as either a "generalist" or a "specialist" in terms of food preference.

1 Food Sources and Feeding Habits

1.1 Carbohydrate Sources

Carbohydrates are obtained from tree sap, honeydew secreted by hemipterans, flower nectar, ripe fruit, mushrooms and manufactured sweets such as syrup juice and soft drinks left in empty tins and bottles, etc. Such sources observed so far which are utilized by *Vespa* are listed in Table 5.1 (Matsuura 1984). The most important carbohydrate source common to all species of *Vespa* is the sap which exudes from borer-made apertures in living trees, especially *Quercus* spp., including *Q. acutissima* Carrath. and *Q. serrata* Thunb., which are widely distributed in great numbers within the same habitat as *Vespa* (Fig. 5.1 D). The relative frequency of visits to tree sap sources by the five sympatric *Vespa* species in Honshu and southward in Japan occurs in the following order: *mandarinia* > *crabro* > *analis* > *tropica* > *simillima*. Within the distribution area of *V. mandarinia*, however, apertures rich in tree sap are generally occupied solely by this species. As will be explained later, this can be explained by the fact that *Vespa* has a remarkable habit of monopolizing a food source by chasing off all visitors, not only other *Vespa* species but any other kind of insect, including beetles, butterflies and moths, and especially because *V. mandarinia* occupies the highest position in terms of interspecific relationships. Tree sap-dependency seems to be rare in *Vespula* and *Dolichovespula* in comparison with *Vespa*, since only one instance of sap-taking has been observed in *Vl. flaviceps* workers (Matsuura unpubl.)

Dependency on flower nectar is low in *Vespa* species and flowers available for them are limited to only a few species, probably due to the small amount of nectar collected per flower, which is insufficient for the nutritional requirements of the large body of the *Vespa*, in addition to the broad head and short tongue, which

Table 5.1. Carbohydrate sources of *Vespa* species (Matsuura 1984)

Sources	Species[a]	*Vespa* species[b]				
		Sim	Cra	Ana	Man	Tro
Tree sap	Fam. Salicaceae					
	1. *Salix gracilistyla* Miq.	○	○	○	○	○
	Fam. Fagaceae					
	2. *Quercus serrata* Thunb.	◎	◎	◎	◎	◎
	3. *Q. acutissima* Carrath.	◎	◎	◎	◎	◎
	4. *Q. glauca* Thunb.	◎	◎	◎	◎	◎
	5. *Shiia cuspidata* Makino	○	○	○	○	○
	Fam. Ulmaceae					
	6. *Ulmus parvifolia* Jacq.	○	○	○	○	○
	7. *Selkova serrata* Makino	○	○			
	Fam. Rutaceae					
	8. *Citrus unshiu* Marcov.	○			○	
	Fam. Euphorbiaceae					
	9. *Mallotus japonicus* Muel. Arg.	○				
Flower nectar	Fam. Berberidaceae					
	10. *Mahonia fortunei* Fedde	◎		◎	○	
	Fam. Vitaceae					
	11. *Cayratia japonica* Gagn.	○	◎	◎	◎	◎
	12. *Ampelopsis brevipedunculata* Trantv.	◎	◎	○	◎	◎
	Fam. Theaceae					
	13. *Camellia japonica* L.	○				
	14. *C. sasanqua* Thunb.	◎		◎	◎	
	Fam. Umbelliferae					
	15. *Foeniculum vulgare* Mill.					◎
	16. *Angelica polyclada* Franch.	◎	◎	◎		
	Fam. Gentianaceae					
	17. *Gentiana scabra* Bunge	○				
Ripe fruit	Fam. Moraceae					
	18. *Ficus carica* L.	◎	◎	◎	◎	◎
	Fam. Anonaceae					
	19. *Asimina triloba* Dual	○			○	○
	Fam. Rosaceae					
	20. *Prunus persica* Batsch.		○		○	○
	21. *Pyrus serotina* Rehder	○		○	○	○
	Fam. Vitaceae					
	22. *Vitis vinifera* L.	◎			◎	◎
	Fam. Ebenaceae					
	23. *Diospyros kaki* Thunb.	○		○	○	○
	Fam. Solanaceae					
	24. *Lycopersicon esculentum* Mill.		○		○	
	Fam. Cucurbitaceae					
	25. *Citrullus battich* Forskål.	○				
	Fam. Bromeliaceae					
	26. *Annas comosus* Merr.				○	

Table 5.1. *(Continued)*

Sources	Species[a]	*Vespa* species[b]				
		Sim	Cra	Ana	Man	Tro
	Insecta					
	Fam. Coccidae					
	27. *Ceroplastes rubens* Maskell			○		
	Fam. Psyllidae					
Honeydew	28. *Psylla tobirae* Miyatake			○		
	Fam. Aphididae					
	29. *Lachnus tropicalis* Goot			◎		
	30. *Toxoptera citricidus* Kirkaldy	○		○		
	31. *T. odinae* Goot			○		
	Gasteramyctes					
Mushroom	Hymenogastrales					
	32. *Kobayasia nipponica* Imai et Kawam.			◎	◎	◎
Manufactured sweets	33. Syrup juice	○	○	○	○	○

[a] Scientific names of plant species from Makino's new illustrated flora of Japan (Makino et al. 1961).
[b] Sim: *V. simillima,* Cra: *V. crabro,* Ana: *V. analis,* Man: *V. mandarinia,* Tro: *V. tropica.*
◎ :important food sources.

make it difficult for them to suck up nectar. *Camellia* spp., which have flowers large enough to allow nectar collection, however, are used by workers and males in the fall and also by overwintered *Vespa* queens as well as those of *Vespula* and *Dolichovespula.* Flowers of *Cayratia japonica* Gagn. in summer and *Angelica polyclada* Franch. in the fall are frequently visited by vespine wasps for nectar collection, since the flowers are dish-shaped, with plenty of exposed nectar. The latter flower is visited especially frequently by males of *Vespula* and *Dolichovespula* as well as by the workers, which not only suck up the nectar but also prey on various flower-visiting dipterans.

The juices of well-ripened fruits are also consumed in great quantities, not only by *Vespa* but also by *Vespula* and *Dolichovespula.* As will be described later, figs, grapes, peaches, etc., the pericarps of which are soft and have a high sugar content, are sometimes damaged severely by *Vespa* (Fig. 5.1B). The wasps gnaw the sarcocarp of these fruits in situ, and the juices are sucked up and carried back to the nests.

Honeydew, which is produced by hemipterans as a sweet secretion, is sometimes consumed by overwintered queens as a food source in the case of *Vespa* (Fig. 5.1A), but it is used more frequently by queens, workers and males of *Vespula* and *Dolichovespula* as their most important source of carbohydrate (Duncan 1939; Sk. Yamane and Kamijo 1976; Edwards 1980).

As for mushrooms, *Kobayasia nipponica* Imai et Kawam. in the fall serves as an important food source for various *Vespa* species. This mushroom grows in a bun-like shape (3–6 cm in diameter) on the ground surface in woods and attracts

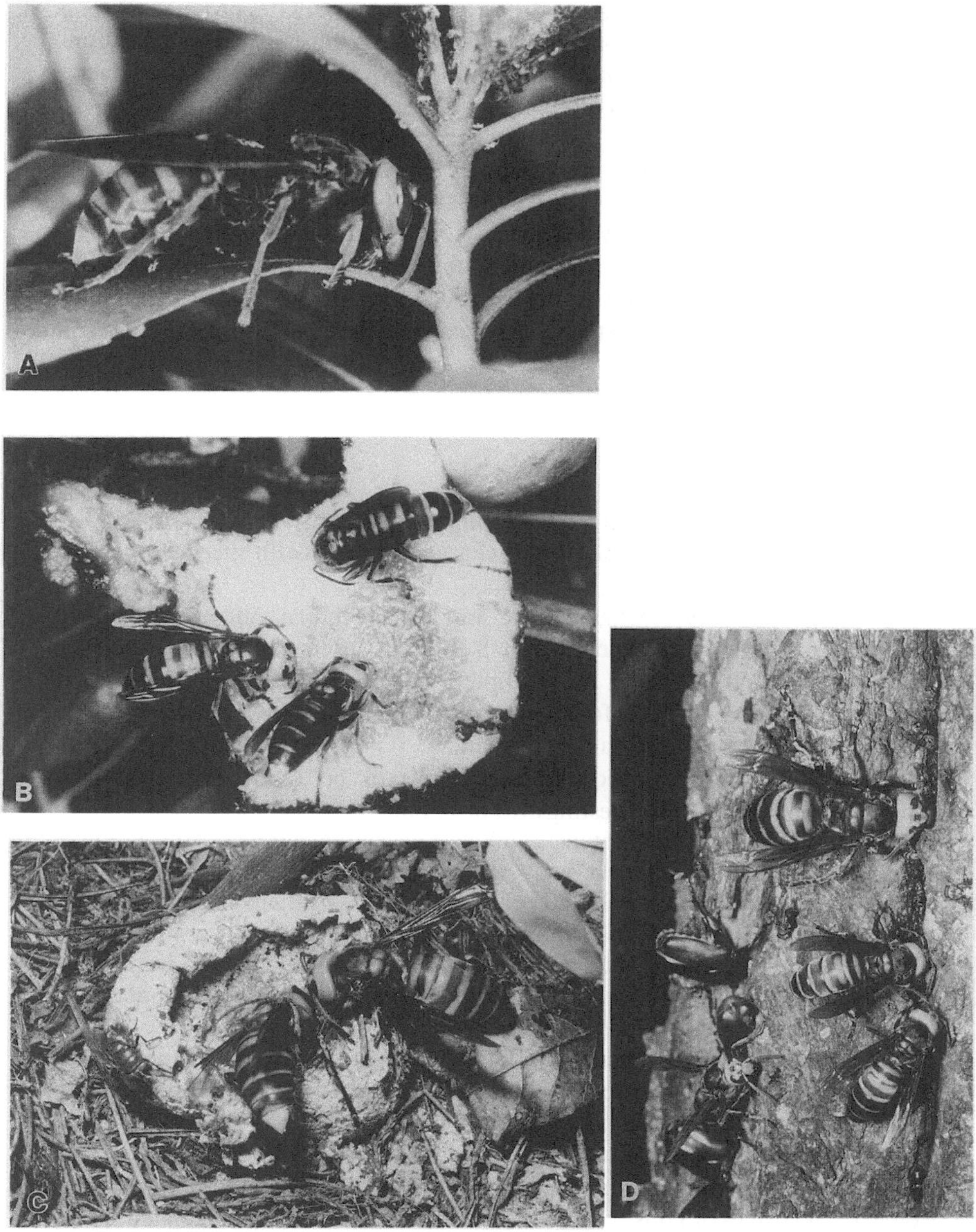

Fig. 5.1. **A** *Vespa analis* foundress licking aphid honey dew; **B** *V. mandarinia* injuring figs; **C** *V. mandarinia* feeding on a mushroom, *Kobayasia nipponica*; **D** uninseminated queen, and workers of *V. mandarinia* visiting tree sap

not only workers but also males of *Vespa* (Fig. 5.1C). No males of *V. mandarinia* have been observed feeding on any foods other than this mushroom (Matsuura unpubl.). Wasps feed on the mushroom's spores, which have nutritional value as protein rather than as carbohydrate (Naohiko Sagara pers. comm.).

Syrup juice and traces of soft drinks in empty tins and bottles left on the ground are used frequently as food sources by *Vespa, Vespula* and *Dolichovespula*. In Japan these beverages started to attract vespine wasps in 1970 when synthetic sweeteners, such as sodium cyclohexylsulfamate, were replaced by natural ones including sugar. Since then, however, gatherings of vespine wasps for juice collection around refuse heaps in such locations as holiday resorts have become a social problem, and the wasp has come to be regarded as a nuisance and an insanitary insect.

1.2 Protein Sources

Most of the protein food collected by vespine wasps consists of insects and spiders (Table 5.2), but prey preference differs characteristically with species, especially in *Vespa*.

V. simillima is a typical "generalist" and tends to hunt a wide range of insects and spiders. More than 44 species from 8 different orders have been observed as its prey, about 60% of them being Diptera adults. Of these, the frequency of captured adults occurs in the following order: *Platypleura kaempferi* Fabricius > *Apis mellifera* L. > Muscidae > Syrphidae > Sarcophagidae > *Sympetrum frequens* Selys. Winged insects are higher in rank because they are the most frequent inhabitants of the nesting area of *V. simillima*. The prey also include the larvae and pupae of Lepidoptera, adults of Locustidae and larvae of Mantidae. *V. simillima* has the smallest body size among the Japanese *Vespa* species. However, its movement is swift and agile, and by catching them with its mandibles by direct body attack, it kills various prey which either suck flower nectar or fly about. Its hunting area is wide and three-dimensional, since it also preys in the same manner on the leaf-feeding beetle, the cicada, etc., by flying about among the branches of trees in woods and thickets.

V. analis also tends to take various insects as a "generalist". More than 53 species of insects from 8 orders have been observed as its prey. The predominant prey at the order level are characteristic and different from those of *V. simillima*, since 32.2% of its prey are various species of wasps and bees belonging to the Aculeata of Hymenoptera in addition to Diptera (37.9%). In particular, it preys on various species of Polistinae belonging to the Vespidae, on the related species *Vl. flaviceps* and even on a species of the same genus, *V. simillima*. For these species, therefore, *V. analis* is not only a competitor for common food sources but also a predator. These social wasps which are potential prey, however, are preyed upon in the field after leaving their nests, and no observation has been reported of a direct attack on their nests or predation on their larvae and pupae, although this is common in *V. tropica* and *V. mandarinia*, as will be described later. *V. analis*, like *V. simillima*, also flies actively around various flowers which are rich in

Table 5.2. Prey of *Vespa* species (Matsuura 1984)

Prey species	Stage of prey[a]	Frequency of observation
V. simillima		
Insecta		
Orthoptera		
Fam. Locustidae		
1. *Parapodisma mikado* Bolivar	L,A	6
2. *Atractomorpha bedeli* Bolivar	A	12
Fam. Tettigoniidae		
3. *Phaneroptera fakata* Poda	L,A	9
4. *P.* sp.	L,A	8
Mantodea		
Fam. Mantidae		
5. *Paratenodera aridifolia* Stoll	L,A	5
6. Gn. sp.	L,A	12
Odonata		
Fam. Libellulidae		
7. *Orthetrum triangulare* Selys	A	2
8. *Sympetrum frequens* Selys	A	45
9. Gn. spp.	A	31
Hemiptera		
Fam. Cicadidae		
10. *Graptopsaltria nigrofuscata* Motschulsky	A	3
11. *Platypleura kaempferi* Fabricius	A	102
12. *Meimuna opalifera* Walker	A	25
Lepidoptera		
Fam. Noctuidae		
13. Gn. spp.	L,A	38
Fam. Nymphalidae		
14. *Vanessa indica* Herbst	A	2
Coleoptera		
Fam. Scarabaeidae		
15. *Popillia japonica* Newman	A	25
Fam. Chrysomellidae		
16. Gn. spp.	A	8
Fam. Attelabidae		
17. *Macorhis ursulus* Roelofs	A	1
Hymenoptera		
Fam. Vespidae		
18. *Eumenes* spp.	A	11
19. *Vespula flaviceps* Smith	A	14
Fam. Apidae		
20. *Apis mellifera* L.	A	> 100
21. *A. cerena* Fabricius	A	24
Diptera		
Fam. Stratiomyidae		
22. *Ptecticus aurifer* Walker	A	34
23. *Hermetia illucens* L.	A	15
Fam. Tabanidae		
24. Gn. spp.	A	67
Fam. Asilidae		
25. *Choerades* sp.	A	34

Table 5.2. *(Continued)*

Prey species	Stage of prey[a]	Frequency of observation
Fam. Bombyllidae		
26. Gn. spp.	A	26
Fam. Syrphidae		
27. *Chrysotoxum* spp.	A	32
28. *Eristalis cerealis* Fabricius	A	18
29. *Lathyrophthalmus* sp.	A	7
30. *Eristalomyia tenax* L.	A	88
31. *Megaspis zonata* Fabricius	A	6
32. *Eumerus* spp.	A	32
33. *Epistrophe balteata* de Geer	A	7
34. *Syrphus* spp.	A	43
Fam. Muscidae		
35. *Muscina* spp.	A	95
36. Gn. spp.	A	87
Fam. Anthomyiidae		
37. *Polietes lardaria* Fabricius	A	15
38. Gn. spp.	A	38
Fam. Calliphoridae		
39. *Calliphora lata* Coquillet	A	41
40. *Lucilia caesar* L.	A	19
41. Gn. spp.	A	54
Fam. Sarcophagidae		
42. *Boettacherisca* sp.	A	28
Fam. Scatophagidae		
43. *Scathophaga stercorarium* L.	A	12
Arachnida		
Araneae		
44. Gn. spp.	A	55
V. analis		
Insecta		
Orthoptera		
Fam. Locustidae		
1. *Atractomorpha bedeli* Boliver	L	3
Mantodea		
Fam. Mantidae		
2. *Paratenodera aridifolia* Stoll	L,A	18
3. *Acromantis japonica* Wastwood	A	15
Odonata		
Fam. Libellulidae		
4. *Sympetrum frequens* Selys	A	25
Hemiptera		
Fam. Cicadidae		
5. *Meimuna opalifera* Walker	A	19
Fam. Coreidae		
6. *Riptortus clavatus* Thunberg	A	2
7. Gn. spp.	A	4
Lepidoptera		
Fam. Sphingidae		
8. *Sphinx ligustri* Butler	L	1

Table 5.2. *(Continued)*

Prey species	Stage of prey[a]	Frequency of observation
Fam. Noctuidae		
9. Gn. spp.	L	33
Coleoptera		
Fam. Harpalidae		
10. *Chlaenius* sp.	A	2
11. Gn. spp.	A	3
Fam. Scarabaeidae		
12. Gn. spp.	A	18
Fam. Tenebrionidae		
13. *Plesiophthalmus nigrocyaneus* Motschulsky	A	3
Hymenoptera		
Fam. Tiphiidae		
14. *Tiphia* spp.	A	5
Fam. Colletidae		
15. *Colletes* spp.	A	5
Fam. Andrenidae		
16. *Andrena* spp.	A	7
Fam. Halictidae		
17. *Halictus* spp.	A	8
18. *Lasioglossum* spp.	A	9
Fam. Megachilidae		
19. *Megachile xanthothrix* Yasumatsu et Hirashima	A	3
20. *M. tsurugensis* Cockerell	A	2
21. *M.* spp.	A	5
Fam. Apidae		
22. *Tetralonia* sp.	A	3
23. *Apis mellifera* L.	A	33
24. *A. cerana* Fabricius	A	12
Fam. Sphecidae		
25. *Larra amplipennis* Smith	A	2
26. *Sceliphron inflexum* Sickmann	A	1
27. *Cerceris* spp.	A	3
Fam. Vespidae		
28. *Eumenes* spp.	A	7
29. *Orancistrocerus drewšeni* Saussure	A	5
30. *Polistes snelleni* Saussure	A	2
31. *P. japonicus* Saussure	A	3
32. *P. mandarinus* Saussure	A	3
33. *P. jadwigae* Dalla Torre	A	20
34. *P. chinensis* Fabricius	A	17
35. *Parapolybia indica* Saussure	A	6
36. *Vespa simillima* Smith	A	3
37. *Vespula flaviceps* Smith	A	23
Diptera		
Fam. Stratiomyidae		
38. *Ptecticus aurifer* Walker	A	45
39. *Hermetia illucens* L.	A	26
Fam. Asilidae		
40. *Choerades* sp.	A	13
41. Gn. sp.	A	15

Table 5.2. *(Continued)*

Prey species	Stage of prey[a]	Frequency of observation
Fam. Syrphidae		
42. *Eristalomyia tenax* L.	A	18
43. *Syrphus* spp.	A	12
44. Gn. spp.	A	19
Fam. Muscidae		
45. *Musca* spp.	A	23
46. Gn. spp.	A	10
Fam. Calliphoridae		
47. *Calliphora lata* Coquillet	A	9
48. *Lucilia caesar* L.	A	4
49. Gn. spp.	A	18
Fam. Sarcophagidae		
50. *Boettcherisca* spp.	A	3
51. Gn. sp.	A	3
Fam. Scatophagidae		
52. *Scathophaga stercorarium* L.	A	2
Arachnida		
Araneae		
53. Gn. spp.	L,A	28
V. crabro		
Insecta		
Orthoptera		
Fam. Tettigoniidae		
1. *Holochlora* sp.	A	3
Odonata		
Fam. Gomphidae		
2. *Sieboldius albardae* Selys	A	1
Hemiptera		
Fam. Cicadidae		
3. *Cryptotympana japonensis* Kato	A	145
4. *Graptopsaltria nigrofuscata* Motschulsky	A	34
5. *Platypleura kaempferi* Fabricius	A	107
6. *Meimuna opalifera* Walker	A	88
Hymenoptera		
Fam. Vespidae		
7. *Polistes mandarinus* Saussure	A	4
Fam. Apidae		
8. *Apis mellifera* L.	A	10
V. mandarinia		
Insecta		
Mantodea		
Fam. Mantidae		
1. *Paratenodera aridifolia* Stoll	L,A	12
2. *Hierodula patellifera* Servile	L,A	9
3. Gn. spp.	A	6
Lepidoptera		
Fam. Papilionidae		
4. *Papilio xuthus* L.	L	13

Table 5.2. *(Continued)*

Prey species	Stage of prey[a]	Frequency of observation
Fam. Sphingidae		
5. Gn. spp.	L	18
Coleoptera		
Fam. Scarabaeidae		
6. *Anomala cuprea* Hope	A	107
7. *A. albopilosa* Hope	A	15
8. *A.* spp.	A	132
9. *Mimela splendens* Gyllenhal	A	54
10. *Mimela* spp.	A	32
11. *Lachnosterna* spp.	A	24
Hymenoptera		
Fam. Apidae		
12. *Apis mellifera* L.	L,P,A	> 1000
13. *A. cerena* Fabriculus	L,P,A	50
Fam. Vespidae		
14. *Polistes mandarinus* Saussure	L,P,A	22
15. *P. jadwigae* Dalla Torre	L,P,A	54
16. *P. rothneyi* Cameron	L,P,A	27
17. *Vespa simillima* Smith	L,P,A	352
18. *V. crabro* L.	L,P,A	38
19. *V. analis* Fabricius	L,P,A	129
20. *V. tropica* L.	L,P,A	11
21. *Vespula flaviceps* Smith	L,P,A	178
Arachnida		
Araneae		
Fam. Argiopidae		
22. *Argiope amoena* Audouin	L,A	38
23. *A. bruennichii* Scopoli	L,A	25
V. tropica		
Insecta		
Hymenoptera		
Fam. Vespidae		
1. *Polistes snelleni* Saussure	L,P	20
2. *P. japonicus* Saussure	L,P	26
3. *P. mandarinus* Saussure	L,P	12
4. *P. jadwigae* Dalla Torre	L,P	358
5. *P. rothneyi* Cameron	L,P	34
6. *P. chinensis* Fabricius	L,P	54
7. *Parapolybia indica* Saussure	L,P	42
8. *P.* sp.	L,P	23

[a] L: larva; P: pupa; A: adult.

visiting insects, makes a direct body attack on a prey and catches it with its mandibles. It also preys on the larvae of noctuid and notodontid moths by flying about in bushes and woods. It seems that many more species of insect than those shown in the table are preyed upon by *V. analis*, since many lepidopteran larvae are carried into its nest as crushed flesh pellets and they cannot, from their appearance, be identified by scientific name.

According to an examination carried out in the southwest district of Japan, *V. crabro* has a characteristic prey preference which is not found in any other species of the genus; more than 95% of its prey consists of cicadas (Fig. 5.2). This

Fig. 5.2. *Vespa crabro* worker hunting a cicada, *Graptopsaltria nigrofuscata* (*A*), and a pellet of a killed cicada brought in the nest (*B*)

species has a special technique for catching large-sized cicadas, such as *Cryptotympana japonensis* Kato, which cannot be used as prey by any other *Vespa* species (Matsuura 1977b). This species can be considered to be the most cicada-preferring type among the *Vespa* species, because its subspecies, *V. c. germana,* introduced in North America, is also known to capture large-sized cicadas (*Tibicen lyricen* and *T. chloromera*; Davis 1925). This species can be called a "semi-specialist" because it has the ability to change its prey preference to large-sized dragonflies and grasshoppers during seasons when cicadas are lacking.

As an example of cicada utilization as a food source of *V. crabro,* one observation of a nest of this species with about 120 workers revealed that 37 loads of cicada flesh pellets (estimated to be equivalent to at least 12 adult individuals of *Cryptotympana japonensis*) had been brought to the nest by 5 marked workers within 3 h on August 10, 1973, in Kibi-chô, Wakayama Prefecture. In another nest with about 240 workers, observed in the same district on August 6, 1968, 53 (71.6%) of 74 loads examined at the nest entrance were cicada flesh pellets, mainly *C. japonensis* (Matsuura unpubl.). Since cicadas are large in size and muscular, and the population density is high in certain regions, this makes them an important food source for certain vespine wasps which are particularly good at capturing them. The active nesting period of *V. crabro* (July - September) corresponds to the occurrence of Japanese cicadas, the cycle for the breeding of new queens in August and September being synchronized with the occurrence of the cicadas.

V. tropica is a typical "specialist" which depends almost exclusively on obtaining the larvae and pupae of various polistine species for its protein food by attacking their nests. In Japan, *Polistes, Parapolybia* and *Ropalidia* are preyed upon by this species. However, it does not attack hives of the honeybees, *Apis* spp., in the field. It sometimes visits apiaries, but only when attracted by the odor of honey stocks. The attacking of polistine wasps is a unique habit, and a detailed account will be given later.

V. mandarinia is considered to be a "semi-specialist", because its main protein food is the adults of some beetles such as *Anomala* spp.. These beetles sometimes comprise 60–70% of the flesh pellets brought to the nest (Matsuura unpubl.). This species is not very skillful at capturing agile insects in comparison with *V. simillima* and *V. analis,* because of its large body, but it preys on slow-moving arthropods including large larvae of Lepidoptera and large web-weaving spiders such as *Argiope amoena* (Fig. 5.3A). Medium-sized beetles having a hard exoskeleton, such as *Anomala* spp., are used hardly at all by other *Vespa* species. *V. mandarinia,* however, turns these beetles into flesh pellets in situ easily, because of its ability to crush to pieces, with its well-developed mandibles, the exoskeleton along with the muscles. However, the most remarkable trait of this large-sized vespine wasp is not only its solitary hunting but also a special hunting strategy which is not found in any other social wasp. In this behavior, this species attacks nests of other species of vespine wasps, including *Vespa* and *Vespula,* and those of the honeybee *Apis* spp., in a group, annihilates the inhabitants of the nest and carries back most of the larvae and pupae present

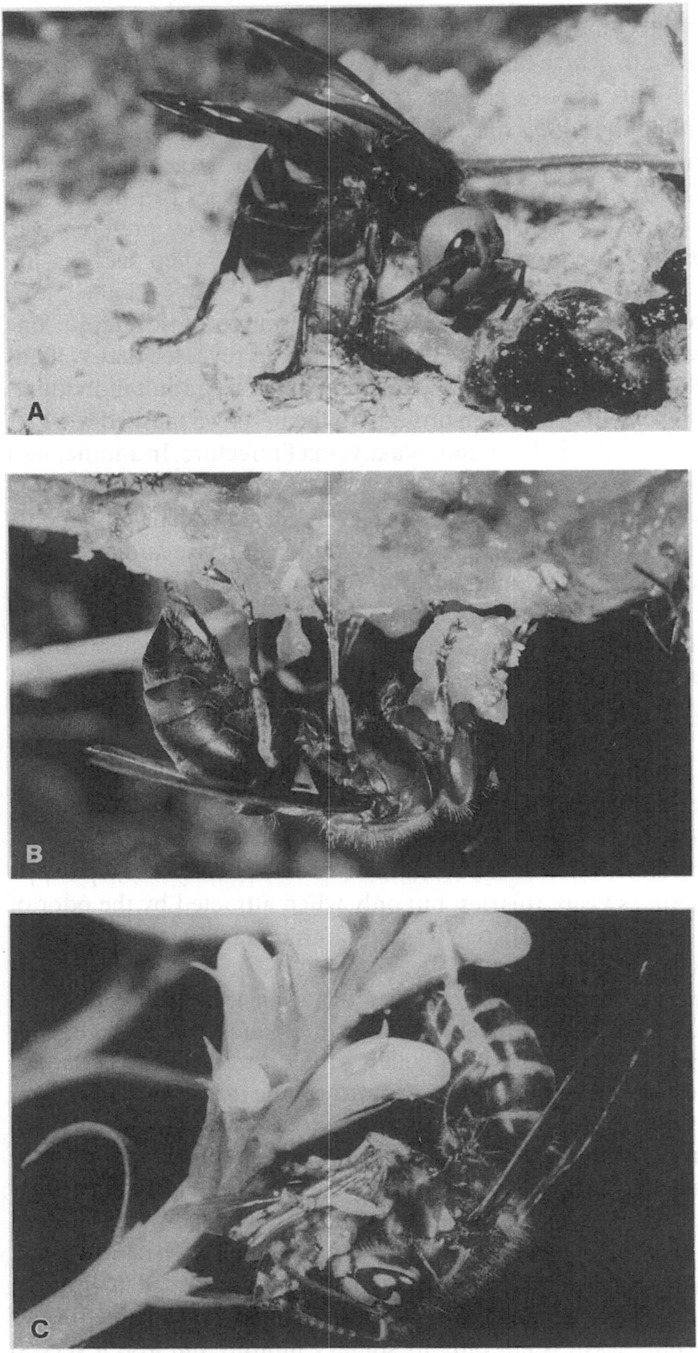

Fig. 5.3. **A** *Vespa mandarinia* worker hunting a lepidopterous larva; **B** *V. dybowskii* worker handling frog flesh; **C** *Dolichovespula media* worker having captured a dipteran

in the nest to its own nest. Details of this hunting strategy will be described later.

Information on the prey of *V. dybowskii* is rare, and only two observations on a species of dragonfly (*Sympetrum frequens* Selys) and a tettigoniid (*Ducetia japonica* Thunberg) have been recorded. The most remarkable habit of this wasp is that it carries back even the muscles of frogs as a food source, and this has rarely been observed for other *Vespa* species (Fig. 5.3B; Matsuura unpubl.). Consequently, it is possible that this species has a wider variety of prey menu in comparison with those of *V. simillima* and *V. analis.*

As for the prey of *Vespula,* Shida (1963) has confirmed 37 species belonging to the Hemiptera and Lepidoptera which were preyed upon by *Vl. flaviceps,* and 17 more species were added later by Iwata (1971a), including Tettigoniidae (Orthoptera), Coenagrionidae (Odonata), Formicidae (Hymenoptera) and spiders. Among these prey, middle- and small-sized flies and horseflies belonging to the Diptera are captured most frequently throughout the nesting period (Matsuura unpubl.). With regard to the developmental stage of the prey, adults and larvae are captured in the case of Lepidoptera (Okutani and Tomomatsu 1951), but the prey seem to be more numerous in various orders in the case of small-sized insects and spiders, which have a relatively soft exoskeleton. An observation has been reported of a worker of *Vl. shidai* collecting flesh from an injured barnacle (Sk. Yamane et al. 1980b). A remarkable habit of the subgenus *Paravespula,* including these two *Vespula* species, is that they use not only the above-mentioned arthropods but also the muscles of vertebrates, such as frogs, snakes and fish, as food sources. They have been known to carry back not only raw meat but also grilled fish (Iwata 1971a) and boiled egg (Duncan 1939) served at table. In some coastal districts, they are disliked as vermin because they tear the meat off fish and squid hung out to dry. With regard to *Vl. vulgaris* of the same subgenus, Broekhuizen and Hordijk (1968) have observed more than 30 species of insect and 8 species of spider as its prey in Holland, including lepidopterans, dipterans and sawfly larvae. The dominant prey were adults of dipterans, occupying 83.5% of collected prey in one of the nests observed. In one nest, however, spiders have been observed to comprise 44.1% of the prey. According to Kemper and Döhring (1962), analysis of prey obtained from a nest of *Vl. germanica* showed that 63% of the prey were dipterans, mainly *Musca domestica* L., while 14% were lepidopterans.

In *Vl. schrenckii,* only one example has been observed of predation on larvae of a moth of the noctuid subfamily Plusiinae. When the flesh of a dead insect or frog is given artificially, this species neither makes a flesh pellet from it nor brings it to the nest (Matsuura unpubl.). *Vl. acadica* and *Vl. atropilosa,* which belong to the *Vl. rufa* group in North America, prey only on live arthropods, including lepidopteran larvae, dipterans and hemipterans (MacDonald et al. 1974; Akre et al. 1981). It seems that the range of prey preference in this subgenus is narrow in comparison with the subgenus *Paravespula.*

In *Dolichovespula,* only two observations have been made in Japan of *D. media* preying on the adult forms of a small noctuid moth and muscid fly, while *D. saxonica* preys on houseflies (Matsuura unpubl.). Neither species, however.

carries the meat of frogs, etc., back to its nest as food (Matsuura unpubl.). In contrast, *D. maculata* in North America is known to feed on the meat of a dead snake (MacDonald et al. 1974) and to attack the hummingbird adult, *Selasphorus rufus* (Grant 1959), and *D. sylvestris* in England has been observed to gnaw meat off a chick just after it had hatched in the nest of a blackcap, *Sylvia atripilla* (Wild 1927).

1.3 The Search for Prey

Generally, *V. simillima* and *V. analis* search for prey while flying actively about, making a zig-zag course around dish- or brush-shaped flowers which insects including Diptera, etc., are visiting, and where they are sucking nectar. *V. mandarinia* and *V. crabro*, in contrast, search for prey by flying upward and downward in a grove or along the trunk of a big tree, but the former species starts to fly about along the eaves of houses after mid-August in search of the nests of *V. simillima*, etc. *V. tropica* also often flies about near eaves, presumably because polistine wasps frequently build their nests under the eaves, and these nests are a favorite food source for this species.

Vl. flaviceps moves in search of prey by flying quickly upward and downward along the stems and leaves of plants. *D. media*, on the other hand, stops for a moment in the air around dish- or brush-like upright flowers in search of prey and then moves to the next hunting place. Such a difference in prey-searching behavior between two species has also been observed between *Vl. vulgaris* and *D. sylvestris* in England (Edwards 1980). *Vl. schrenckii* and *Vl. vulgaris* generally search for prey by flying about slowly in a thicket close to the ground or along plant stems.

1.4 Capture and Treatment of Prey

When a species of *Vespa* (except for *V. tropica*), *Vespula* or *Dolichovespula* finds a prey, it flies at the prey directly from a distance of a few to a little over 10 cm and bites the prey with its mandibles. When *V. mandarinia* hunts a scarab (*Anomala cuprea*) on a leaf, it lands on the leaf and then bites the head of the prey directly. The cicada hunters, *V. simillima* and *V. crabro*, usually fly at a cicada standing still on a tree trunk and bite part of the prey's wings or body, but sometimes they chase a fleeing cicada and capture it in the air. Both the predator and prey drop to the ground, but *Vespa* bites the wings off the cicada immediately by biting the base of the forewings. It seems that *Vespa* does not use its poisonous sting even for the hunting of large-sized prey like cicadas.

After a winged insect has been captured, every wasp bites off the prey's wings first then each leg in turn, throwing these away at the capture site and usually carrying back only the thorax, which is muscular. When the prey are small, they are sometimes carried back to the nest with the remaining parts of the wings and/or legs still intact. Generally, every vespine wasp gives its prey this treatment after transferring the prey to another place distant from the hunting site, such as a tree branch. The wasp hangs on to the branch with its hind legs, turning its head downward, and deals with the prey by rotating it with the middle and fore-legs (Fig. 5.3C). When a large prey is captured, the wasp visits the hunting site several times to carry back the remaining parts of the prey in small pieces. The prey is always transferred to the nest by flying, holding the prey with the mandibles. The size of the prey carried back by a *Vespa* worker in one flight is usually 5–7 mm in diameter. The heaviest prey carried by *V. analis* and *V. mandarinia* workers was 0.215 g (48.1% of the weight of the *V. analis* worker) and 0.495 g (47.7% of the *V. mandarinia* weight) respectively (Matsuura 1973b). In other words, every worker of *Vespa* species can carry a load of nearly half its own body weight. Among the vespine wasps, only *V. tropica* transports its prey as liquid to its nest, by sucking up the body fluid from the larvae and pupae of polistine wasps and holding the fluid in its crop. In such cases, each worker can carry a maximum amount of fluid about 60% of its own body weight in one flight (Matsuura 1973b).

2 Dependence of *Vespa tropica* on Polistine Wasps

2.1 Objects of Attack

The following description is based mainly on a report by Matsuura (1984).

As described above, *Vespa tropica* can be designated as the most highly specialized "specialist" among the Vespinae, since it depends entirely on the larvae and pupae of Polistinae for its protein food. Three genera of Polistinae serve as prey in temperate regions, i.e., *Polistes* (Sakagami and Fukushima 1957b), *Parapolybia* (Sekijima et al. 1980) and *Ropalidia* (de Ruitter 1916). Stenogastrinae can be added to the list of prey in the tropical regions of Southeast Asia (Williams 1919; Ward 1965), which is the main distribution area of *V. tropica*. In this area, it is probable that the prey menu consists of all the species belonging to the Polistinae, including *Polybioides,* which inhabits the same area as *V. tropica*. Although no observation has been reported of any attack by this species on the nests of other vespine wasps, there has been an example of *V. tropica* pulling pupae from a nest of *V. simillima*, from which adults had already been evacuated due to an attack by *V. mandarinia*. The *V. tropica* took advantage of the occupant's (*V. mandarinia*) state of unreadiness, transferred the pupae to a nearby place, sucked up their body fluids and carried them back to its nest (Matsuura unpubl.). This species feeds readily on any pupa of other *Vespa, Vespula* or *Dolichovespula* species when given artificially. *V. tropica*, however, hardly shows any interest if a comb bearing an *Apis* spp. brood is left in the field after removal of the adults. Consequently, its food preference is limited to the easily attackable lower social wasps among the Vespidae.

2.2 Collection of Prey

When *V. tropica* attacks an active nest of polistine wasps, the majority of the adults escape from the nest to places a little over 10 cm distant from the nest site, such as a tree branch or under the eaves of a house. However, the queen and, occasionally, a few workers remain behind at the back of the comb. *V. tropica* pays little attention to these remaining adults and first opens a cocoon with its mandibles, pulling out the pupa or prepupa. *V. tropica* seems to be able to judge the growth stage of the pupa in the cocoon by external examination with its antennae, because it first selects a cocoon which contains a prepupa or a young pupa, the body of which has not yet hardened, then a cocoon with a hard pupa and finally a larva. *V. tropica* hangs on to the comb with its two pairs of hind legs, retaining the prey in its fore-legs and mandibles. Then the wasp sucks all the fluid from the prey, turning it counter-clockwise with the fore-legs, and stores the fluid in her crop. After the contents have been swallowed, the victim is molded into a small pellet and thrown to the ground under the nest. A nest of polistine wasps invaded by *V. tropica* can be recognized at a glance, since the ground directly under such a nest is scattered with the ground-up remains of cocoons, together with pieces of flesh from the victims. In the case of large *Polistes* species in Japan, such as *P. jadwigae* and *P. rothneyi*, a single prepupa or pupa provides enough fluid to be carried away, but in smaller species such as *P. snelleni* and *P. chinensis*, the contents of two or three individuals are successively swallowed. The time spent in sucking up a victim's fluid in one attack has been estimated to be from 130 to 250 s (m = 185, n = 72). The remaining *Polistes* adults, the owners of the nest, stare at the activities of the invader with no sign of resistance. On very rare occasions, an owner wasp may move toward the *V. tropica* in the process of handling its victim, receive a piece of the victim's flesh and distribute it to other larvae in the nest (Matsuura unpubl.). Such unusual behavior seems to be related to the fact that *V. tropica* does not attack *Polistes* adults under any circumstances.

The polistine wasps, having escaped from the attacked nest, return after *V. tropica* has flown away, and resume their nesting activities. Once found, however, the nest is continuously visited by the same invader for several days until all the larvae and pupae in the nest are eaten up (Table 5.3) and, finally, the invader eats even the fourth and fifth instar larvae and slightly hardened pupae. The polistine wasps, on the other hand, continue nursing activities as long as eggs and young instar larvae remain in the nest.

Occasionally, a nest in which a *V. tropica* worker is handling its prey is visited by another individual of the same species but of a different nest. In such a case, combat starts between them, and the winner remains on the site. In all cases so far observed (> 400), polistine nests have always been attacked by a single individual of *V. tropica*, though Sakagami and Fukushima (1957b) state that a *V. tropica* worker often comes back accompanied by one or two nest-mates after the first visit. In addition, observations of the interiors of *V. tropica* nests have shown no evidence that a worker which has carried back prey has informed other workers of the location of the polistine nest, or that it has been accompanied by other workers in the next attack.

2.3 Phenology of Attack

Polistes starts its nest foundation in and after mid-April, in the warm districts of southwest Japan. The relationship between the developmental process of the nest and attack on the nest by *V. tropica* is shown in Fig. 5.4. *V. tropica* queens, on the other hand, emerge from their hibernacula during late May and early June, about one month later than other *Vespa* species. A *V. tropica* queen starts her first attack, prior to her nest foundation, on a *Polistes* nest in which a few workers are just emerging. The queen visits *Polistes* nests only one to two times a day, and her object of attack is the prepupa or young pupa which, presumably, enhances development of her ovaries as a protein source before nest foundation. The number of visits increases as the *V. tropica* nest is founded and eggs in the nest begin to hatch, and the number of attacks reaches its maximum when workers emerge in mid- to late July (Fig. 5.4). The foraging activities of the queen, however, decrease gradually from late July to mid-August, and then the activities are handed over to the workers. Since *V. tropica* workers engage in extranidal activities 2 to 3 days after emergence, their attacks on *Polistes* nests begin in late July and continue until rearing of the brood in *Polistes* nests is completed in late August to early September. In *V. tropica* nests, on the other hand, males and new queens are reared from early July, and these reproductives emerge in late August to mid-September. After mid-September, food sources of *V. tropica* change from polistine brood to carbohydrate, including tree sap, mushrooms and ripe fruit, in order to feed newly-emerged males and new queens. Thus the life cycle of *V. tropica* is synchronized with that of polistine wasps, which are *V. tropica*'s principal food source and, consequently, the cycle is the shortest among those of the Japanese *Vespa* species.

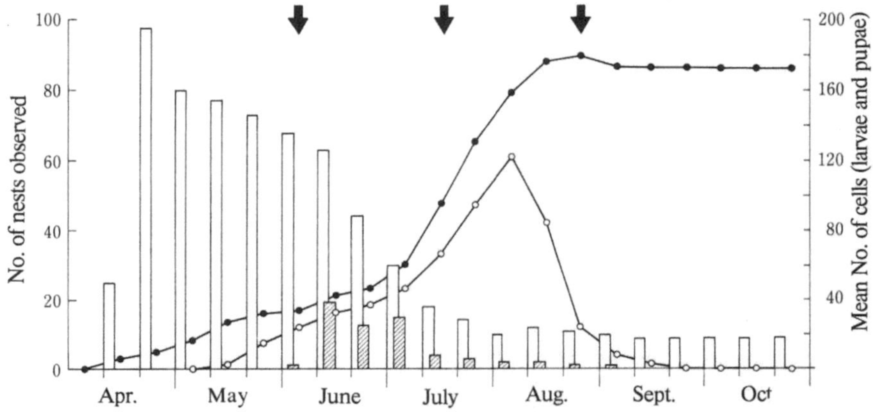

Fig. 5.4. Changes in number of active nests of *Polistes jadwigae (blank)* and those of nests attacked by *Vespa tropica (hatched)* in a 20 ha horticultural section in Wakayama. ● Mean cell number in *P. jadwigae* colonies; ○ mean number of larvae and pupae in *P. jadwigae* colonies (After Matsuura 1984)

2.4 Ratio of Polistes Nests Attacked by V. tropica

Table 5.3 shows the number of *Polistes jadwigae* nests attacked by *V. tropica*, examined over 7 years within an area of about 20 ha in Wakayama Fruit Tree Experimental Station (Kibi-chô, Wakayama Prefecture). Developmental steps of *P. jadwigae* nests were classified into three periods for discussion: (I) before the emergence of first workers, (II) thereafter until the emergence of males and new queens and (III) thereafter until disintegration of the nest. Only one nest (0.4%) was attacked during period (I) but the frequency reached 45.5% during period (II), although the frequency varied from 10.9% to 93.2% depending on the year examined. During period (III), 46.9% of the nests, on average, were attacked, but annual fluctuation in the frequency was also remarkably high. Such a fluctuation was considered to be related to the number of *V. tropica* nests, because a tendency for the frequency of attack on *Polistes* nests by *V. tropica* was found to be high in years when the density of *V. tropica* nests was high.

Table 5.3. Number and developmental period of *Polistes jadwigae* nests attacked by *Vespa tropica* within confines of Wakayama Fruit Tree Exp. St. (Matsuura 1984)

Year	Period[a] I No. nests attacked/observed (%)	II No. nests attacked/observed (%)	III No. nests attacked/observed (%)	Total No. nests through periods I-III attacked/observed (%)	No. nests[b] of *V. tropica*
1969	0/25 (0)	4/13 (30.8)	13/17 (76.5)	17/45 (37.8)	4
1970	0/27 (0)	5/46 (10.9)	2/16 (12.5)	7/58 (12.1)	1
1971	0/88 (0)	8/38 (21.1)	12/15 (80.0)	20/121 (16.5)	5
1972	0/18 (0)	16/31 (51.6)	9/34 (26.5)	25/47 (53.2)	2
1973	1/48 (2.1)	41/44 (93.2)	3/9 (33.3)	45/58 (77.6)	4
1974	0/22 (0)	9/14 (64.3)	7/12 (58.3)	16/39 (41.0)	2
1975	0/18 (0)	7/12 (58.3)	7/10 (70.0)	14/30 (46.7)	3
Total	1/246 (0.4)	90/198 (45.5)	53/113 (46.9)	144/398 (36.2)	21

[a] I: before emergence of first workers; II: thereafter till emergence of sexuals; III: thereafter till disintegration.

[b] Nests discovered within about 500 m from Wakayama Fruit Tree Exp. St.

2.5 Period of Predation on Polistes Nests

Once *V. tropica* finds a *Polistes* nest, it visits the same nest daily until all the larvae and pupae available are exhausted. An attacking queen visits the same *P. jadwigae* nest consecutively for 4.5 days on average, while workers take 7.2 days on average (Fig. 5.5; Table 5.4). The difference between predation periods of queens and workers is due to the difference in colony size of *P. jadwigae* at the time when they are attacked by *V. tropica*. That is, attack by *V. tropica* queens takes place mainly in colonies during the period of slow increase, when there is only a small brood available for prey. In contrast, workers take more time than queens to exhaust the prey, even though they prey upon 8.8 to 13.3 mature larvae and pupae per day in late July to early August, since the size of a *Polistes* colony in this period reaches its maximum for the year.

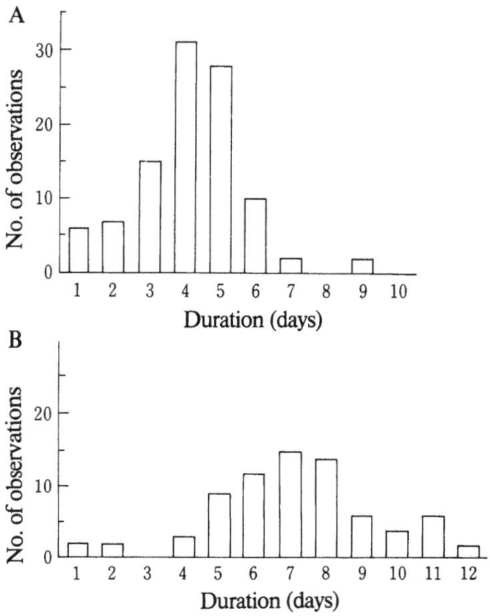

Fig. 5.5. A Duration (in days) of attacks by *Vespa tropica* queens, and **B** workers upon the same colonies of *Polistes jadwigae* (After Matsuura 1984)

Table 5.4. Some records of attack by queens and workers of *Vespa tropica* on *Polistes jadwigae* nests in 1973 (Matsuura 1984)

Nest code of *P. jadwigae*	No. of cells[a]	*V. tropica* No.[b]	Date of first attack	Date of final attack	Duration of attack (in days)	Total no. of brood eaten by *V.t.*[c]	No. of brood eaten/day
Pj 3007	27	Q 01	June 9	June 13	5	12	2.4
Pj 3012	28	Q 01	" 13	" 16	4	9	2.3
Pj 3015	31	Q 01	" 17	" 20	4	14	3.5
Pj 3017	32	Q 01	" 18	" 21	4	13	3.3
Pj 3019	28	Q 01	" 25	" 27	3	10	3.3
Pj 3026	30	Q 02	" 6	" 10	5	11	2.2
Pj 3027	36	Q 02	" 22	" 25	4	16	4.0
Pj 3028	52	Q 03	" 26	" 29	4	18	4.5
Pj 3035	159	W 01	July 27	Aug. 2	7	71	10.1
Pj 3041	207	W 02	" 29	" 4	7	93	13.3
Pj 3047	142	W 03	Aug. 2	" 7	6	53	8.8

[a] Number of cells in *P. jadwigae* nest on day when first attack by *V. tropica* was observed.
[b] Q: queen hornet; W: worker hornet.
[c] Number excluding those of eggs and I-III instar larvae.

2.6 *Fate of Attacked Polistes Colonies*

As the attacks of *V. tropica* are repeated until all the eggs, larvae and pupae are consumed, the nests finally disintegrate. When a polistine nest is attacked before emergence of males and new queens, the adult wasps finally abandon their original nest, and 66.7% (*P. jadwigae*) and 30% (*Parapolybia indica* Saussure; Sekijima et al. 1980) of the attacked colonies construct new nests near their original nests and start brood-rearing again (Table 5.5). The behavior of *P. jadwigae* and *Parapolybia indica* (Sekijima et al. 1980) during this period is as follows.

Table 5.5. Fate of *Polistes jadwigae* nests attacked by *Vespa tropica* (Matsuura 1984)

Period[a]	No. of nests attacked	Reconstruction at another site	Abandonment or disappearance	Aggregation around nest site
I	4	2(50.0%)	2(50.0%)	0
II	192	171(89.1%) (159)(82.8%)[b]	21(10.9%)	0
III	75	8(10.7%)	1(1.3%)	66(88.0%)
Total	271	181	24	66

[a] See Table 5.3.
[b] Number of nests reattacked by *V. tropica* after reconstruction.

After the start of attacks by *V. tropica,* the queen and workers stop their building activities, though they continue resting on the comb. After two to three attacks by *V. tropica,* some workers leave the nest and fly about in search of another nesting site nearby. When a suitable site is found, the scout workers settle at that place and spend most of their time resting on it, with occasional returns to the original nest. With the search for a new nest site starts, the workers in the original nest stop building activities and the queen also stops oviposition. The queen and workers then leave the nest individually in search of the new nest site, which is usually situated in a sheltered place, or in a bush where the new nest cannot be discovered easily, either by *V. tropica* or by polistine wasps. About one day after the arrival of the queen, the workers usually start building a new nest, and the queen begins to lay eggs as soon as the cells are constructed. The distance between the two nests is within 2 to 3 m, but reaches 4 m at maximum, both in *P. jadwigae* and *Parapolybia indica.* After they have constructed a new nest, most of the workers frequently visit the original nest over several days, though they never reuse the abandoned nest even if *V. tropica* stops attacking it. However, 93.0% (159/171 nests) of the nests reconstructed by *P. jadwigae* are discovered and reattacked by *V. tropica* before males and new queens have emerged, and the nests finally disintegrate. On the other hand, 88.0% of *P. jadwigae* nests attacked by *V. tropica* after emergence of males and new queens do not construct a new nest, and all the adults including workers, males and new queens remain on or near the empty nest and then leave the nest. In the case of *Parapolybia indica,* nine of 30 nests attacked by *V. tropica* from July to early August were reconstructed, and new queens emerged from some of the new nests (Sekijima et al. 1980).

2.7 Restriction of V. tropica Colony Size in Temperate Regions

V. tropica in temperate regions constructs the smallest nests among the species of Vespinae. The average number of cells in one nest is 165.8, and even the biggest nest so far observed contained only 313 cells (Matsuura 1984; cf. Chap. 6). Since Vespinae usually construct larger nests with many more workers than do Polistinae in the temperate regions, including *Polistes* and *Parapolybia*, such a dwarfing of colony size in *V. tropica* can be considered to be a specialized strategy related to its life cycle, and, from the viewpoint of adaptation, its food source is considered to be one aspect of this strategy. As can be seen clearly in Table 5.2, all the other *Vespa* species capture adults of polistine wasps and use them as food for their larvae, but differ from *V. tropica* in that they do not depend on larvae and pupae of polistine wasps as their sole protein source.

For *V. tropica,* polistine wasps as a food source can be rated highly on the basis of the following points: (1) larvae and pupae, a protein source with a high nutritional value, are located in one nest and at high density, (2) prey can be obtained with no loss of predators, since adults of the prey make hardly any counterattack and (3) the predation process, from the searching for prey to carrying it back to the nest, can be undertaken exclusively and without any competition, since attacks on the nests of polistine wasps for food by other *Vespa* species are quite rare. Although polistine wasps provide these advantages as a food source for *V. tropica,* the duration of brood-rearing by polistine wasps in temperate regions is short because they are originally inhabitants of the tropics; their brood rearing becomes maximal in July to mid-August and is completed in late August in most species of polistine wasp inhabiting southwestern Japan (Matsuura 1976b). As a consequence of the use of the brood of polistine wasps as a food source, the predator has to shorten its nesting period greatly by synchronizing it with the short rearing period of the polistine wasps. *V. tropica* in temperate regions does this by lowering its colony size to the level of that of polistine wasps. Competition with other *Vespa* species does not occur in respect to larval food sources, because emergence of new queens of *V. tropica* has already been completed when other vespine wasps are engaged in maximum brood rearing in and after September (cf. Chap. 6). At the same time, since lowering of the colony size increases predators' flexibility of response to food density, this small colony size will result in a reduction in danger when depending entirely on polistine wasps as the food source, the density of which is not high in temperate regions. In a species with a large colony size and a long life cycle, like *V. simillima,* it is impossible to obtain the same food source throughout the entire nesting period. However, it seems possible to become a "specialist" when the colony size is small with a short life cycle. On the other hand, such a synchronization of the life cycle becomes unnecessary in the tropics, where polistine wasps continue their nesting activities throughout the year and their density is high. In fact, though *V. tropica leefmansi* in equatorial Sumatra has an annual life cycle, it contrasts markedly with its fellow subspecies in temperate regions, because, backed by abundant food sources, including Stenogastrinae, *Ropalidia* and *Polistes,* in these districts (Matsuura 1983b), its colony size in terms of the numbers of cells, workers and emerged new queens is more than ten times larger than those of the same species in Japan and Taiwan. Consequently, it can be considered that this species in temperate regions maintains its population density by adaptation to food sources through miniaturization of the colony size. However, data on geographical changes in the colony size are not yet available for this species, which widely inhabits various areas of Southeast Asia, ranging from the tropics to temperate regions.

3 Group Predation by *Vespa mandarinia*

It is well known that *Vespa mandarinia* attacks honeybee hives en masse and causes serious damage to honeybee colonies (Okada 1961; Matsuura and Sakagami 1973). Attacks by *V. mandarinia* are not limited to honeybee colonies but also extended to nests of other species of the same genus and of the genus *Vespula* (Matsuura 1967a). From a comparative viewpoint, such habits of the species are remarkably specialized, and attacks by all other vespine wasps so far observed have been restricted to those occurring in the course of individual hunting by a queen or a worker.

3.1 Phenology of Attacks

In southwestern Japan, *V. mandarinia* attacks nests of other vespine wasps en masse from late August to early November, with a peak in late September to mid-October (Table 5.6), coinciding with the period of its attacks on honeybee colonies (Fig. 5.6). In the nests of many vespine wasps, the numbers of workers and cells reach a maximum in September to October, and the lower combs contain a large number of larvae and pupae of males and new queens, which afford a large quantity of protein food for larvae of *V. mandarinia*. However, the reason why *V. mandarinia* does not visit apiaries or other vespine nests until mid-August is still unknown. However, it may indicate that: (1) a change of protein source is required in late summer, when the numbers of its preferred prey decrease, beetles such as scarabs, for instance (2) a qualitative change in protein requirement as a food source occurs in parallel with the start of the rearing of new queens or (3) the number of emerged workers in the nest of *V. mandarinia* becomes sufficiently large to allow attacks on other social wasps only in late summer. At present, there is little concrete evidence to test these assumptions.

Table 5.6. Time of attack and number of nests of social wasps attacked by *Vespa mandarinia*, 1965–1975 (Matsuura 1984)

Species	July			Aug.			Sept.			Oct.			Nov.			Total No. of nests attacked
	E	M	L	E	M	L	E	M	L	E	M	L	E	M	L	
Polistes spp.	0	0	0	0	4	1	1	0	0	0	0	0	0	0	0	6
Vespa simillima	0	0	0	0	1	7	8	11	13	16	14	9	5	0	0	84
Vespa analis	0	0	0	0	2	2	3	8	9	8	4	3	1	0	0	40
Vespa crabro	0	0	0	0	0	0	0	2	2	1	1	0	0	0	0	6
Vespa tropica	0	0	0	0	0	0	0	1	0	0	0	0	0	0	0	1[a]
Vespula flaviceps	0	0	0	0	0	6	5	10	12	18	11	7	8	0	0	77
Total No. of nests attacked	0	0	0	0	7	16	17	32	36	43	30	19	14	0	0	214

[a] Nest kept at vespiary. E, M and L mean early, mid- and late in each month respectively.

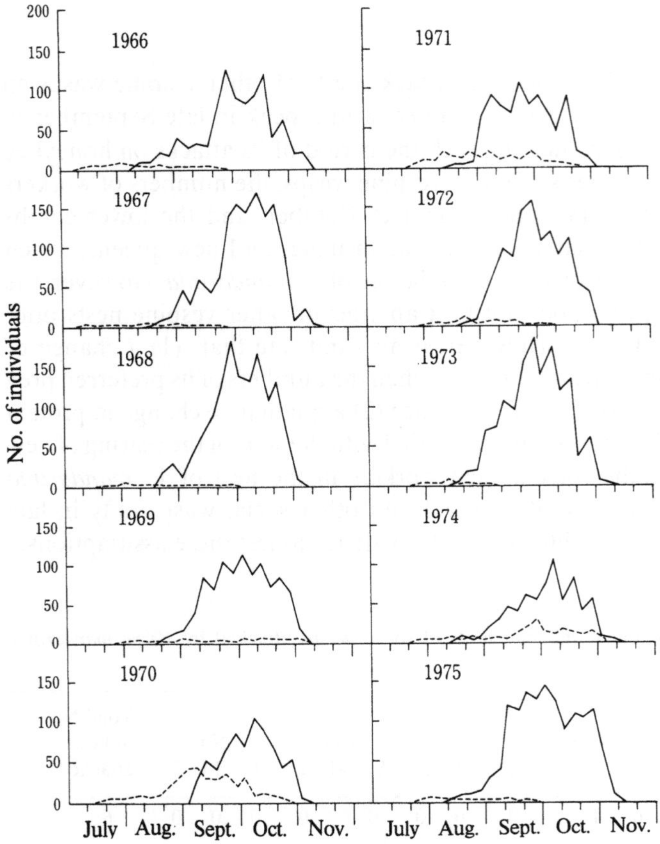

Fig. 5.6. Annual fluctuations of seasonal changes in the activities of *Vespa mandarinia* (—), and the other four *Vespa* species (--- mainly *V. simillima*) in an apiary (Wakayama, Honshu) (After Matsuura 1984)

3.2 Predatory Habits

Group predation by *V. mandarinia* is divided into three successive phases: hunting, slaughter and occupation (Matsuura and Sakagami 1973).

Hunting Phase. Similar to the case in other *Vespa* species, a worker of *V. mandarinia* during this phase individually visits an apiary or the nest of another *Vespa* species or that of *Vespula*, and, using its mandibles and fore- and mid-legs, captures a worker from the nest which is flying about near the entrance. An apiary-visiting *V. mandarinia* worker tends to choose a bee which is leaving the nest, returning to the hive or counterattacking the predator. Once the prey has been captured, the predator hangs on to a nearby tree branch with its hind-legs and gnaws off the prey's gaster, leaving the thorax muscles, which are turned into a flesh pellet and carried back to the nest. The behavior of *V. mandarinia* in the hunting phase is basically the same as that of other *Vespa* species: that is, each worker repeats a chain performance of visiting a hive→waiting around the hive →catching one bee→preparing the prey→carrying it back to the predator's own nest. In apiaries, workers of *V. mandarinia* do not stick to a specific hive, but visit many hives in search of those from which it is easy to catch prey. A few workers of *V. mandarinia* from the same nest sometimes visit an apiary alternately or simultaneously to hunt bees, but each hornet always hunts individually. Although *V. mandarinia* preys solely on adult bees during this phase, the number of prey is small because *V. mandarinia* are less skilful at catching bees than are other quick-moving species such as *V. simillima.* Direct loss from bee colonies, therefore, is generally negligible.

Attacks on nests of other vespine wasps also start with the hunting phase. Once a worker of *V. mandarinia* finds a nest of another vespine wasp, it hovers around the nest for several to a little over ten minutes and suddenly pounces upon a worker which is just leaving or returning to the nest or, sometimes, one which is guarding the entrance. After falling to the ground in combat, *V. mandarinia* bites the victim to death with its mandibles, gnaws off the victim's mesosoma and carries it back to the predator's nest. However, the attacked species occasionally kills the *V. mandarinia* individual by counterattacking with its sting. This tendency is strong in medium-sized species such as *V. analis* and *V. crabro*. In large nests of *V. simillima* and *V. crabro*, many workers rush out of the nest at the arrival of *V. mandarinia* and run about on the envelope, guarding their nest. In most cases, some of them counterattack and chase away the invader. In the fall, one to two corpses of *V. mandarinia,* killed by counterattacks, can be seen almost every day around the nest entrances of vespine wasps. In nests of vespine wasps which have the ability to chase away *V. mandarinia* when it is in its hunting phase, the shift of *V. mandarinia* to the slaughter phase is repressed. In the European honeybee, *Apis mellifera,* however, the hunting phase always shifts to the next phase in the absence of artificial protection.

Slaughter Phase. Once the attack by *V. mandarinia* workers shifts from the hunting phase to the slaughter phase, the behavior changes drastically. The attack in this phase is characterized by a simple repetition of slaughter by several to dozens of workers from the same *V. mandarinia* nest.

In the case of attacks on beehives, *V. mandarinia* workers from the same nest successively visit one particular hive in an apiary, ignoring adjacent hives. It has been shown that if a *V. mandarinia* worker is captured during this phase and then liberated near the apiary, it does not go home but returns immediately to the victim hive to recommence the attack. Each *V. mandarinia* worker stays near the hive entrance and, with its mandibles, bites to death any counterattacking bees, one by one, but usually does not use its venomous sting for killing. If a counterattacking bee bites at and clings to a part of the body of *V. mandarinia*, the invader soars upwards, removes the bee in the air with its legs, bites it to death and returns to the victim hive to recommence the slaughter. The slaughter may start in the early morning and last until evening, but usually starts in the forenoon. Once the slaughter phase develops, no participating *V. mandarinia* worker leaves the victim hive until all counterattacking has stopped or until it has been killed in the fighting. When the struggle cannot be settled within the day, all *V. mandarinia* workers return to their nest at around sunset, but reattack the victim hive early on the following morning.

Individual marking has shown that hornet workers participating in the slaughter at a hive always belong to the same nest. Occasionally, workers of *V. mandarinia* from different nests appear at an apiary in the course of the slaughter, but they are chased away from the site by "legitimate" invaders. On the other hand, other hornets from the same nest visit the victim hive in succession after the slaughter has started. The newcomers participate immediately in the slaughter, and there is no individual hunting behavior at all.

The duration of the slaughter varies according to the number of participating *V. mandarinia* workers and the intensity of the defence. Counterattacking bees, during the initial part of the phase, are entrance guards and relatively old workers. In the subsequent period, however, foraging bees participate in the defence, and almost all worker bees, except for those which had only emerged a few days beforehand, fight against the *V. mandarinia* workers. When more than ten *V. mandarinia* workers make a group attack on a hive containing 10 000 to 30 000 bees, counterattacks cease between 30 min and 3 h after the initiation of the battle, and the periphery of the hive entrance is covered with a layer of bee corpses (Fig. 5.7A). Many of the young bees that survive leave the hive, but some remain and show no resistance. During the battle, food is supplied for the invaders by newly arrived hornets, which can be seen performing frequent trophallaxis on nearby tree branches, at hives and on the ground. Some hornets, however, starve to death during the battle when the number of newcomers is insufficient, or if the attack becomes prolonged, causing a shortage of food supply.

The number of bees killed by *V. mandarinia* attacks varies according to the size of the victim colony, the number of hornets and the intensity of the counterattack. In southwestern Japan, each of the hives managed under the non-migratory apiculture system contains 15 000 to 30 000 bees from August to October, but visits by 20 to 30 *V. mandarinia* workers usually result in the deaths of between 5000 and 25 000 bees in the space of 30 min to 3 h. In contrast, *V. mandarinia* workers dying during the slaughter phase usually amount to only one or two. In one instance, about 25 000 of 30 000 bees were killed during a 3-h attack involving 30 *V. mandarinia* workers, while only two hornets were killed by the bees. This means that each hornet killed approximately one bee every 14 s (Matsuura and Sakagami 1973).

In *Vespa* nests, combat in this phase is naturally much more severe than in attacks on bees. One or a few workers of the victim species confront each *V. mandarinia* individual in flight. The two opponents fall to the ground and fight violently, using their mandibles, legs and stings. The slaughter phase lasts several hours or, sometimes, continues until the following day. In 22 of 24 examples so far observed, victim colonies of *Vespa* species were exterminated in this phase. In the remaining two examples, however, the battle was suspended with no further attacks when 34 and 29 *V. mandarinia* individuals were killed by *V. simillima* and *V. crabro* respectively. In these cases, the victim colonies were saved from

Fig. 5.7. **A** *Vespa mandarinia* attacking a hive of the European honeybee. **B** Workers of the Japanese honeybee, in a body, having killed a *V. mandarinia* worker that attempted to invade the hive. **C** *V. mandarinia* worker predating a callow wasp of *Polistes jadwigae*

extermination because of a shortage of reinforcements for the *V. mandarinia*, but their colony viability was greatly reduced after each battle.

In one instance of an attack on *Vl. flaviceps,* 10 *V. mandarinia* workers occupied the victim nest after about 1 h of fighting, losing only one attacker. As in the case of attacks on honeybees, *V. mandarinia* attacks *Vl. flaviceps,* using its mandibles as its sole weapon, and bites the victim to death, staying near the entrance of the nest. The counterattack by *Vl. flaviceps* is not as severe as that made by honeybees. In a nest occupied by about 20 *V. mandarinia* workers, several tens of *Vl. flaviceps* were still alive in the nest, and the departures and returns of a few *Vl. flaviceps* workers were observed (Matsuura and Sakagami 1973).

The causal mechanism underlying the shift from the hunting to the slaughter phase is as yet unknown, but it is obvious that a behavioral change is manifested in *V. mandarinia,* generally after it has experienced intense counterattacking. However, the shift develops only when more than one *V. mandarinia* worker from the same nest is engaged in individual hunting near the victim nest. Therefore, it is plausible that an interindividual relationship between workers of the same nest stimulates initiation of the slaughter while they are visiting a victim nest. In this case, the slaughter phase starts with two to three *V. mandarinia* workers, but the number of participating hornets increases rapidly thereafter. From this, many beekeepers assume that in honeybees the information carried by homing *V. mandarinia* workers during the hunting phase favors the arrival of other nest-mates. However, observations of *V. mandarinia* nests have shown no evidence of foraging site-informing behavior inside the nest, such as is common in honeybees, and this would seem to preclude such an assumption (Matsuura unpubl.). The mechanism for leading a fellow attacker to a nest during the slaughter phase may be explained by assuming that pheromone-like substances are emitted by each fighting *V. mandarinia* worker during the shift to the slaughter phase, and that this substance attracts nest-mates flying around the periphery of the battlefield and stimulates them to participate in the slaughter (Matsuura 1975b). Such an assumption seems to be substantiated by the fact that when several *V. mandarinia* workers are kept in a cage and left in the field, their nest-mates visit the cage one after the other, and by the fact that when some workers, freshly killed, are left at a particular point, a remarkable attraction is exerted on nest-mates flying about nearby (Matsuura unpubl.). This problem, however, is still open to further critical study.

Occupation Phase. After the decline or cessation of defence by the victims, *V. mandarinia* immediately start to enter the nest one after the other. They pay little attention to any bees still surviving in the colony or leaving the nest. The invader principally seeks out pupae or mature larvae, pulls them out and carries them back to its own nest, retaining the game in its mandibles. When the victim is small, as in the case of a honeybee, it is pulled out and carried away as it is. When the victim is large, such as with other *Vespa* species, the invader at first carries the prey inside the victim's nest or to a nearby tree branch and sucks up part of the body fluid into its crop. When the prey is a pupa, the legs and wings are gnawed off, and mostly only the thorax muscles are carried back to the invader's nest. At first, the invaders prefer developed pupae, then gradually shift to younger ones and finally to mature larvae. Soon after the slaughter phase, the corpses of victims and the dead invaders' own bodies are also transformed into flesh pellets after removal of the head, gaster, legs and wings, and transported back to their nest. However, most of these corpses are left behind on and after the following day, probably because of putrefaction, which disqualifies them as food.

In this way, the occupied beehive or hornet nest is continuously visited by *V. mandarinia* workers from one nest, and the pupae and larvae inside the hive or nest are carried away until none remain. This transportation process lasts for a few days or up to two weeks, and is performed by ten-odd to 80 workers, including both the survivors of the battle and newcomers. When a hive is rich in honey reserves, the honey is transported continuously for 1 to 2 months after the pupae and larvae have been exhausted. During this period, several to ten-odd hornets remain at the hive, not even returning home at night.

Once a hive or nest is occupied, its entrance is guarded by several to ten-odd *V. mandarinia* workers, which stand immobile, facing toward the outside. The approach of large animals, such as men or dogs, and vespine wasps of different species or of the same species but belonging to alien nests, precipitates violent attacks by these guards just as if they were defending their own nest. However, they do not make such attacks during the hunting and slaughter phases, so that at these times bee-attacking hornets can be beaten to death with wooden sticks, etc., in apiaries. In the occupation phase, however, this operation involves a degree of danger. Before attacking, each guard flies around the men or animals approaching the occupied hive or nest and threatens them with a peculiar loud clicking noise produced by a rapid opening and closing of the mandibles. This threatening behavior is common to all *Vespa* species at or near their nests, but is not known in *Vespula* and *Dolichovespula*. In *V. mandarinia*, this action is performed not only at the hornets' own nests, but also at the hives or nests occupied after the slaughter phase and near its food sources such as tree sap. This is not noted in any other species, and indicates its well developed territorial defense of abundant food sources.

3.3 Foraging Distance

According to tracking observations made between nests and food sources, the flight radius of *V. mandarinia* is 1-2 km on average, reaching a maximum of 8 km. Daily visits by several to a dozen *V. mandarinia* workers of the same nest continue when the distance between the nest and apiaries is 1-2 km (max. 2.5 km), but the most serious damage appears mainly at apiaries located less than 1 km from the nest (Matsuura unpubl.). Other vespine nests attacked are also mainly located less than 1 km from a *V. mandarinia* nest. In the case of attacks on vespine nests by a colony of *V. mandarinia*, observed in 1972 at Kibi-chô, Wakayama Prefecture, 8 nests of *Vespa* and *Vespula* species were found within 1 km from the nest of *V. mandarinia*, and five of the eight nests were attacked and destroyed one by one between September and October. This indicates that vespine nests in the neighborhood of *V. mandarinia* nests are always exposed to the danger of attack by invaders (Matsuura 1984).

3.4 Ratio of Nests Attacked by V. mandarinia

Attacks on *Vespula* and *Vespa* nests are made only by *V. mandarinia*. Table 5.7 shows the ratio of attacks on nests of other species by *V. mandarinia* observed in and near Kibi-chô, Wakayama Prefecture, during an 11-year period from 1965 to 1975. The highest ratio was found for *Vl. flaviceps*, the proportion of nests attacked reaching 50% of total nests of this species observed continuously during this period. On the other hand, nests of *V. analis*, *V. simillima* and *V. crabro* were attacked by *V. mandarinia*, but the ratios were low compared with that of *Vespula*. Generally, cases in which attacks shift to the slaughter phase seem to be rare in large-colony species having a large number of workers, because such nests can resist the attack and inhibit it at the stage of the hunting phase. The order of relative resistance ability of vespine wasps against *V. mandarinia* is: *V. crabro* > *V. simillima* > *V. analis* >>> *Vl. flaviceps*. Attacks on *V. tropica* are quite rare under natural conditions. This may be due, as described before, to the short

Table 5.7. Number of vespine nests attacked by *Vespa mandarinia*, 1965–1975 (Matsuura 1984)

Species	No. nests observed (A) / No. nests attacked (B)[a]	1965	1966	1967	1968	1969	1970	1971	1972	1973	1974	1975	Total	Nests attacked (%)
V. simillima	A	2	10	5	9	6	24	17	21	9	7	8	118	
	B	1	0	0	2	0	3	3	3	2	1	1	16	13.6
V. analis	A	5	2	15	13	2	0	7	10	8	3	13	78	
	B	3	0	0	0	0	0	1	5	1	1	1	12	15.4
V. crabro	A	1	4	5	4	0	4	3	3	4	5	7	40	
	B	0	1	0	0	0	0	1	0	0	0	0	2	4.8
Vl. flaviceps	A	7	6	5	7	6	5	5	6	8	8	5	68	
	B	5	2	3	3	3	2	2	3	5	4	2	34	50.0
Total	A	15	22	30	33	14	33	32	40	29	23	33	304	
	B	9	3	3	5	3	5	7	9	8	6	4	62	20.4

[a] Number of nests observed only during attacking period from mid-August to early November.

colony cycle of this species, which reaches the phases of completion of brood rearing and dissolution of the nest before the attacks by *V. mandarinia* intensify in September to October. However, once its nests are discovered, this least ferocious of the *Vespa* species is easily conquered by *V. mandarinia*. For example, in mid-September 1973, two nests of *V. tropica* which had been reared by a window in an open room of a house in Kibi-chô, Wakayama Prefecture, were each attacked and occupied by only three workers of *V. mandarinia*.

No intraspecific attack of *V. mandarinia* has so far been observed. Such a possibility would hardly seem likely, since no instance has yet been discovered, either of numerous dead hornets near a nest of this species or of a vacant nest, indicating such an intraspecific attack.

3.5 Defense Strategy in Japanese Honeybee Apis cerana

Among the vespine wasps, ability to resist *V. mandarinia* attacks can be seen to a certain degree in *V. simillima* and *V. crabro*, which have a large colony size. Among other social wasps and bees, the Asiatic honeybee, *Apis cerana*, occasionally has the ability to inhibit invasion by *V. mandarinia*. Since a report by Tokuda (1924), it has been thought that the Japanese honeybee *A. c. japonica*, the easternmost and largest subspecies of Asiatic honeybee, avoids damage by retreating into its hive and showing no resistance upon the arrival of *V. mandarinia*. This honeybee, however, defends itself well against invasion, not only by this passive retreat behavior but also by a particular form of defense strategy against the invader, which cannot be found in the introduced species, *A. mellifera*.

The following observations were made mainly with *A. cerana* reared in and near Kibi-chô, Wakayama Prefecture. In this district, several hundreds of colonies are still being reared, using hives, wooden boxes which were designed by Mitsuichi Sada, the father of Japanese Apiculture, who lived in Arida near Kibi. This hive is characteristic of this district, having two to three small entrances (7–8 mm in height and width) on its front side, so that only *A. cerana* can enter the box.

On the arrival of *V. mandarinia*, departures from the hive abruptly cease, and a warning fizzing-like sound, which is characteristic of this species, is emitted from the hive. This sound, called "shimmering", is due to the simultaneous folding of the wings by all bees in the hive. Homing foragers keep their distance from *V. mandarinia* and quickly enter the hive through holes situated away from the enemy. Since, unlike *A. mellifera*, *A. cerana* does not make solitary counterattacks, *V. mandarinia* waits for a while at the entrance but then gives up and leaves. If the attacker attempts to enter the hive, however, numerous bees appear from the out-of-the-way entrances. Without flying or counterattacking, they form a circle around the enemy, pointing their heads towards the intruder, slightly raising their gaster and vibrating their wings. The hornet tries to catch a bee with its mandibles, but the attempt seldom succeeds, for the movements of *A. cerana* workers are incomparably more agile than those of *A. mellifera*. Meanwhile, one of the worker bees surrounding the hornet pounces rapidly upon the enemy's head, and other bees also rush upon the hornet at almost the same time. Within about 10 to 15 seconds of this brave attack by the first bee, the hornet is thickly covered with several tens to several hundreds of bees, this black ball of bees sometimes reaching the size of a clenched fist (Fig. 5.7B). The continuously emitted shimmering sound in the hive then ceases, and the departure of foragers recommences. The bees covering the enemy maintain their position for more than 20 min, with little motion. The corpse of the hornet, left behind near the entrance, shows no external injury; it is apparently killed by stings in the softest part of the neck, which connects the head and thorax. The only occasional existence of dead or injured bees at the scene indicates that their counterattack against the hornet is quick and accurate.

Thus, the attack by *V. mandarinia* on *A. cerana* is interrupted in the initial part of the hunting phase and rarely proceeds to the subsequent slaughter phase. However, if a colony of *A. cerana* is kept in a hive with a wide entrance space like that used for *A. mellifera,* the colony cannot use the effective defense strategy described above. For example, a colony of *A. cerana* reared by Matsuura in a hive with a wide entrance space was attacked by *V. mandarinia* on October 22, 1972, and damaged by the attack as follows: When three *V. mandarinia* workers were discovered flying about the hive, and the inside of the hive was inspected, the corpses of four *V. mandarinia* workers and about 80 *A. cerana* bees were found. Since the nest of the attackers was located only about 100 m away, it was thought that the colony of *A. cerana* would have been totally destroyed by the attack if the attackers had not been killed by the beekeeper. This suggests the possibility that even colonies of wild *A. cerana* are not able to inhibit invasion by *V. mandarinia* simply through the colony's nest site structure. However, it is known among beekeepers in the Arida district that when *A. mellifera* and *A. cerana* are reared in the same location *V. mandarinia* principally attacks the former type of hive, while damage by *V. mandarinia* attack is negligible when only *A. cerana* is reared in an apiary (Matsuura and Sakagami 1973). This indicates that *A. cerana* is an unsuitable target for attack by the hornet, when compared with *A. mellifera* and other vespine wasps.

This coexistence of *A. cerana* and *V. mandarinia* in the field in Japan is related to the fact that *A. cerana* has developed two effective defense strategies which are not found in *A. mellifera;* one being the absence of solitary counterattacks by workers and the other an effective form of mass counterattack. These skillful defensive strategies contrast sharply with its weakness against other species of the same genus; that is, *A. cerana* possesses no means of resisting pillage of its honey reserves by the introduced species, *A. mellifera.* For example, when colonies of both species are kept in the same location, in most cases *A. mellifera* invades the hive of *A. cerana,* pillages its honey reserves and finally exterminates the *A. cerana* colony (Sakagami 1959). Thus, the introduction of *A. mellifera* to Japan produced, within a century, a curious three-cornered relationship among the Japanese and European bees and *V. mandarinia* in the fields of Japan.

3.6 Characteristics of Group Predation

With the largest body size among the Vespinae and its enormously developed genae giving support to the muscles of its strong mandibles, *V. mandarinia* possesses a formidable ability to attack and slaughter other insects and animals. This ability is reinforced by its violent aggressiveness and the potency of its sting. Because of these attributes, *V. mandarinia* occupies an unrivalled position among insects and small animals, having the highest rank in the food web, at least in Japan. On the other hand, however, its relatively sluggish movement compared with *V. simillima* and *V. analis,* because of its large body size, makes it difficult for it to capture fleeing agile prey. Consequently, *V. mandarinia* and other vespine wasps have come to occupy slightly different ecological niches. *V. mandarinia* did not choose specialized behavior like that practiced by *V. tropica,* but began to seek prey which was easy to catch in large quantities. For this purpose, *V. mandarinia* set its sights on colony-forming social insects, since their nests contained large amounts of concentrated protein food. Consequently, dependence on the *Polistes,* which form small colonies, as in *V. tropica,* would be impossible for *V. mandarinia,* from the viewpoint of securing a sufficient quantity of prey, because not only do its nests reach a large size, but it also has a large body. On the other hand, nests of other vespine wasps are very attractive to *V. mandarinia* as food sources, irrespective of the inevitable sacrifices occurring due to counterattacks. Moreover, the presence of counterattacking individuals has a somewhat favorable result for the overpowering invader in battle. Thus, ecologically, *V. mandarinia* affects other congeneric species in two ways: as a competitor in seeking common food sources, and as a predator. Attacks on nests of other species by *V. mandarinia,* first initiated as individual hunting forays, would be rewarded with a nest containing an abundant supply of larvae and pupae as a protein source for the invader's own larvae, once the attack had been continued until resistance by the victim was overcome. Even if the attack was accompanied by a loss of several to several tens of nest-mates, some of them could be fed back to the colony as a food source, and damage to the colony from such a loss would be relatively light. The slaughter phase, the most interesting stage in the attack by *V. mandarinia* on other social wasps, is considered to be a later insertion between the hunting and occupation phases. This phase is seemingly quite light or absent during an attack on a species such as *Vespula,* whose counterattacking behavior is not well developed, but it is distinct when attacks are made on victims offering severe resistance, i.e., other *Vespa* species and *A. mellifera.* Its functional significance is to facilitate the occupation of a victim's nest. Although it is obvious that the initiating mechanism of this phase is related to the intense counterattacks and the presence of nest-mates of the invader near the attack site, further studies will be required in order to elucidate the minute details of the mechanism.

The habit of defending the occupied nests of other wasps, as seen in *V. mandarinia,* has been observed not only in *V. orientalis* (Ishay et al. 1967) but also in cleptobiotic stingless bees of the genus *Lestrimelitta* (Nogueira-Neto 1970). Compared with the professionalized attacks on other stingless bees and the resulting behavior of nest occupation by *Lestrimelitta,* attacks by *V. mandarinia* are far less efficient and, moreover, the latter still retains the habit of preying on various insects individually. *V. mandarinia*'s habit of attacking social wasps, therefore, can be regarded as a form of incomplete specialization.

4 Foraging Distance

Vespine wasps have a habit of visiting the same carbohydrate or protein source repeatedly until the source is exhausted. Therefore, the location of a vespine nest and its distance from a food source can be identified by placing in a given location an artificial feeder which is replenished with fresh honey or sugar solution daily, so that various vespine wasps will visit the feeder and continue their food-foraging activities, allowing tracking of each individual to its nest. Such investigations have been performed by Matsuura for 20 years, by setting these feeders in hilly country in Kibi-chô, Wakayama Prefecture, and Anô-chô, Mie Prefecture. Table 5.8 shows the results of investigations of the distances between the food source and the nests of six *Vespa* species. For *V. dybowskii,* however, the investigation was performed in 1978 in Iida City, Nagano Prefecture, Central Japan. In each species, most of the nests were found within 1 km of the food source. However, a significant number of individuals showed further extended flight radii: 2–3 km for *V. analis* and *V. crabro* and more than 3 km for *V. mandarinia* and *V. simillima,* reaching a maximum of 4.8 km for *V. mandarinia* and 5.2 km for *V. simillima.* These results indicate that the flight radius for food-foraging is wider in large-colony than in small-colony species, because the former need a larger amount of food, and each individual in a nest collects food independently within a wide range. Similar flight radii can be applied in the case of predation, although reports on prey-foraging distance are as yet insufficient. When *V. mandarinia* workers visiting an apiary were marked with paint and tracked to their nests, 64.6% (31/48) of their nests were found within 1 km, 22.9% within 1 to 1.5 km, 10.4% within 1.5

Table 5.8. Foraging distances for *Vespa* workers (Matsuura unpubl.)

Species	Distance between feeding station and nest (m)									
	0–100	100–200	200–400	400–600	600–800	800–1000	1000–1500	1500–2000	2000–3000	>3000
V. analis	16	27	28	18	15	9	3	3	1 (2700)	0
V. crabro	4	9	7	7	5	4	2	2	1 (2400)	0
V. tropica	8	11	25	14	10	6	1 (1400)	0	0	0
V. mandarinia	3	15	18	17	11	10	8	7	3	5 (4800)
V. simillima	5	18	14	19	16	18	14	16	5	5 (5200)
V. dybowskii	1	1	2	1	1	1 (900)	0	0	0	0

to 2 km and 2.1% within 2.7 km (Matsuura unpubl.). A maximum distance of 8 km has also been reported (Matsuura and Sakagami 1973).

The flight radii of *Vespula* and *Dolichovespula*, whose body sizes are smaller than that of *Vespa*, are narrow. It is well known that in Nagano and Gifu Prefectures, Central Japan, nest collectors set a lump of frog or fish flesh attached to a strand of floss silk at a certain location in the field and trace workers of *Vl. flaviceps* or *Vl. shidai* which have taken the bait, using the "streamer" as a flying marker in order to find the nest. The distances between such bait-sites and the wasp nests observed by such nest collectors and by Matsuura are shown in Fig. 5.8. In such cases, the tracking attempt was usually abandoned when a wasp took too much time to carry the bait back to its nest and to revisit the trapping site, because it was assumed that the nest was too far away to be tracked down. Data on long foraging distances, therefore, are insufficient to allow detailed discussion, but such cases only amounted to less than 10% of the total. The flight radius in the above two species as a linear distance was, on average, 100 to 400 m (72% of the total), reaching a maximum of 980 m. According to an investigation on *Vl. pensylvanica* performed in the United States by setting a bait of fish flesh marked with a piece of metal, nests of half of all marked workers were found within 160 m, 90% within 0–400 m, indicating that the food-foraging activities of this species were undertaken within relatively short distances from the nest (Akre et al. 1975). However, it has been reported that this species flew back to its nest, which was as much as 1 km distant from a honey-containing feeder (Akre et al. 1973). *Vl. germanica*, which has been introduced and has become naturalized in New Zealand, generally undertakes predatory activities within 50 to 250 m of its nest.

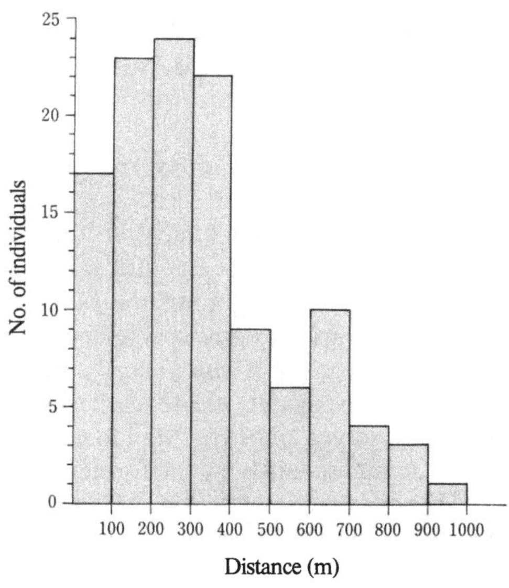

Fig. 5.8. Frequency distribution of distances from nest to foraging site in two Japanese *Vespula* species (Matsuura unpubl.)

except for a record of 1200 m by one individual (Perrott 1975). *Vl. rufa* in England generally also performs its extranidal activities within a range of 200 to 300 m from the nest (Arnold 1966 cited in Spradbery 1973a).

As for *Dolichovespula*, two observations that have been made for *D. media* are that a worker preyed on a fly entering a house 8 m distant from its nest, while another preyed on a moth at a point about 120 m distant from its nest (Matsuura unpubl.). Also, Japanese *D. saxonica* and *D. norvegicoides* in the process of prey treatment were observed on a plain about 150 m distant from each worker's nest (Matsuura unpubl.). In England, *D. sylvestris* has a flight distance of 200 to 300 m around the nest (Arnold 1966 in Spradbery 1973a).

Judging from the above observations, it can be concluded that *Vespula* and *Dolichovespula* undertake food-foraging activities within a distance of several 100 m of their nests. In particular, the flight radii of *Vl. schrenckii* and *D. media*, both of which have a small colony size, will be narrower than that of *Vl. flaviceps*, whose body size is small but whose colony size is several times larger than those of the former two species.

5 Intra- and Interspecific Relationships at Tree Sap Apertures

Various species of *Vespa* visit borer-made apertures in the trunks of *Quercus* and other tree species in order to take in tree sap, their favorite carbohydrate source, which is secreted from these apertures. In early spring, post-hibernating queens visit tree sap locations before devoting themselves to nest-founding, and, after the summer, such tree sap is used by workers as food for larvae and adults in their nest. At sites of tree sap exudation, certain intra- and inter-specific dominance relationships exist among post-hibernating *Vespa* queens (Matsuura 1969), and *Vespa* workers from a particular nest assert among themselves their territorial rights to the food source (Matsuura 1980a).

5.1 Dominance Relationships Among Post-Hibernating Queens

Post-hibernating *Vespa* queens visit tree sap-exuding apertures of *Quercus serrata, Q. acutissima, Salix gracilistyla,* etc., and continue their feeding activities the whole day, before beginning nest-building activity. An aperture rich in tree sap is visited by various *Vespa* species including *V. mandarinia.* In southern Honshu in Japan, queens of *V. simillima, V. mandarinia, V. crabro* and *V. analis* emerge from their hibernacula in late April to May, but the activities of *V. tropica* queens are not observed until late May to early June. Consequently, tree apertures are visited simultaneously by the former four *Vespa* species.

The results shown in Fig. 5.9 and Table 5.9 were obtained by marking tree aperture-visiting pre-nesting queens with colored oil-paint on their mesonotum, in order to distinguish each individual, and by observing their activities during a 3-day period in early May.

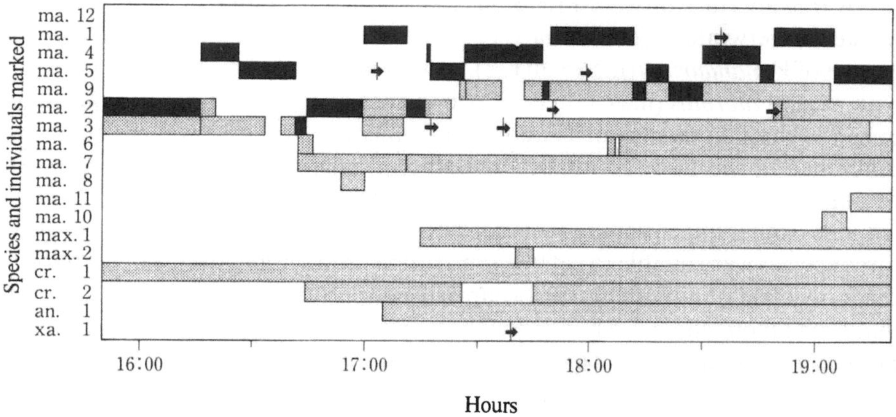

Fig. 5.9. Activities of tree sap intake observed among 18 marked individuals of four species of *Vespa* that visited a tree aperture in the daytime of May 5, 1967. *Black rectangles* indicate occupation of tree aperture. *Spotted rectangles* indicate staying around it without feeding. *Arrows* indicate flying away soon after flying around the tree aperture. *ma* Queens of *V. mandarinia*; *max* stylopized workers of *V. mandarinia*; *cr* queens of *V. crabro*; *an* queen of *V. analis*; *xa* queen of *V. simillima* (After Matsuura 1969)

Table 5.9. Length of time (min) spent taking tree sap by each individual visiting tree aperture over 3 days at beginning of May, 1967 (Matsuura 1969)

Species and individual marked[a]	Frequency of visits	Frequency of feedings	Total time for feeding	Average each feeding time	Time for non-feeding
ma 12	5	5	103	21.0	0
ma 1	11	9	158	17.5	0
ma 5	13	9	79	8.8	1
ma 4	9	8	82	10.0	1
ma 9	11	8	31	4.0	128
ma 2	10	6	42	7.0	46
ma 3	18	5	28	5.6	126
ma 6	9	10	59	3.9	286
ma 11	6	2	4	2.0	239
ma 7	1	1	1	1.0	153
ma 8	2	0	0	0	6
ma 10	7	0	0	0	29
max 1	4	0	0	0	129
max 2	3	0	0	0	153
cr 1	8	0	0	0	198
cr 2	2	0	0	0	132
an 1	3	0	0	0	233
xa 1	2	0	0	0	0

[a] ma 1–12: queen of *V. mandarinia*; max 1–2: stylopized worker of *V. mandarinia*; cr 1–2: queen of *V. crabro*; an: queen of *V. analis*; xa: queen of *V. simillima*.

A tree aperture observed for 3 days was visited by a total of 18 queens and workers (overwintered individuals) of four *Vespa* species, which consisted of 12 queens of *V. mandarinia* (ma 1 ~ ma 12), two workers of the same species stylopized with *Xenos crabronis* (max 1, max 2), two queens of *V. crabro* (cr 1, cr 2) and one queen each of *V. analis* (an 1) and *V. simillima* (xa 1). There was a significant difference in the tree sap-intake time among individuals. For example, each feeding time for a particular *V. mandarinia* queen, named ma 12, reached 21 min, while two other queens of the same species (ma 8 and 10) and queens of the other three species were unable to get tree sap at all in the daytime. It is clear, therefore, that only certain individuals can occupy a tree aperture and take in tree sap, and that even among these, each individual occupies the tree aperture according to its place in a given hierarchy. That is, in the daytime, a tree aperture exuding sap was always occupied by a single individual surrounded by others that were unable to get the sap. Whenever they tried to approach the aperture, these individuals were driven back by the occupant, which threatened them with mandibles widely opened. No counterattack efforts were made.

Feeding individuals changed places when the first became fully engorged and flew away, or when it was driven from the tree aperture by a superior individual arriving from another place. When a tree aperture already occupied by an individual (A) is visited by a queen of the same or different species, a visiting individual (B) shows one of the following four reactions:

1. Alighting at a position 5 to 6 cm distant from the original occupant after flying around the tree aperture ($A >> B$);
2. Flying away soon after flying around the tree aperture ($A > B$);
3. Severe combat, using mandibles and stings, between the two individuals soon after alighting at or near the tree aperture; the victor begins to take tree sap and the defeated individual flies away, with death occurring only rarely ($A = B$);
4. An immediate changeover from the occupant to the visitor without any combat; the occupant gives up its position and stays around the site or flies off ($A < B$).

Such "inferior" and "superior" types of behavior have been observed not only among occupants and visitors as described above, but also among non-occupants surrounding the tree aperture. The inferior individuals retreated when they perceived the approach of a superior, which frequently threatened them with opened mandibles, so that each individual took up a position away from the others, forming an approximately uniform spatial distribution around the aperture. Table 5.9 indicates the frequencies of dominance and subordination among the above-mentioned 18 marked individuals.

On the basis of these observations given in Fig. 5.9 and Table 5.9, it is clear that there is a distinct linear dominance relationship among the tree sap-visiting queens of the four *Vespa* species including *V. mandarinia*. That is, the following order was observed among 12 queens and 2 stylopized workers of *V. mandarinia*:

$$\text{ma } 12 \gg \left\{\begin{array}{ccc} \text{ma } 4 & > & \text{ma } 1 \\ \wedge & & \llcorner \\ \text{ma } 5 & & \end{array}\right\} \gg \text{ma } 9 > \text{ma } 2 > \text{ma } 3 > \text{ma } 6 > (\text{ma } 7, \text{ma } 8, \text{ma } 11 > \text{ma } 10) > \text{max } 1 > \text{max } 2.$$

The interspecific dominance relationship is as follows: *V. mandarinia* $\gg$ *V. crabro* $>$ *V. analis* $>$ *V. simillima.*

Each individual occupies an aperture and takes tree sap according to this order. The queen ma 12 of *V. mandarinia* with the top ranking repeatedly took her fill of tree sap and flew away at any given time when she visited the tree aperture during the observation period, meeting with absolutely no resistance from the other individuals. The next position was a little complex, but was occupied by three queens of *V. mandarinia*, ma 1, ma 4 and ma 5, among which a three-cornered relationship existed as shown above. That is, each of these three queens took tree sap in the absence of superiors, not giving up the aperture to inferiors, but giving up the aperture when a superior individual arrived, and flying away immediately to seek other sources of tree sap.

Individuals under the above four queens in the hierarchy did not usually migrate to other trees but spent much time near the tree aperture awaiting their opportunity, which came when the occupant had taken her fill of tree sap and flown off. Occasionally, these individuals took tree sap for a brief period, watching for a chance when the aperture was momentarily empty during the frequent changes made by the four superior queens. Even among these waiting individuals, however, lower rankers were hardly ever able to utilize such an opportunity. The hierarchical relationship among the four queens at the bottom of the ladder, ma 8, ma 7, ma 11 and ma 10, was not clear, although they were always superior to the two stylopized workers, max 1 and max 2. These stylopized *V. mandarinia* workers were always superior to other *Vespa* species.

In the case of bottom-ranking individuals, however, opportunities for taking tree sap were quite rare, and most of these individuals seemed to spend the entire day without any chance of food-taking, in spite of their long vigil. Such a situation is evident not only from these data but also from many other observations. However, even individuals at the bottom of the hierarchy can take tree sap in the evening as long as they keep waiting near a tree aperture which secretes abundant tree sap, since, unlike the situation in the daytime, new arrivals at the aperture cease in the evening and the occupant leaves the spot after taking its fill of tree sap and does not return. In the evening, each individual with a full crop walks to the top of the tree exuding the sap and settles on the underside of a branch 7 to 8 m high. The waiting individuals near the tree aperture then finish their tree sap-intake one by one, irrespective of differences in species, but according to the dominance relationship existing among them.

5.2 Behavior Toward Other Insects

Tree sap is an important food source in the springtime, not only for post-hibernating *Vespa* queens but also for various other insects. In the daytime, a tree aperture exuding sap is visited by various insects, such as flies, butterflies, cryptophagid beetles and polistine wasps, but is usually occupied by *Vespa* species. On such occasions, these insects stay within 2 to 5 cm of the tree aperture and try to take in tree sap whenever it is unoccupied by a *Vespa* individual. However, opportunities for these insects to take tree sap are rare, as is the case for other vespine wasps which are waiting near the tree aperture, because the aperture-occupying *Vespa* individual chases away any visitor which tries to steal the sap.

In the evening, beetles, such as cryptophagids, still remain near the tree aperture, and the sap is newly visited by several moths of the subfamily Catocalinae, *Arcte coerulea, Dinumma deponens* and others. Each of these species walks restlessly around the sap-taking *Vespa* and tries to steal the sap by frequently stretching out its proboscis. These moths, however, usually retreat or fly away from the tree aperture because, once they are discovered, the occupant *Vespa* pounces upon them with its open mandibles.

5.3 Occupation of Tree Apertures by Workers in Summer

The tree aperture-occupying behavior by post-hibernating queens ceases gradually as their nesting activities start. Meanwhile, worker emergence starts in the nests of *V. mandarinia* from around mid-July, and tree apertures are again occupied by tree sap-seeking workers. Whereas the sap-intake of a post-hibernating queen is used for her own nutriment, the tree sap is carried back by workers to their nests and distributed to larvae and adults over a long period as a stable carbohydrate source. Especially in the case of the large-colony species *V. mandarinia*, the tree sap is an important food source, because it constitutes a large proportion of the carbohydrate food carried back to the nest, and substitute sources for tree sap are quite rare.

During and after the summer, one can sometimes find a tree aperture occupied by a group of several vespine workers, mainly *V. mandarinia*, taking in tree sap. In such cases, several workers take the tree sap at the same aperture simultaneously, pressing their bodies together and showing none of the dominance behavior common among post-hibernating queens in spring. These workers always belong to the same colony, with no exceptions being made to allow stray individuals from other nests to enter the main group, even members of the same species (Matsuura 1980a, 1984). Each individual with a full crop goes back to its nest in turn, gives the sap to individuals in the nest by regurgitating it from its crop, and then returns to the tree aperture to repeat the tree sap intake. The number of *V. mandarinia* workers visiting a tree aperture varies according to the condition of tree sap secretion and the degree of development of the nest, but a tree aperture with abundant sap is usually visited by many workers, reaching a maximum of 20 or more. According to distinguishing experiments involving the marking of each individual with paint, it seems that tree aperture-visiting workers are engaged mostly in tree sap transportation work, attending to hardly any intranidal activity.

A tree sap-exuding aperture is always occupied by *V. mandarinia* workers belonging to the same colony, which claim a territorial right to it by putting up ferocious resistance to any visitor from a different nest. Colonies of workers which visit the same tree aperture vary from day to day. When *V. mandarinia* workers belonging to a different nest arrive near a tree aperture which is already occupied by another colony's workers, a severe battle sometimes starts between them, involving mandibles and stings. One to 2 dead bodies of *V. mandarinia* workers in particular can always be found below a tree aperture exuding abundant tree sap, and these bodies can be regarded as the victims of such a battle.

When a tree aperture is exuding a great deal of tree sap, several workers remain at the site, even in the evening, and stay on until the following morning. Contrary to the tree sap-intake behavior of queens in the evening in spring, these workers take small amounts of tree sap and never drink their fill. These workers seem to be watching for visitors from alien colonies in order to secure their territorial rights.

In Taiwan also, many *Vespa* workers have been observed to visit the same tree seeking sap, and *V. mandarinia* is the dominant species among these, dominating *V. basalis* and *V. velutina* as well. However, *Polistes gigas,* the largest species of the genus *Polistes* in the world, is obviously superior to *V. mandarinia*. Numerous corpses of *V. mandarinia* beaten in battle with *P. gigas* can sometimes be found scattered about on the ground (Sk. Yamane unpubl.).

Chapter 6

Population Dynamics

This chapter will treat the vespine wasp population on two levels: individually and on colony. On the individual level, survivorship of colony members is discussed mainly here in relation to life tables obtained from workers in normally developed colonies. On the colony level, the survival of nests, colony development and annual nest fluctuations will be discussed, and an appraisal of some of the factors affecting them will be offered. These are mainly based on the observations reported by Matsuura (1984).

1 Survivorship Curves of Workers

1.1 Length of Developmental Stage

Table 6.1 shows the length of each developmental stage (in days) from egg to emergence in workers of five Japanese *Vespa* species, mainly during the polyethic period. Total duration of brood development from egg to adult varied on average from 30.8 days in *V. simillima*, the species with the smallest body size, to 40.1 days in *V. mandarinia*, the largest species. This shows a tendency for workers with larger bodies to have a longer duration of development. Comparing the data on each developmental stage, the lengths of development of the fifth instar larval stage vary widely, while the lengths for the egg, and immature larval stages, (especially first to third instars), and the cocoon stages are fairly constant with each species. Such wide variability in the fifth instar larval period seems to be caused by the difference in the provision of food for each individual. Larval growth is most active and a large supply of food is required at this stage.

To date, no studies have been reported on Japanese species of *Vespula* and *Dolichovespula*. Archer (1980b) examined the variability of total duration of brood development in relation to the enlargement of cells with colony development. He reported that, in *Vl. vulgaris* in England, the duration from egg to adult was 24–39 days in the case of small cells, and 34 days for large cells. He also reported that such a duration in *D. sylvestris* was 35–37 days for small cells and 28 days for large cells.

Generally, the total duration of brood development in vespine wasps is longer in the early stage of nesting. This is caused by the small number of workers present and, more particularly, because (1) larvae cannot be cared for fully by workers, since their nursing burden is high and their number insufficient, and (2) minimal nest temperature needed for brood development cannot be regulated or maintained by so small a number of adults. The duration of brood development

Table 6.1. Length of each developmental stage in five *Vespa* workers (days) and size of adult workers (Matsuura 1984)

Species	Egg	Larval instar[a] I	II	III	IV	V	I-V	Co-coon	T.	No. of individuals observed	Size of adult workers[b] (mm)
V. simillima	4.3 (0.2)	1.0 (0.0)	1.0 (0.0)	1.0 (0.2)	1.5 (0.4)	6.7 (1.2)	11.2 (1.2)	15.3 (0.4)	30.8 (2.5)	115	21.3
V. crabro	5.0 (0.2)	1.0 (0.0)	1.1 (0.1)	1.2 (0.2)	2.2 (0.5)	6.8 (2.1)	12.0 (2.5)	15.5 (0.3)	32.4 (4.7)	73	24.1
V. analis	5.0 (0.2)	1.0 (0.0)	1.0 (0.0)	1.3 (0.1)	2.1 (0.5)	6.5 (1.8)	11.9 (2.1)	15.6 (0.4)	32.4 (4.1)	60	24.5
V. mandarinia	6.0 (0.1)	1.0 (0.0)	1.5 (0.0)	2.0 (0.1)	3.5 (1.6)	7.8 (1.9)	15.9 (2.2)	18.2 (0.4)	40.1 (4.6)	84	34.2
V. tropica	6.2 (0.1)	1.0 (0.0)	1.0 (0.0)	1.3 (0.1)	2.1 (0.6)	7.2 (2.2)	12.6 (2.5)	16.8 (0.5)	35.6 (4.7)	71	30.1

[a] Parenthesized numbers show standard deviations (±).
[b] Mean of 10 workers for each species.

becomes short as the number of workers increases, as the ratio of nurse-workers increases and the nest temperature becomes stable.

1.2 Survivorship Curves

Life tables, or survivorship curves, are used for making ecological comparisons of the lifespan of various animals. The life table is a diagram of periodical profiles of decrease in the number of simultaneously-born individuals. It consists of several parameters, including the number of survivors (lx), mortality factors (Fx), number of deaths (dx) and mortality rate (qx). The survivorship curve is made by plotting changes in the survival rate of the initial individuals (ordinate axis) against time (transverse axis). These diagrams have been widely used as a general means of analyzing the dynamics of animal populations (Southwood 1967; Itô 1978) ever since a review by Deevey (1947). With regard to social Hymenoptera, reports have been published on *Lasioglossum* (Sakagami 1977), *Apis* (Fukuda and Sakagami 1968a,b; Sakagami and Fukuda 1968; Fukuda and Ohtani 1977), *Bombus* (Brian 1951), *Myrmica* (Brian 1951), *Mischocyttarus* (Jeanne 1972) and *Polistes* (Matsuura 1977a; Miyano 1980).

Figures 6.1–6.5 present survivorship curves for all developmental stages of workers of five *Vespa* species, prepared from well developed colonies in the polyethic period. From the results shown in these figures, it can be seen that the types of survivorship curves among the five *Vespa* species are all quite similar, and show a remarkably low mortality rate during the period between the egg stage and the few days immediately after emergence of workers. There is a constantly changing profile in the mortality rate after the initiation of extranidal activities.

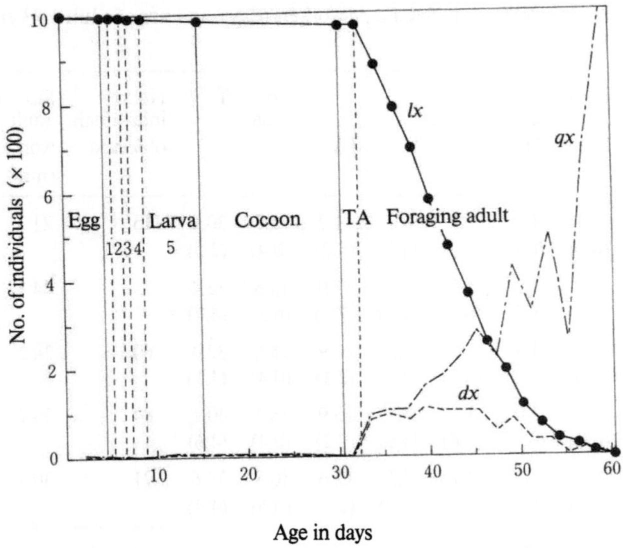

Fig. 6.1. Survival curve for *Vespa simillima* workers. *TA* Teneral adult. For *lx*, *dx* and *qx*, see text (After Matsuura 1984)

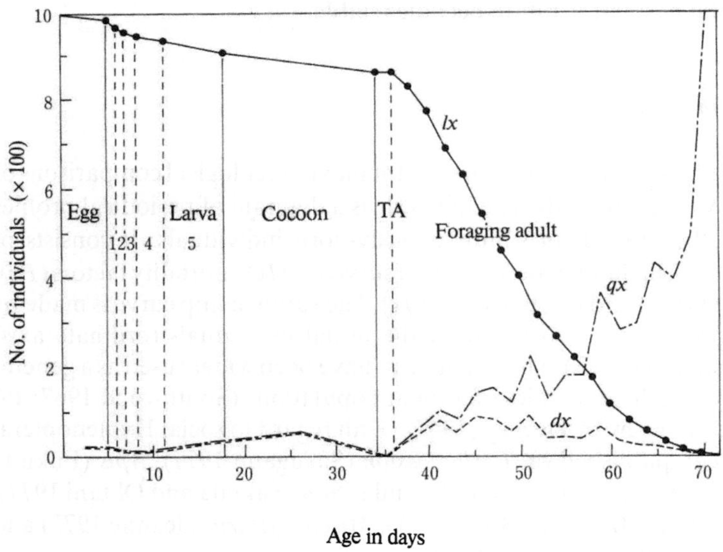

Fig. 6.2. Survival curve for *Vespa crabro* workers. *TA* Teneral adult. For *lx*, *dx* and *qx*, see text (After Matsuura 1984)

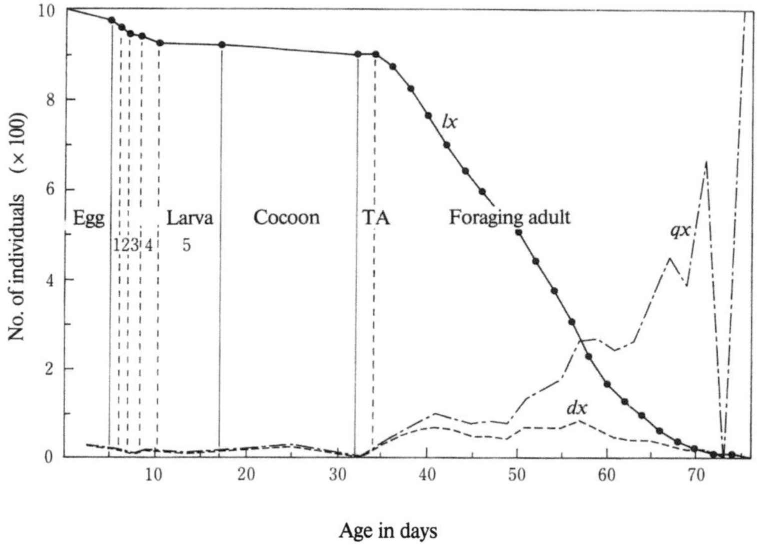

Fig. 6.3. Survival curve for *Vespa analis* workers. *TA* Teneral adult. For *lx, dx* and *qx*, see text (After Matsuura 1984)

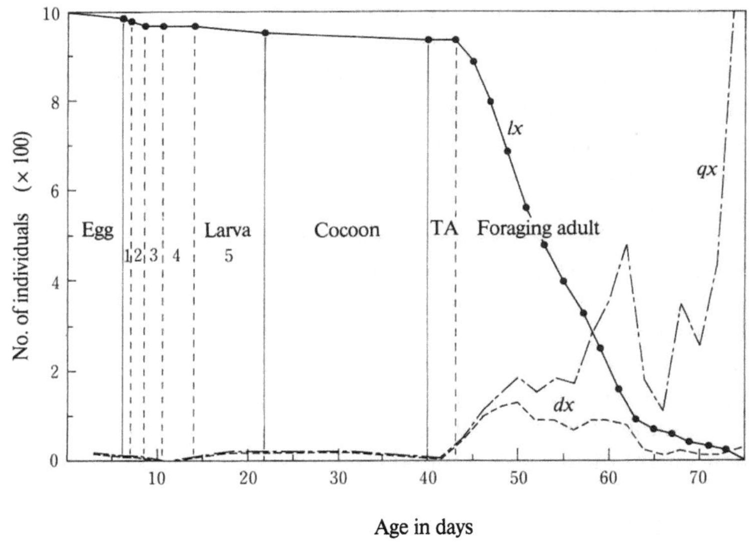

Fig. 6.4. Survival curve for *Vespa mandarinia* workers. *TA* Teneral adult. For *lx, dx* and *qx*, see text (After Matsuura 1984)

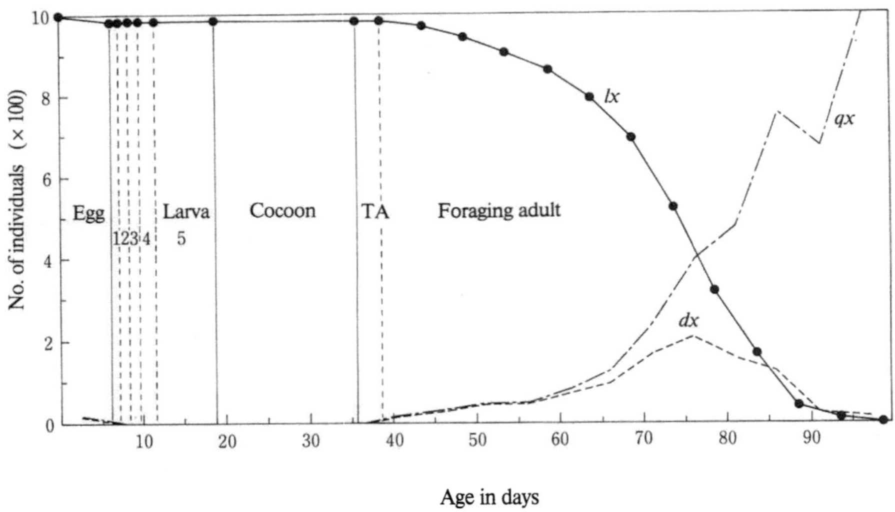

Fig. 6.5. Survival curve for *Vespa tropica* workers. *TA* Teneral adult. For *lx*, *dx* and *qx*, see text (After Matsuura 1984)

Therefore, the survivorship curve is quite close to the type seen in large mammals, including humans, which has been named Type I by Deevey (1947) and is considered one of the three basic types of curve. It has been indicated by Itô (1978) that such a feature originates in the remarkably low mortality during non-adult stages from egg to pupa inside the nest. This is as a result of the advanced brood-rearing habits practised by social adult Hymenoptera. This is opposed to the quite high mortality among general insects at corresponding stages of egg to first instar larva, final instar larva and aged adult. A similar example has also been reported in the *Apis mellifera* (Sakagami and Fukuda 1968), which shows a convex survivorship curve, rare in insects.

Figure 6.6 shows the survivorship curves of five *Vespa* species for the same cohorts given in Figs. 6.1–6.5. These deal at daily intervals with adults alone. Common features among these species are that hardly any individual deaths occur during the first 2–3 days after worker emergence and that a gradual decrease in the number of individuals begins at an almost constant rate with the commencement of the first extranidal activity. Although mean longevity in *Vespa* varies according to species, the maximum longevity is usually about twice the mean. There is a general tendency in the five *Vespa* species for the longevity to become shorter with increase in colony size: the longevity is shortest in *V. simillima* (12.9 days; s.d. ± 6.2) and becomes longer in the order of *V. mandarinia* – *V. crabro* – *V. tropica*, although the longevity of *V. tropica* (34.9 days; s.d. ± 11.5) is outstandingly long in comparison with the other four species. Such an interspecific difference seems to reflect the various ecological niches occupied, including feeding habits, foraging distances and intra- and interspecific relationships, in addition to differences in colony activities which are overtly size-dependent.

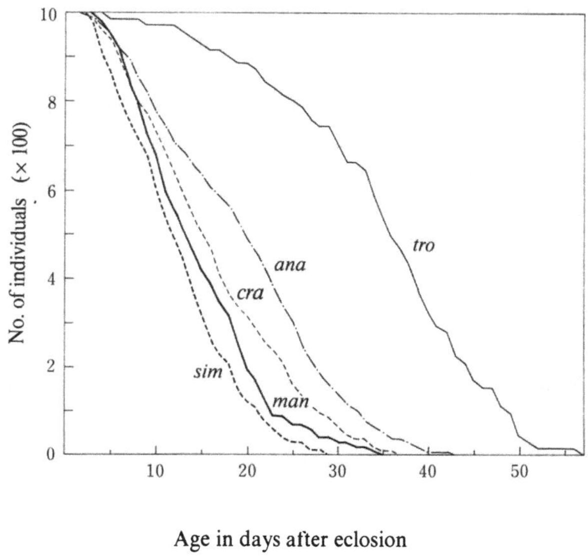

Fig. 6.6. Comparison of survival curves of the five *Vespa* species in southwestern Honshu. *sim, V. simillima*; *man, V. mandarinia*; *cra, V. crabro*; *ana, V. analis*; *tro, V. tropica* (After Matsuura 1984)

According to a study on *Vespula vulgaris* in Europe (Archer 1980b), the mean longevity of workers is 10.3–22.3 days, whereas the longevity of workers in early nests is generally shorter because of the high burden of their nursing. The longevity becomes greater, however, as the developmment of the colony proceeds. The number of workers increases and, therefore, the total level of activity carried out by each individual is reduced. The number of larvae per worker in a colony becomes minimal and the longevity of workers reaches its maximum at a stage just before the commencement of nursing of new queens. Life span decreases again as the nursing of new queens starts.

In *Apis mellifera*, mortality of adult workers before extranidal activity commences is remarkably low, although the survivorship curve after the commencement of extranidal activities shows a similar profile to that of vespine wasps. The workers become engaged in intranidal activities, such as nest-construction and nursing, almost 3 weeks after their emergence and then start their extranidal activities such as food collection (Sakagami and Fukuda 1968).

1.3 Fate of Brood at Emergence of Reproductives

In every species, the males and new queens are produced toward the end of the seasonal cycle. They feed on the secretion of the fifth instar larvae and on the food brought in by workers during the time the young remain in the nest. Commencement of the emergence of males and new queens means that the colony enters a disorganized state at the same time. The first sign of disintegration in the colony takes the form of larvae being taken out of their cells and abandoned outside the nest. Such abandonment of larvae is a habit which can be seen in all vespine wasps at the end of the seasonal cycle, though hardly at all in other periods (Duncan 1939).

Table 6.2 summarizes the life table data for all members of the brood of a *V. tropica* colony from nest foundation to colony disorganization, for the purpose of comparing the mortality of the brood accompanying the emergence of males and new queens with those during other nesting periods. The life table is classified into three periods according to the time when the eggs are laid: namely, the preemergence period; the co-operative periods before emergence of reproduc-

Table 6.2. Life table for *Vespa tropica* brood in nest (T7401) at different nesting periods (Matsuura 1984)

Period	Duration of oviposition (1970)	Developmental stage		Fdx[b]	lx[b]	dx[b]	qx[b]
Eggs laid before emergence of workers (n=39)[a]	June 9 – July 13	Egg		–	39	0	0
		Larval instar	I	–	39	0	0
			II	–	39	0	0
			III	–	39	0	0
			IV	–	39	0	0
			V	–	39	0	0
		Cocoon		–	39	0	0
		Adult		–	39		
Eggs laid before emergence of sexuals (n=269)[a]	July 14 – Aug. 23	Egg		Cannibalism by adults	269	7	2.6
		Larval instar	I	Cannibalism by adults	262	1	0.4
			II	–	261	0	0
			III	–	261	0	0
			IV	–	261	0	0
			V	Cannibalism by adults	261	77	29.5
		Cocoon		Cannibalsim by adults	184	16	8.6
		Adult		-	168	-	-
Eggs laid after emergence of sexuals (n=98)	Aug.24 – Sept. 25	Egg		Cannibalism by adults	98	28	28.6
		Larval instar	I	Starvation	70	2	2.8
			II	Starvation	68	3	4.3
			III	Starvation	65	6	9.2
			IV	Starvation	59	7	11.8
			V	Starvation and unknown	52	47	90.2
		Cocoon		Cannibalism by adults	5	5	100.0
		Adult			0	–	–

[a] No. of eggs laid by the queen during each period.
[b] See text.

tives; and the period after their emergence. In this nest all the eggs laid during the preemergence period developed into adults (workers), but high mortality occurred during the fifth instar larval and cocoon stages, after the emergence of reproductives (after mid-August). Mortality during the later period was almost entirely due to cannibalism by workers, the dead larvae and pupae being either provided for males and new queens as food, or abandoned outside the nest. Cannibalism in *V. tropica* is assumed to be caused by lack of food protein, which almost entirely depends on the polistine brood, because the colonies of polistine wasps cease their brood rearing after mid-August. During the period after the emergence of males and new queens, higher mortality was observed throughout all developmental stages from egg to pupa, particularly the egg and fifth instar larval stages. Mortality of eggs and pupae during this period was mainly due to cannibalism by workers, while death among larvae, especially the fifth instar, was mainly due to starvation caused by the termination of provisioning. This was unlike the factors prevalent before the emergence of reproductives.

A tendency similar to that of *V. tropica* is also recognized in the life table of a *V. analis* brood in the later stages of its seasonal cycle (Table 6.3). About half of the eggs laid before the emergence of reproductives reached the adult stage (36 new queens and 27 males). Mortality was high, especially at the fifth instar larval and cocoon stages, mortality being again caused by worker cannibalism. On the other hand, the brood hatching from eggs laid by the foundress after the

Table 6.3. Life table for *Vespa analis* brood oviposited by queen before and after emergence of reproductives in nest (T7207) (Matsuura 1984)

Period	Duration of oviposition (1972)	Developmental stage		*Fdx*[a]	*lx*[a]	*dx*[a]	*qx*[a]
Eggs laid before emergence of sexuals (n = 124)	Aug. 12 – Sept. 24	Egg		Cannibalism by adults	124	7	5.6
		Larval instar	I	Cannibalism	117	2	1.7
			II	–	115	0	0
			III	Cannibalism	115	1	0.9
			IV	Cannibalism	114	2	1.7
			V	Cannibalism	112	35	31.2
		Cocoon		Cannibalism	77	14	18.0
		Adult (36 ♀♀ 27 ♂♂)			63	–	–
Eggs laid after emergence of sexuals (n = 82)	Sept. 25 – Oct. 10	Egg		Cannibalism	82	36	43.9
		Larval instar	I	Cannibalism	46	2	4.3
			II	Cannibalism	44	1	2.2
			III	–	43	0	0
			IV	Cannibalism	43	1	2.3
			V	Starvation and unknown	42	39	92.8
		Cocoon		Cannibalism	3	3	100.0
		Adult			0	–	–

[a] See text.

emergence of reproductives died from starvation or were fed upon by adults during the period from egg to cocoon. No adults emerged in this period.

Judging from these observations, together with observations of many other nests, it is probable that *Vespa* workers devote their attention mainly to the provisioning of the adult reproductives which have emerged earlier; thus most of the larvae fail to be fed because of decrease in attention given to them when the reproductives emerge and, as a result, growth-terminated and under-nourished larvae are abandoned outside the nest. It is also considered that a rapid decline in nursing activity occurs because of the fact that eggs and pupae are used as food.

2 Seasonal Changes in Population Structure

It is quite difficult to observe events inside the nests of vespine wasps, because they are covered with a thick envelope. Some species also construct their nests in covered locations. Therefore, states of seasonal change in colony development have been examined by collecting nests periodically, breaking them up and counting the number of cells, determining the constitution of each developmental stage, changes in the nest, etc. (for example, Spradbery 1971; Archer 1972). In this section, seasonal changes in the numbers of cells and individuals at different growth stages, including eggs, larvae, pupae and adults, are compared and discussed based on data for five *Vespa* species (*V. tropica, V. analis, V. crabro, V. mandarinia* and *V. simillima*) gathered over a period of 10 years in the field in and near Kibi-chô, Wakayama, in southwestern Japan (Matsuura 1984).

2.1 Number of Cells

Seasonal changes in the numbers of cells in the colonies of the above-mentioned five *Vespa* species are shown in Fig. 6.7. Interspecific differences can clearly be recognized between termination time of nest activities and colony size. In *V. tropica,* the mean number of cells per colony was only 165.8 (s.d. ± 77.3) even in September, a period during which nest development reached its maximum; nest activities ceased completely in early October. In *V. analis,* the colonies lasted about 6 months, or until late October to early November, and the mean number of cells reached 516.7 (s.d. ± 190.1) in early November. This was the smallest colony size next to *V. tropica.* In *V. crabro,* nest activities continued until early to mid-October with a rapid increase in development after August. The mean number of cells was 767.6 (s.d. ± 962.5) in the nests collected in August, and 1 984.8 (s.d. ± 1116.2) in October. In *V. mandarinia*, nest activity continued until November and the mean number finally reached 2 712.4 (s.d. ± 984.7), although after August the number fluctuated widely among nests. In *V. simillima,* the colony size of every nest was already larger than those of other species by July, and each colony subsequently showed more rapid development with the colony's relocation of the nest site to an open situation. Similar to *V. mandarinia*, the nesting activity of *V. simillima* lasted for a long time, occasionally until late

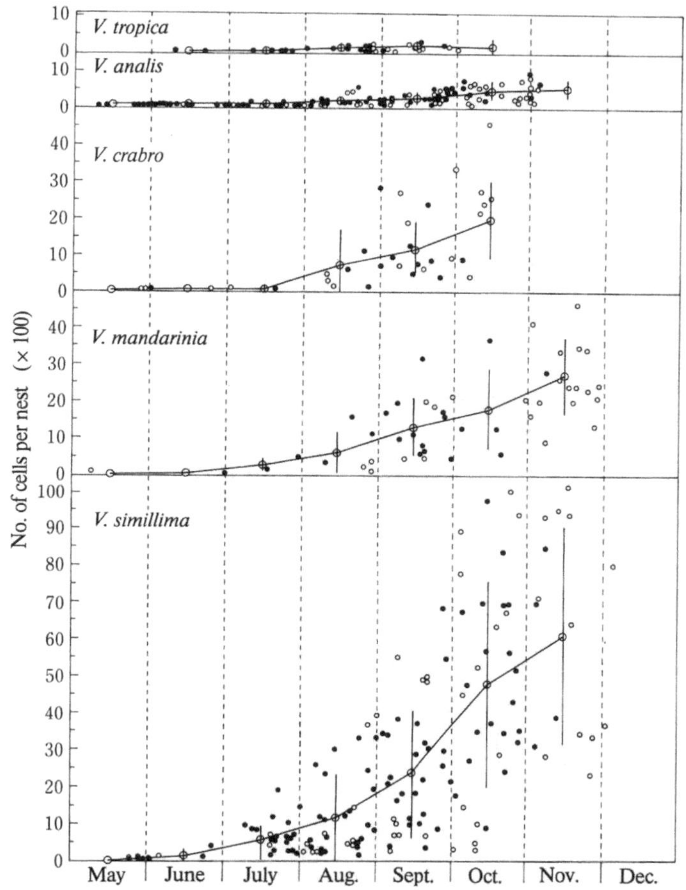

Fig. 6.7. Seasonal changes in number of cells per nest in five *Vespa* species inhabiting southwestern Honshu. ● Colonies with foundress; ○ colonies having lost foundress; ⊙ mean cell number per nest in each month (After Matsuura 1984)

November to early December. The mean number of cells per colony changed from 575.7 (s.d. ± 435.6) in July to 6087 (s.d. ± 2,869) in October, although the number fluctuated widely from nest to nest as in *V. mandarinia*. For example, in the case of foundress-occupied nests in mid-October, the number of cells in the largest nest was 9936, which was 13.3 times as large as the number in the smallest nest (746).

2.2 Number of Individuals According to Developmental Stage

Figures 6.8–6.17 show seasonal changes in the numbers of individuals in colonies of five *Vespa* species according to developmental stage, including eggs, larvae,

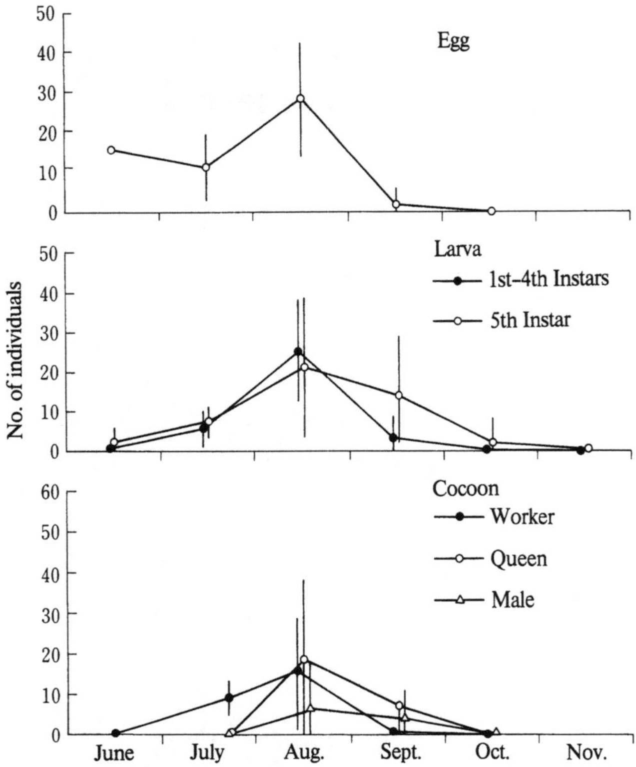

Fig. 6.8. Numbers of eggs, larvae and cocoons in colonies of *Vespa tropica*; monthly means (After Matsuura 1984)

cocoons and adults. The mean number of individuals and standard deviation (s.d.) were calculated by gathering data for periods of 10 days in the case of *V. analis* and *V. simillima* since the number of collected nest samples was sufficient or by gathering data monthly in the cases of the other three species for which insufficient nest samples were available.

Eggs, larvae and cocoons:

In *V. tropica,* the mean number of individuals at each immature stage reached a peak in August, decreased rapidly in September, except for the fifth instar larvae and cocoons of males and new queens. It became negligible in October (Fig. 6.8). In *V. crabro,* the mean number of individuals at each stage reached a peak in September, except for worker cocoons, which reached a maximum in August and whose numbers in each nest in September were lower than those of cocoons of males and new queens (Fig. 6.9). In *V. mandarinia,* the mean numbers of eggs and larvae reached their peaks during September and October, and the number of fifth instar larvae maintained a high level even in November. The number of worker cocoons reached a peak in September and declined rapidly in October.

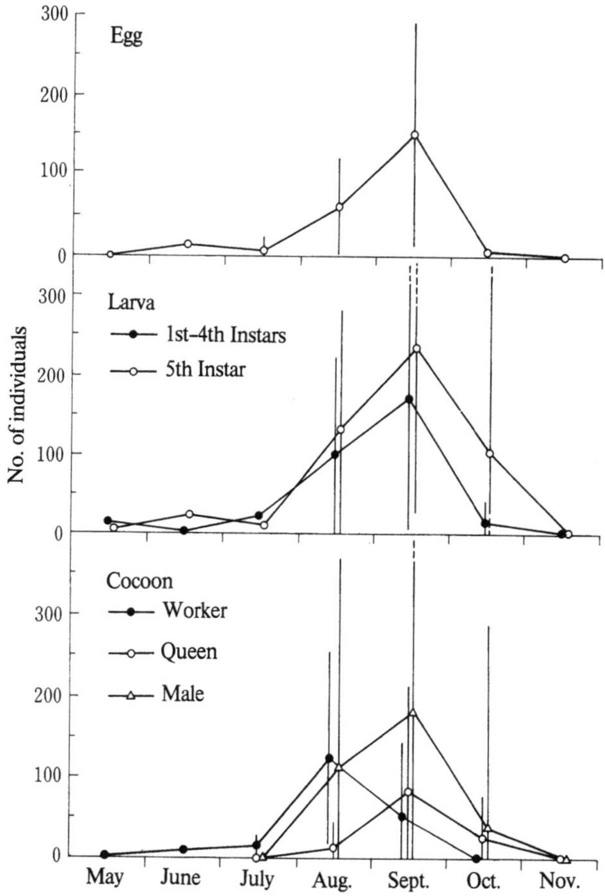

Fig. 6.9. Numbers of eggs, larvae and cocoons in colonies of *Vespa crabro*; monthly means (After Matsuura 1984)

With the onset of the decline in the number of worker cocoons, those of males and new queens started to increase and the number of new queen cocoons reached a peak in November (Fig. 6.10). In *V. analis*, the mean numbers of eggs and larvae, especially the first to fourth instars, reached their peaks in mid-August, continued almost unchanged until early October and then started to decline rapidly in mid-October. The number of worker cocoons reached a peak during late August and early September, but hardly any could be found after late September. Male cocoons, on the other hand, were found on two occasions during early and mid-August and during late September and mid-October. The male cocoons found during the former period, however, can be regarded as individuals that had originated from oviposition by workers in an orphaned colony which had lost its foundress during the polyethic period from June to July (cf. Chap. 4). Cocoons of new queens of this species were found over a long period from mid-September to early November (Fig. 6.11). In *V. simillima,* the mean numbers of eggs and first

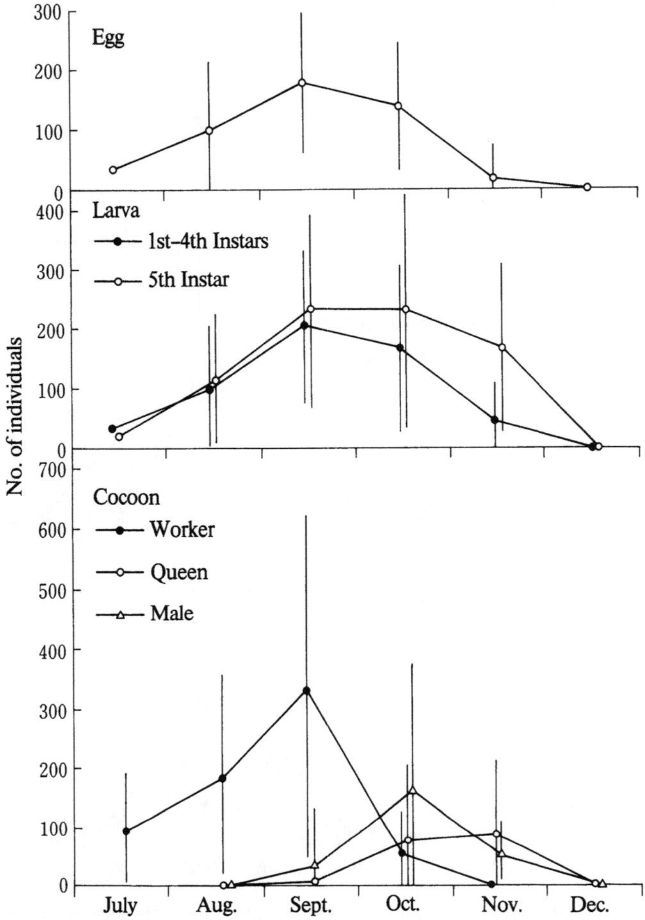

Fig. 6.10. Numbers of eggs, larvae and cocoons in colonies of *Vespa mandarinia*; monthly means (After Matsuura 1984)

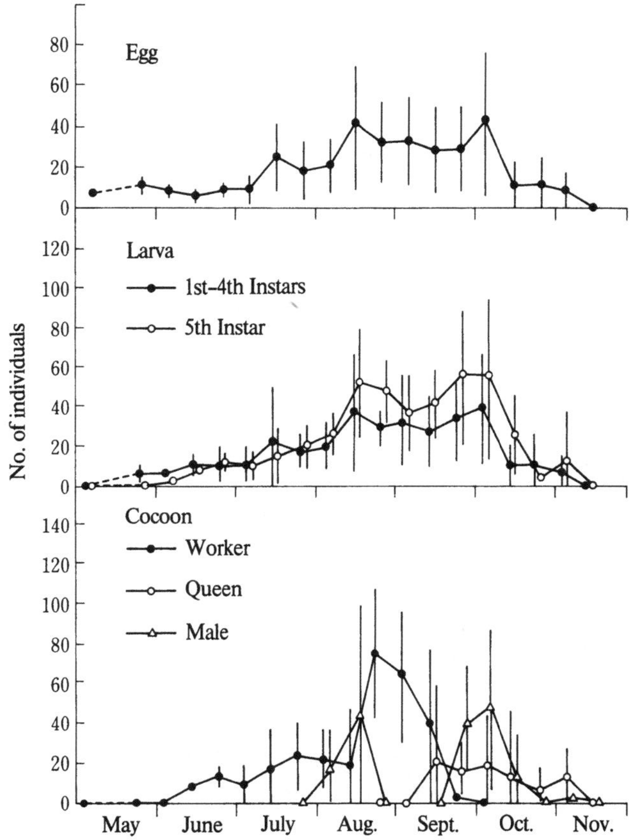

Fig. 6.11. Numbers of eggs, larvae and cocoons in colonies of *Vespa analis*; 10-day means (After Matsuura 1984)

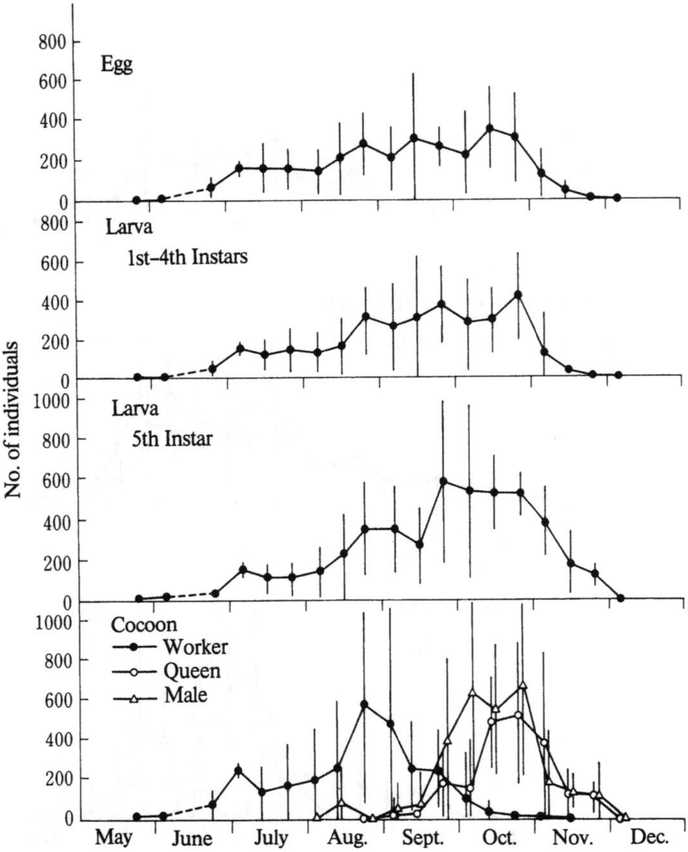

Fig. 6.12. Numbers of eggs, larvae and cocoons in colonies of *Vespa simillima*; 10-day means (After Matsuura 1984)

to fourth instar larvae increased rapidly after late August, continued to increase or remained at the same level until late October and then declined rapidly in November. The number of fifth instar larvae, on the other hand, increased rapidly after late September, was maintained at a constant level higher than the number of first to fourth instars until late October, and then started to decline in early November, although some could still be found in late November. Peaks of the mean numbers of cocoons of workers, males and new queens were observed in late August-early September, October and mid-October-early November respectively (Fig. 6.12).

Adults:

In *V. tropica,* workers were observed from late July to mid-October, but most frequently during late August and late September. Males and new queens were observed from late August to early October, with their numbers peaking in September, although a number of disorganized colonies were found before the

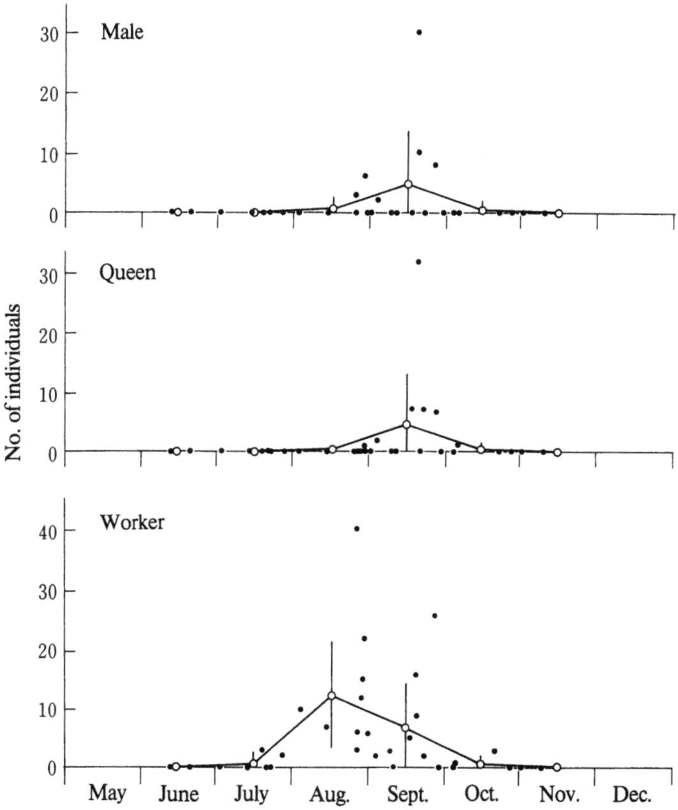

Fig. 6.13. Seasonal change in numbers of workers, males and new queens in colonies of *Vespa tropica* (After Matsuura 1984)

emergence of any males or new queens (Fig. 6.13). In *V. crabro*, workers were observed from mid-June to late October with a peak in about September. Males and new queens were observed from late August to mid-October and from mid-September to mid-October respectively, although the number of emerged individuals of either sex fluctuated widely from one nest to another (Fig. 6.14). In *V. mandarinia*, workers were observed from mid-July to late November, but most frequently during September and October. Males were usually observed from mid-October to late November and occasionally from late August to September. New queens were observed only during November, except for one nest with eight queens collected in mid-October (Fig. 6.15). In *V. analis*, workers were observed from late June to late November, with a mean number per colony during mid-August and early October maintained at a level of around 50. Males were first found in a few queenless colonies collected from early to mid-August. However, no emergence of males in colonies with queens could be observed until mid-September, although emergence was observed after that until early November with a peak during early and mid-October. New queens began to emerge from mid- to late September and their mean number per colony reached a peak during

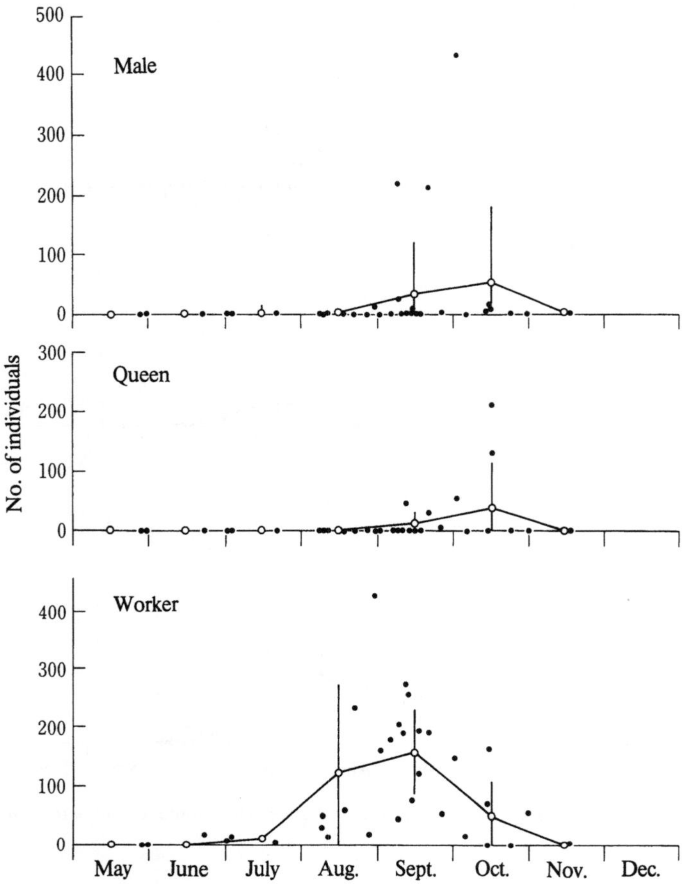

Fig. 6.14. Seasonal change in numbers of workers, males and new queens in colonies of *Vespa crabro* (After Matsuura 1984)

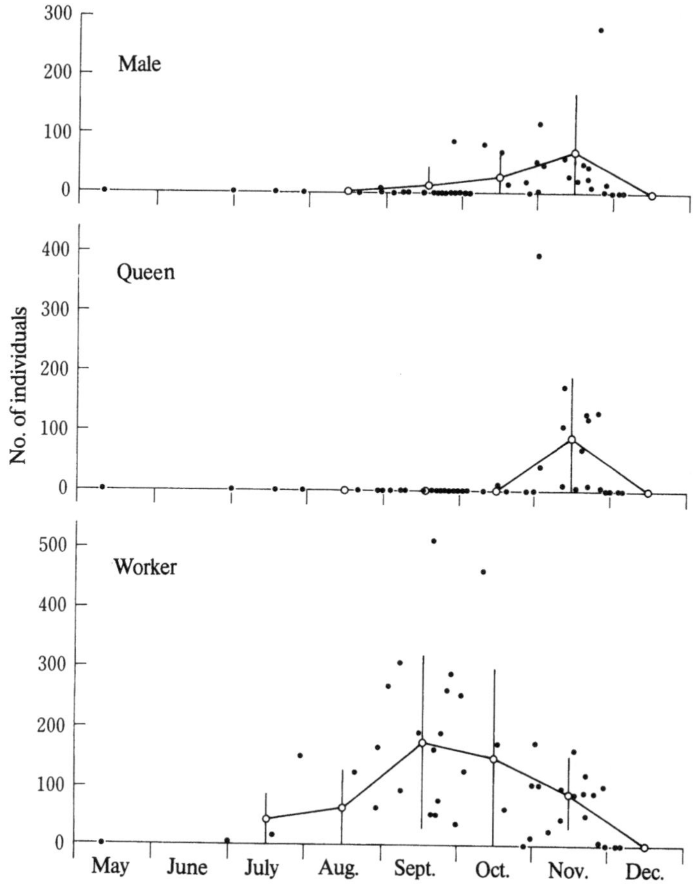

Fig. 6.15. Seasonal change in numbers of workers, males and new queens in colonies of *Vespa mandarinia* (After Matsuura 1984)

mid-October and early November. This was a little late compared with the peak for males (Fig. 6.16). In *V. simillima,* workers began to emerge from mid- to late June and their mean number per colony increased at a higher rate than those of the other species, reaching a peak in late September and starting to decline gradually thereafter, with a rapid decline in November. Males and new queens emerged at almost the same time during early and late September, the males being observed throughout a period during early October and early November and the new queens during mid-October and mid-November (Fig. 6.17).

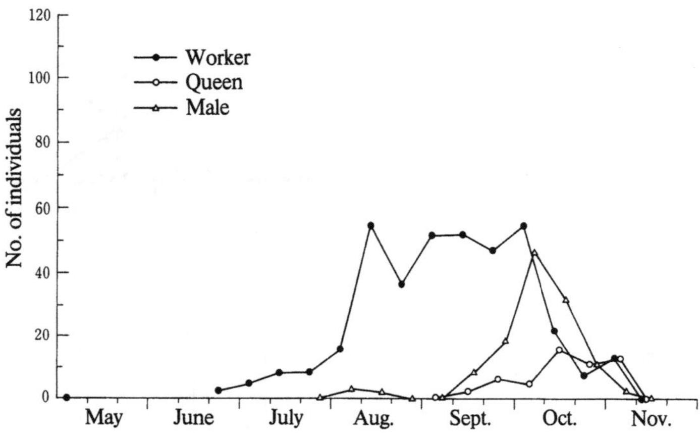

Fig. 6.16. Seasonal change in numbers of workers, males and new queens in colonies of *Vespa analis*; 10-day means (After Matsuura 1984)

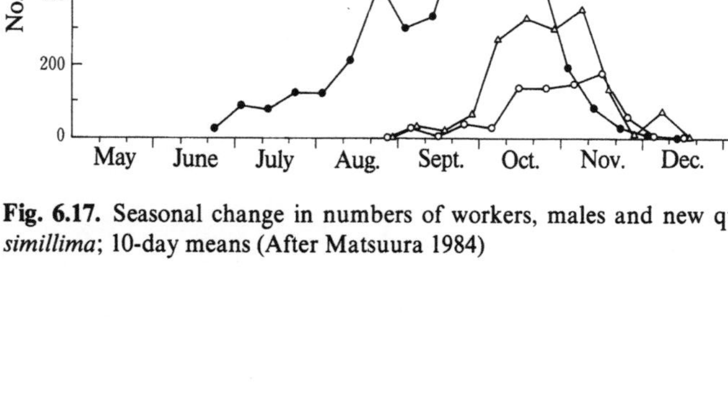

Fig. 6.17. Seasonal change in numbers of workers, males and new queens in colonies of *Vespa simillima*; 10-day means (After Matsuura 1984)

2.3 Colony Size

Vespine wasps develop huge colonies compared with those of the Polistinae (*Polistes, Parapolybia*) in temperate regions, and the colony size shows clear interspecific differences. Colony size can be classified roughly into three types, in accordance with the final numbers of cells and combs at the later stage of the seasonal cycle, based on records so far reported on the colony size of Vespinae in Japan (e.g., Iwata 1971a; Matsuura 1974a, 1976b, 1980b, 1984; S. Yamane 1977; Sk. Yamane et al. 1980b; Makino et al. 1981; Makino 1982) and the data on *Vespa* described above:

1. Small-colony type:
   The numbers of cells and combs per colony are generally about 1000 or below and 3–4 respectively. *Vespa*: *V. tropica, V. analis; Vespula*: *Vl. schrenckii; Dolichovespula*: *D. media, D. saxonica, D. norvegicoides.*
2. Medium-colony type:
   The numbers of cells and combs per colony are generally 2000–4000 and 4–12 respectively. *Vespa*: *V. crabro, V. dybowskii, V. mandarinia, V. simillima simillima, V. affinis.* No corresponding species of *Vespula* or *Dolichovespula.*
3. Large-colony type:
   The numbers of cells and combs per colony are generally 4000–10 000 and 6–12 respectively. *Vespa*: *V. simillima xanthoptera; Vespula*: *Vl. vulgaris, Vl. flaviceps, Vl. shidai.*

As described already, the colony size of vespine wasps differs remarkably even among nests of the same species in the same region, and this tendency is strong in such species of the large-colony type (cf. Fig. 6.7). In *V. simillima,* such a difference can be seen clearly even between its subspecies, such as the subspecies in Hokkaido (*V. s. simillima*) and the subspecies in South Honshu (*V. s. xanthoptera*). This can be regarded as a result of climate-dependent restriction of colony size, because the seasonal cycle is shortened in a cold district like Hokkaido. However, in a species whose colony size is small and the seasonal cycle relatively short, such as *D. media,* differences in the number of cells, etc., between its subspecies in Hokkaido and Honshu are small. Any difference originating from their geographical distribution can easily be neutralized.

Table 6.4 summarizes a comparison of nests of the largest colony size in each species of vespine wasps so far observed, based on the above-described data on *Vespa* and other reports so far available. The nests shown in this table are quite rare examples, because the numbers of cells and combs in each nest exceed the ranges of numbers seen in usual colony-size nests as described above, although it is necessary to bring the biological characteristics of a population of a species into full play under optimum environmental conditions for the achievement of such large numbers. In any case, difference in colony size as a species-specific property can be seen clearly in the table. For example, in *Vespa* species, the number of cells in the largest nest of a small-colony-type species, *V. tropica,* is only about 1/46 of that of the largest nest of a large colony-type species, *V. simillima xanthoptera.*

Table 6.4. Number of cells in largest nest for each Japanese species of Vespinae

Species	Locality	Date	No. of combs	No. of cells	Literature
Vespa					
V. tropica	Kibi, Wakayama	Sept. 18, 1971	3	313	Matsuura (1984)
V. analis	Kibi, Wakayama	Oct. 29, 1975	4	752	Matsuura (1984)
V. crabro	Kibi, Wakayama	Oct. 22, 1973	6	4563	Matsuura (1984)
V. mandarinia	Kibi, Wakayama	Nov. 21, 1964	6	4661	Matsuura (1984)
V. simillima simillima	Sapporo, Hokkaido	Sept. 11, 1979	8	4209	Makino et al. (1981)
V. s. xanthoptera	Sasayama, Hyogo	Dec. 15, 1955	10	14 272	Iwata (1971a)
Vespula					
Vl. schrenckii	Saku, Nagano	Aug. 29, 1980	4	1425	Takamizawa (1981a)
Vl. flaviceps	Musashino, Tokyo		12	16 800	Iwata (1971a)
Vl. shidai	Saku, Nagano	Nov. 4, 1980	7	16 850	Takamizawa (1981b)
Dolichovespula					
D. media media	Otaru, Hokkaido	Aug. 11, 1980	3	1987	Makino (1982)
D. norvegicoides	Nagawa, Nagano	Aug. 31, 1980	5	2251	Wakabayashi, (unpubl.)
D. saxonica	Otaru, Hokkaido	Aug. 2, 1979	3	1521	Makino (1982)

A relation can be seen between such species-specific colony size and phylogenetic positions. Although interspecific differences in colony size are remarkable in *Vespa*, species of the medium- and large-colony types are basically the norm for this genus, and the two small colony-type species, *V. tropica* and *V. analis*, can be regarded as exceptions. These two exceptional species are distributed widely from the tropical lowlands to the subtropics, but their colony sizes in the tropics are several to a little over ten times as large as those in temperate zones. This indicates that reduction in colony size has been proceeding in temperate zones as a result of secondary adaptation to climate and to the amount of available food resources (Matsuura 1983b; cf. Chap. 7). *V. basalis, V. velutina,* and *V. affinis* in Taiwan are also of large-colony types, and their colony sizes are equivalent to or larger than that of *V. simillima* in Japan (Matsuura 1970a, 1973c; S. Yamane 1977).

In the genus *Vespula,* a bidirectional relation can be seen between two subgenera: *Vespula* as a small colony type and *Paravespula* as a large colony type. *Vl. rufa* in Europe, which belongs to the subgenus *Vespula,* and *Vl. acadica* and *Vl. consobrina* in North America, which also belong to the same group, have similar small colony sizes (Gaul 1948; L.H. Taylor 1939; Spradbery 1973a; MacDonald et al. 1975; MacDonald and Matthews 1976; Akre et al. 1981), but *Vl. atropilosa* and *Vl. vidua* in North America have a slightly larger colony size, sometimes containing more than 2000 cells per nest (MacDonald et al. 1974; MacDonald and Matthews 1976; Akre et al. 1981). In the subgenus *Paravespula,* on the other hand, the number of cells per colony has been known to reach several thousand or even exceed 10 000, including not only the above three species in Japan but also *Vl. germanica* in Europe and *Vl. maculifrons, Vl. pensylvanica* and

Vl. flavopilosa in North America (MacDonald et al. 1974; MacDonald and Matthews 1976; Spradbery 1971; Akre et al. 1981). Overwintered colonies of *Vl. germanica*, which naturalized recently into the Southern Hemisphere, are especially noteworthy (Thomas 1960). In New Zealand, the colony size of annual nests of this species is similar to those in Europe and North America, but overwintered colonies of the naturalized species sometimes construct huge nests containing 3–4 million cells per nest (Spradbery 1973b). In the genus *Vespula, Vl. squamosa* and *Vl. sulphurea* in North America are considered to have different phylogenetic positions from the above two subgenera, and the mean number of cells per colony in the former species, of which colony size has been examined, is about 5000, with the largest annual nest containing nearly 10 000 cells (Akre et al. 1981).

As for the genus *Dolichovespula,* the above-mentioned three species in Japan are the typical small-colony type. The same species in Europe also show the same tendency (Janet 1903; Kemper and Döhring 1967), although a huge nest of *D. norwegica* containing 4078 cells has been found in England (Archer 1980b). The number of cells in typical colonies of *D. arenaria* and *D. maculata* in North America is 300–1500, but the colony size of either species sometimes reaches 3000–4000 cells per colony (Greene et al. 1976; Akre et al. 1981).

2.4 Duration of Life Cycle and Type of Life History

The life cycle of vespine wasps is in principle annual. There are examples of perennial colonies, in *Vl. vulgaris* (California, USA) and *Vl. squamosa* (Georgia and Florida, USA; 65 000 and 95 000 cells per colony respectively) (Duncan 1939; Ross and Matthews 1982) in addition to the famous case of the naturalized population of *Vl. germanica* in New Zealand and Tasmania (Spradbery 1973b). In Taiwan, *Vl. flaviceps karenkona* develops biennial colonies, but the number of cells per colony is below 15 000 (Sk. Yamane and S. Yamane 1975). These colonies do not last for more than two years.

Duration of life cycle is restricted not only by changes in the geographic environment in association with altitude and latitude, but also by specific and attributive properties and the size of the colony.

Among Japanese *Vespa* species, *V. tropica,* a small-colony type and *V. simillima,* a large-colony type, have the shortest (ca. 3.5–4 months) and longest (7–7.5 months) life cycle durations respectively, and the other three species are placed between these two extremes in the order of *simillima-mandarinia-analis-crabro-tropica. V. simillima* and *V. tropica* can be regarded as a long-cycle species and a short-cycle species respectively.

The life history type of the long-cycle species is characterized by: (1) long duration of life cycle because of the early initiation and late termination of nesting activity, (2) large colony size, (3) high production numbers of males and new queens, (4) high threshold values for the numbers of workers and cells related to the rearing of new queens and (5) a generalist on feeding habit. In contrast, the life history of short-cycle species is characterized by: (1) late initiation and early termination of nesting activity, (2) small colony size, (3) low production numbers of males and new queens, (4) low threshold values for the numbers of workers and cells related to the rearing of new queens and (5) a specialist on feeding habit. However, it is difficult to classify the other three *Vespa* species in either of the above two categories, since each of them has its own trend with regard to the above five items. For example; *V. analis* is a long-cycle species with a generalist feeding habit, but is a species of the small-colony type.

In *Vespula*, on the other hand, *Vl. schrenckii*, which belongs to the subgenus *Vespula*, is a typical short-cycle species, while *Vl. flaviceps, Vl. shidai* and *Vl. vulgaris* of the subgenus *Paravespula* are regarded as long-cycle species, because they possess the first four of the five characteristic features of the long-cycle species of *Vespa*. With regard to this fifth characteristic, feeding habit, *Vl. schrenckii* preys solely on live insects as a specialist, while these three species of the subgenus *Paravespula* have a broad food preference, including the dead flesh of vertebrates and roasted fish, as a generalist (cf. Chap. 5). As for the subgenus *Vespula*, a similar life history to the Japanese *Vl. schrenckii* as a short-cycle species has been found in every species of the "*rufa* group", such as *Vl. rufa* in Europe and *Vl. acadica, Vl. atropilosa, Vl. consobrina, Vl. vidua* and *Vl. intermedia* in North America (Spradbery 1973a; Edwards 1980; Archer 1981a; Akre et al. 1981). In the subgenus *Paravespula*, on the other hand, *Vl. germanica* in Europe and *Vl. maculifrons, Vl. pensylvanica* and *Vl. flavopilosa* in North America are the long-cycle species and, although the accumulated data on each species are insufficient, it seems that these species have a similar life history type to that of the same subgenus in Japan (Spradbery 1973a; Archer 1981a,b; Akre et al. 1981). Consequently, the genus *Vespula* differentiates itself into two distinct bidirectional life history types: the subgenera *Vespula* and *Paravespula*. Each type can be regarded as a phylogenetic characteristic of the corresponding subgenus.

With regard to *Dolichovespula*, any of the three Japanese species (*D. media, D. saxonica, D. norvegicoides*) is a typical short-cycle species, having the above-described life history characteristics: (1)–(4) and a food preference restricted to live insects, similar to the case of the *Vl. rufa* group. A similar life history type can be seen in *D. sylvestris* in England and *D. arenaria* in North America (Archer 1981a,b; Greene et al. 1976), but *D. maculata* occasionally feeds on the raw meat of vertebrates (Payne and Mason 1971; Greene et al. 1976). Although comparative ecological studies on the species of this genus have not yet been performed in full, it is obvious that the genus possesses a short life-cycle history as a phylogenetic property. Since this genus has many similarities to the subgenus *Vespula* (*Vl. rufa* group) with regard to life history type as well as nesting habits (cf. Chap. 3) and feeding habits (cf. Chap. 4), it may be of great interest to study them from the viewpoint of their phylogenetic relations.

2.5 Seasonal Development of a Single V. tropica Colony

It is very difficult to follow the development of individual colonies of vespine wasps during their active period from nest foundation to disintegration, because

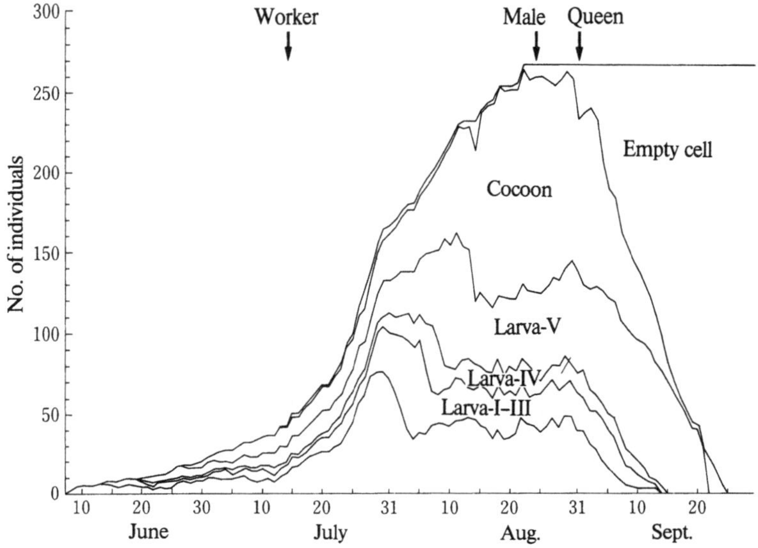

Fig. 6.18. Seasonal change in numbers of eggs, larvae, cocoons and cells in a colony of *Vespa tropica* (After Matsuura 1984)

of the complex nest structure, although this is less so for *V. tropica,* whose colony size is smallest. Since the colony size of vespine wasps varies greatly even among nests of the same species, as mentioned already, it is important to know the changes occurring in the development of a standard colony from nest foundation to disintegration. As an example of such a study, data on a single colony of *V. tropica* (Matsuura 1984) are introduced below (Figs. 6.18, 6.19).

The nest used in this study was founded in an empty box in a hut in Kibi-chô, Arida-gun, Wakayama Prefecture, on June 9, 1974. Examinations were performed every day on the number of cells, the constitution of each developmental stage of growth and the numbers of emerged workers, males and new queens throughout the active season, from nest foundation to disintegration via the emergence of males and new queens (Matsuura 1984).

Cells:

In the period from nest initiation to worker emergence, the mean number of cells constructed per day was 1.1 ± s.d. 1.3 (range: 0–6). After worker emergence, the number of cells increased rapidly with the mean number of cells constructed per day reaching 4.9 ± s.d. 3.4 (range: 0–15) until the nest reached its maximum size on August 21, containing 268 cells and 3 combs. The number of empty cells rarely exceeded 6.0% of the total number of cells until the emergence of sexuals, because the foundress generally laid an egg as soon as a new cell was constructed. The construction of cells was terminated with the emergence of males, and the empty cells increased rapidly after the emergence of new queens, due to the cessation of oviposition in spite of the existence of the foundress.

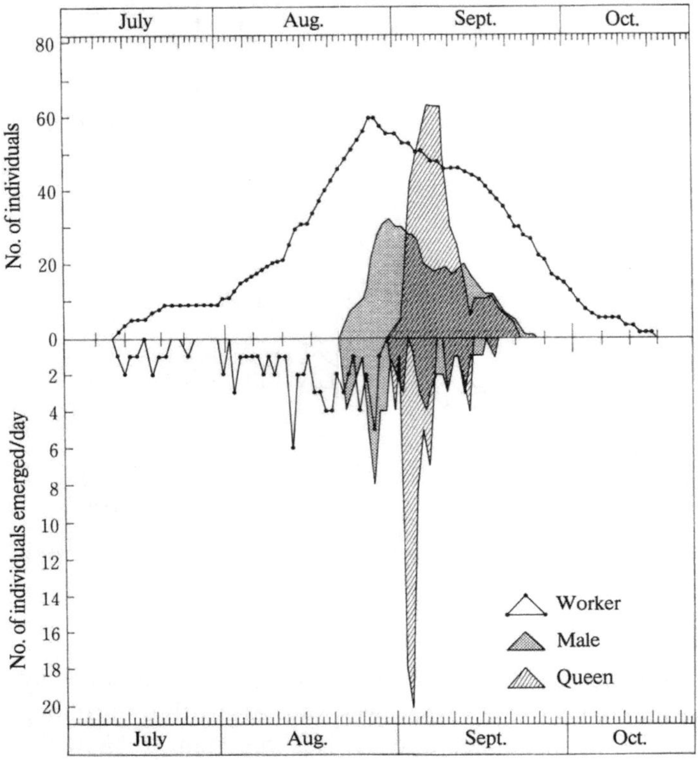

Fig. 6.19. Daily counts of workers, males and new queens in a colony of *Vespa tropica* (After Matsuura 1984)

Eggs, larvae and cocoons:

The foundress laid an average of 1.4 eggs ± s.d. 1.8 (range: 0–4) per day during preemergence period. After worker emergence, the oviposition rate increased rapidly and the highest rate reached 14 eggs per day on July 27 and 28. Thereafter, the oviposition rate started to decline until the final oviposition on September 13.

In the larval stages, numbers of first to fourth instar larvae reached a peak on August 5, resulting in a rise in the number of fifth instar larvae after about one week, when they accounted for nearly 75% of all the larvae present.

The number of cocoons reached a peak on August 25, when they accounted for nearly 52.5% of the total brood, with a maximum of 137 cocoons per day. All the brood stages, including eggs, larvae and cocoons, decreased rapidly with the commencement of the emergence of reproductives and disappeared during mid- and late September, in the order of immature larvae, cocoons and fifth instar larvae.

Adults:

The total number of workers was less than 10 for the first 18 days after first emergence, but reached a maximum of 60 in late August and then stopped. Emergence of workers lasted for 44 days. The total number of workers produced in this colony was 71, and the rate of emergence was 1.7 (s.d. ± 1.2) workers per day up to a maximum of six workers per day. The first males emerged at 67 days, and the first new queens emerged at 74 days after colony foundation. All males emerged within 25 days and all new queens within 18 days. The total number of reproductives produced in this colony was 61 males and 75 new queens. In particular, the emergence of new queens was characterized by a concentrated occurrence within a short period: in this colony, a total of 38 new queens emerged during September 1 and 2, and the number was nearly 50.7% of the total number of adults. Both sexual types stayed for about two weeks in the nest: males began to depart on and after September 1, new queens from September 10, leaving only workers in and around the nest in late September.

Judging from these results, it may be supposed that *V. tropica* in temperate zones achieves maximum colony development when the number of workers reaches a peak, since abundant adult workers and a stock of mature larvae as provisioners of secretion facilitates concentrated production of males and new queens, but the colony starts to decline rapidly when the reproductives leave the nest. Such a tendency seems to be a common phenomenon in species with small colonies and a short life cycle, such as *Vl. schrenckii* and each species of *Dolichovespula*, but can also be regarded as a basically immutable life history type, even in species with large colonies and a long life cycle, including other *Vespa* species and each species of *Paravespula*.

3 Production of Reproductives

3.1 Number of Reproductives Produced per Colony

In vespine wasps in temperate zones, new queens and males are produced when the number of workers reaches a peak. Early emergence of males occurs occasionally from eggs laid by a *V. analis* foundress in a nest at an early stage just after the first emergence of workers (Matsuura 1966a), and their emergence occurs also from eggs laid by workers in an orphaned colony or one which has lost its foundress after worker emergence (cf. Chap. 4). The males produced before the emergence of new queens take no interest in the workers and do not engage in mating activity, because of the absence of new queens with which to copulate. Emergence of males occurs a little before the emergence of new queens in every species of vespine wasp and, in a single colony, it generally continues even after their emergence.

The total production number of males and new queens per colony can be obtained by, for example, the marking of all emerged individuals in the small-colony-size species, like the above-described case of *V. tropica.* Such an examination, however, is quite difficult to perform technically when the colony size is large, because of the long period required for adult emergence. Consequently, the numbers of adults and cocoons of males and new queens were examined for five *Vespa* species in their nests collected in southwestern Japan in order to determine the rough numbers of emerged individuals. The results are shown in Table 6.5.

In *Vespa,* the maximum numbers of both males and new queens varies from several tens in a *V. tropica* nest to over 1000 in a *V. simillima* nest, showing remarkable interspecific differences. The number of reproductives produced per colony is ranked as *simillima* > > *mandarinia* > *crabro* (> *dybowskii*) > *analis* > *tropica*. The production of reproductives, especially new queens, depends greatly upon the colony size, not only among different species but also among conspecific colonies. Figure 6.20 shows the relationship between the number of cells and the total number of new queens and their cocoons in *V. simillima* nests collected in October, a season of full emergence of new queens. In the large successful colonies exceeding about 8000 cells, at least 1000 new queens were produced. In the poor colonies, such as one colony with 746 cells (collected on October 13) and another with 949 cells (collected on October 10), the total number of reproductives was only 2 and 16 respectively. It seems that the number of new queens produced from a poor colony does not exceed 50, even if the number of eggs and larvae being reared are included. Concerning artificially controlled colonies of *V. simillima,*

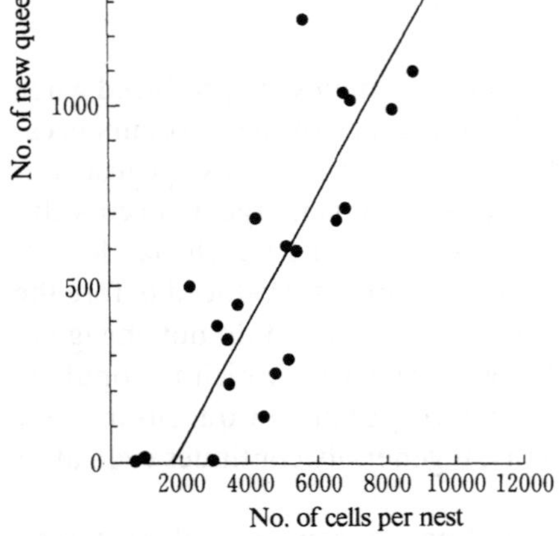

Fig. 6.20. Relation between numbers of new queens (cocoons + adults) and of cells in colonies of *Vespa simillima* collected during October in Wakayama (After Matsuura 1984)

Table 6.5. Number of reproductives produced per colony in five *Vespa* species[a] (Matsuura 1984)

Species		No. of new queens present: Cocoon (A)	Adult (B)	A+B (C)	No. of males present: Cocoon (D)	Adult (E)	D+E (F)	♀/♀+♂ (C/C+F)	No. of workers present	No. of cells (G)	Correlation coefficients between C and G
V. tropica n=6[b]	$\overline{m}$	28.6 (23.6)	8.2 (12.1)	36.8 (26.9)	11.0 (11.0)	4.8 (3.8)	15.8 (10.8)	0.65 (0.30)	18.5 (14.1)	207.0 (71.1)	r=0.348
	min.	0	2	2	15	2	17	0.11	2	126	
	max.	56	0	56	24	3	27	0.68	40	275	
V. analis n=20[b]	$\overline{m}$	27.7 (22.8)	13.0 (21.4)	40.7 (36.8)	31.3 (36.4)	32.7 (26.6)	64.0 (54.6)	0.45 (0.32)	53.4 (32.1)	498.3 (179.9)	r=0.472
	min.	4	8	12	1	4	5	0.71	21	225	
	max.	74	77	151	0	6	6	0.96	24	707	
V. crabro n=13[b]	$\overline{m}$	79.5 (119.8)	27.8 (39.3)	107.4 (122.5)	131.0 (207.8)	58.2 (128.7)	189.2 (316.4)	0.54 (0.33)	141.8 (76.4)	1,378.3 (958.5)	r=0.352
	min.	28	4	32	4	1	5	0.39	53	327	
	max.	420	35	455	188	212	400	0.53	192	2476	
V. mandarinia n=12[b]	$\overline{m}$	101.0 (105.1)	103.7 (107.8)	204.7 (150.5)	117.9 (147.5)	94.3 (99.1)	212.2 (176.8)	0.50 (0.21)	136.2 (111.8)	2546.7 (1,021.7)	r=0.584 ($p < 0.001$)
	min.	0	8	8	14	20	34	0.19	22	822	
	max.	333	0	333	532	85	617	0.35	458	3,690	
V. simillima n=21[b]	$\overline{m}$	558.9 (371.6)	137.2 (111.3)	696.0 (467.8)	567.0 (335.6)	289.6 (206.3)	856.6 (460.6)	0.42 (0.31)	510.2 (315.4)	5,465.9 (2,336.9)	r=0.856 ($p < 0.001$)
	min.	16	0	16	170	3	173	0.09	81	949	
	max.	1380	312	1692	462	661	1123	0.60	471	9323	

[a] Figures in parentheses denote SD (±).
[b] No. of nests observed.

males were produced, but in poor nests with less than 500 cells at the time of colony completion in early November the females were all workers. By contrast, 42 and 61 new queens were produced in two successful colonies with 1200 and 1480 cells respectively (Matsuura unpubl.). Judging from these data obtained in the field and under artificial conditions, it seems that the number of cells in a colony of *V. simillima* needs to reach a colony level of at least 1000 cells in order to attain a sufficient colony size to produce new queens. On the other hand, threshold values of colony size necessary during the production period of new queens are low in species of small- and medium-colony type in comparison with *V. simillima.* The minimum number of cells required for new queen production in each of the species is estimated on the basis of Table 6.5 etc., to be as follows: *V. mandarinia,* ca. 800 cells; *V. crabro,* ca. 300 cells; *V. analis,* ca. 200 cells, and *V. tropica,* ca. 100 cells, showing a tendency for colonies of smaller size to have a lower threshold value among the four species.

In *Vespula,* reports on the number of produced reproductives are few. *Vl. schrenckii* produces the lowest number of reproductives and the number per colony is generally below 200 for either males or new queens, although the nests of the other three species in Japan (*Vl. flaviceps, Vl. shidai* and *Vl. vulgaris*) sometimes contain from several hundred to over a thousand reproductives at the time of nest collection (Shida 1953; Matsuura 1976b; Sk. Yamane et al. 1980b; Takamizawa 1981b; Ikeda and Kodomari 1982). In this genus in Europe and North America, the number of reproductives is high in one subgenus, *Paravespula,* in comparison with the other subgenus, *Vespula,* which is similar in colony size to the Japanese *Vl. schrenckii.* Specially, in *Vl. maculifrons* in North America (Akre et al. 1981), the mean number of cells of new queens per colony was found to be 1622 (in 1974, n = 12) – 2780 (in 1975, n = 10).

In two *Dolichovespula* species in Japan (*D. media* and *D. saxonica*), total numbers of adult males and new queens per colony in the nests collected in August were 34–188 and 47–106 respectively, and those of cocoons of males and new queens were 70–718 and 37–273 (Makino 1982). In *D. sylvestris* and *D. norwegica* in England, the numbers of males and new queens produced per colony of the former species were 282 and 252 respectively, and those of the latter species were about twice as great as those of the former species (Archer 1981a). In *D. arenaria* in North America, the maximum numbers of adult males and new queens at the time of nest collection were recorded as 188 and 100 respectively (Akre et al. 1981). Two nests of *D. norwegica albida* have been collected as the northernmost record of this genus, and the numbers of new queens in those nests were estimated to be 13 and several tens respectively (Sk. Yamane et al. 1980a). Since this genus is a small-colony type as described above, the number of reproductives produced per colony will not exceed 1000, unlike with the *Paravespula.*

3.2 Larva: Worker Ratio (L/W)

The change in the larva:worker ratio in a colony has been shown to be one of the most important factors favoring the production of new queens in various social

Hymenoptera (e.g., *Bombus*: Cumber 1949; Polybiini: Richards and Richards 1951).

Figure 6.21 shows the seasonal change of larva: worker ratio (L/W ratio) in colonies of five *Vespa* species. The L/W ratio remained generally high for a while after emergence of workers, but started to decline gradually with the development of colonies. The ratio approached less than 2:0 before the period when colonies began to produce the queen brood and during the period from queen brood rearing to the cocoon stage. The ratio then started to rise again or become irregular as the new queens began to emerge in the nest. Remarkable variation in the L/W ratio was observed among the nests of each species throughout the whole cycle, and this was thought due to the existence of a significant difference in numbers of larvae and workers depending on each nest, even in the same species. Consequently, the relationship between the L/W ratio and the number of workers in

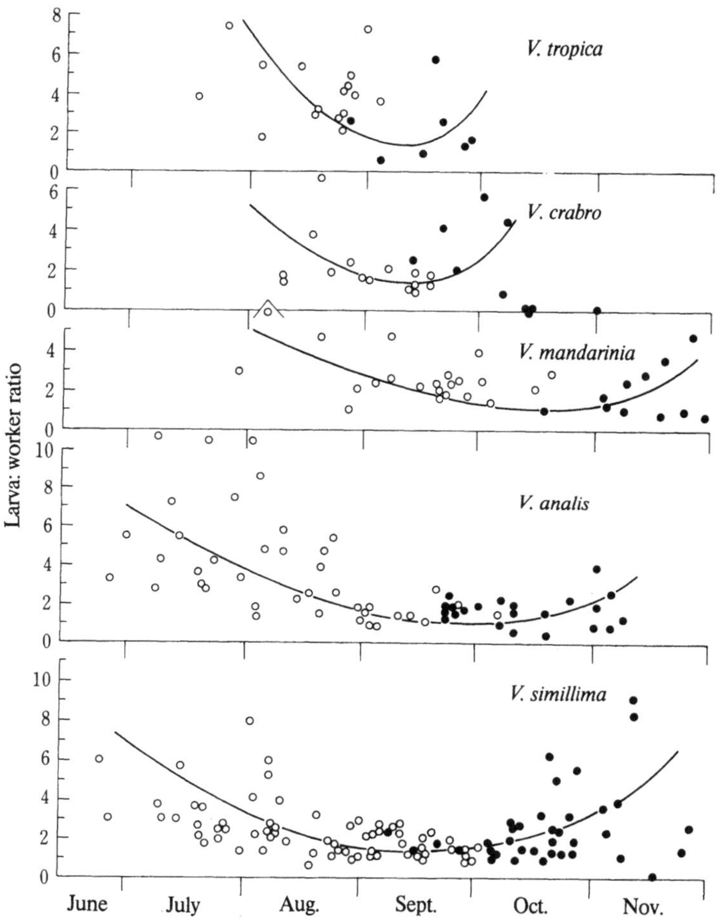

Fig. 6.21. Seasonal change in larva: worker ratio in five *Vespa* species in southwestern Honshu. ○ Colonies not rearing new queens; ● colonies with immatures or adults of new queens (After Matsuura 1984)

the colonies before the emergence of new queens was checked by rearranging the data used in Fig. 6.21. The results are shown in Figs. 6.22 and 6.23. Despite the remarkable difference in colony size between *V. simillima* and *V. analis,* the L/W ratios for the two species showed quite similar values when the number of workers in the colony of each species reached a certain level. That is, the L/W ratios varied only between 2 and 1 when the number of workers exceeded 300 in *V. simillima* and 40 in *V. analis.* In *V. crabro* and *V. mandarinia*, which are very similar in colony size, the L/W ratios also varied between approximately 2 and 1 in colonies with more than about 100 workers during the queen-rearing period. Judging from these data, it seems likely that a reduction in the L/W ratio to a point below 2 is a signal for commencement of new queen production, though the minimum number of workers for rearing the queen brood differs considerably among *Vespa* species, as does the number of cells, as is described above.

In *Vespula*, colonies of *Vl. vulgaris* and *Vl. germanica* in England started construction of cells for new queens when their L/W ratios were reduced to 1 and 0.6 respectively, with more than 1000 workers per colony in either species. This was generally around early August (Spradbery 1971). However, when the L/W ratio in a colony of *Vl. vulgaris* was reduced artificially by introducing more

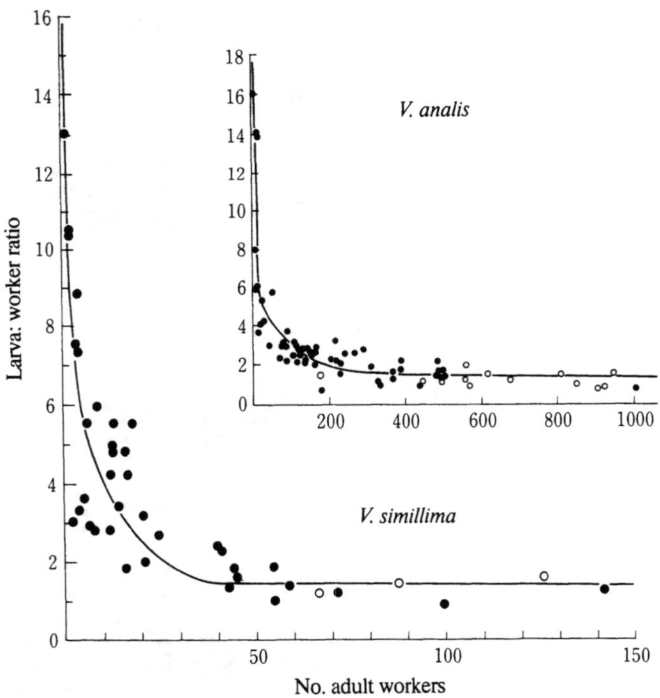

Fig. 6.22. Relation between number of workers and larva : worker ratio in colonies of *Vespa simillima* and *V. analis* prior to new queen eclosion. ● Colonies without new queen cocoons; ○ colonies with new queen cocoons (After Matsuura 1984)

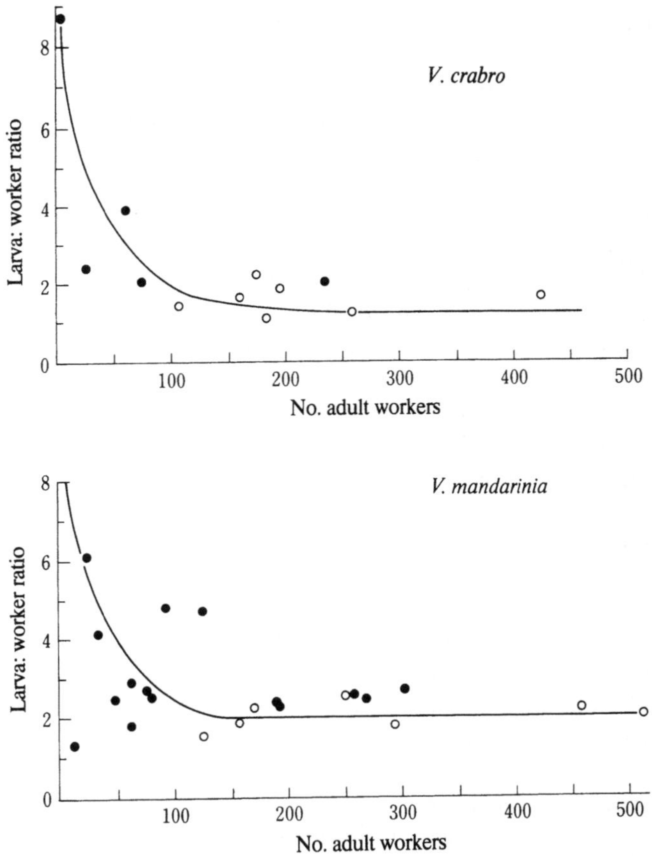

Fig. 6.23. Relation between number of workers and larva: worker ratio in colonies of *Vespa crabro* and *V. mandarinia* prior to eclosion of new queens. *Symbols,* see Fig. 6.22 (After Matsuura 1984)

workers from other colonies, construction of cells for new queens did not begin (Potter 1964; Edwards 1980).

As for *Dolichovespula,* the L/W ratio in the colonies of *D. sylvestris* in England became less than 5 about 10 days after emergence of the first workers, and continued to decrease after that, with the increase in the number of workers reaching its lowest value at the time of new queen production (Archer 1981a).

L/W ratio can also be regarded as an index of the potential labor force in a colony, and the ratio reaches its lowest value when the number of workers reaches the maximum for the whole life cycle of the colony. Such a potential labor force seems to contribute not only to nursing activity for the production of new queens, but also to total colony stabilization by means of improvement of thermoregulation and protection of the nest.

4 Sympatry and Mechanism of Coexistence in Five *Vespa* Species

Each of the four genera of Vespinae is a group showing advanced specialization, having a highly developed social structure. Remarkable diversification of habit can be seen especially among the species of *Vespa*.

The genus *Vespa* is distributed in a wide area from the tropics to temperate zones, but a region between the eastern Himalayas and southern China is regarded as the original site, since many of its species are most frequently found there (van der Vecht 1957). In East Asia, this genus has a tendency to be reduced to few species as it approaches its northern and western limits, and this is in marked contrast to *Vespula* and *Dolichovespula* (Sk. Yamane et al. 1980b). Also, in Japan, *Vespula* and *Dolichovespula* generally inhabit high-latitude districts, such as Hokkaido and Tohoku, or in mountainous areas over 1000 m above sea level in the South of Honshu; Vespa is rare or absent in both cold districts and high mountains, but is rather abundant in hills and open fields in southwestern Honshu. Thus, stable sympatric populations are maintained among the five species: *V. simillima, V. analis, V. crabro, V. mandarinia* and *V. tropica*. Each of these species is a large-sized predator armed with a poisonous sting and well developed mandibles, and, backed up by their mighty colonies, they seem to occupy the highest ecological niche among small animals, including insects in the fields and mountains of Japan. It would be of great interest, from the viewpoint of population ecology, to know the mechanisms responsible for the stable coexistence of such sympatric species.

It can be seen, by comparing the life history characteristics of these five *Vespa* species, as has been done above, that each species has developed a diversified specificity with regard to features such as nesting habits (Chap. 3), feeding habits (Chap. 5) and life history types (Chap. 6).

First, it can be pointed out that each of these five species has its own specific nesting situation which serves as a basis for the various preference for nesting sites, from terrestrial open spaces in the case of *V. analis* to subterranean situations in *V. mandarinia*. This prevents interspecific competition and, as a result, helps in the development of species-specific nest structure specialization and nest site modification, which cannot be found in *Vespula* and *Dolichovespula*. In particular, relocation of colonies in the middle of the seasonal cycle by *V. simillima* and *V. crabro*, whose nest sites differ from each other, is a most characteristic habit which cannot be found in any other species of Vespinae.

The second point is interspecific differentiation in food preference and diversification of foraging behavior. Differences in the source of prey used for provisioning of their larvae differentiate these social wasps into two extremes: specialists and generalists. *V. tropica* depends for its food source solely upon the brood of Polistinae, which have high nutritional quality. Such a special food preference, however, greatly restricts the colony size in temperate zones, where food is insufficient in comparison with the tropics. On the other hand, *V. crabro*, being a semi-specialist, possesses a special ability as a hunter of various cicadas, but potentially has a relatively broad food preference. *V. analis* and *V. simillima* are regarded as generalists, although they have some food preferences which are characterized by the use of adult Hymenoptera by the former species and use of adult Diptera by the latter. These two species utilize various sources of food in a wide area within 1–4 km of their nests, but a time lag between their reproductive periods, during which demand for food is greatest, seems to serve as a means of sharing their food resources. *V. mandarinia* has quite unique foraging habits, because it has the largest body size among social Hymenoptera and, therefore, demands for food during its larval stages are high. The principal food of this species is slow-moving beetles whose body is heavily sclerotized, but it has developed a method for foraging high-quality food in large amounts: it kills all the adults in colonies of other *Vespa, Vespula* and honeybees by group predation and takes all brood away from the nest. As a result, a difference in ecological niches occurs between *V. mandarinia* and other species of the same genus. To other species, *V. mandarinia* is not only a competitor for common food sources but also the most dreaded natural enemy at the colony level. Success in this specific foraging activity of *V. mandarinia* is dependent not only on its unusually large body size and relatively large colony size but also on efficient group predation. Consequently, it can be considered that the five *Vespa* species share their food resources by developing differentiation of food habits and foraging behavior.

The third point is that each species has its own life history type which is closely related to the appearance and disappearance of its prey. *V. tropica,* a specialist, is adapted to food resources as a small-colony type and short-cycle species, while a generalist, *V. simillima,* is a large-colony type and a long-cycle species whose seasonal cycle is about twice as long as that of *V. tropica.* Compared with these two species, which have established contrasting life history types, each of the other three species takes its own standpoint by acquiring characteristics which, with regard to the main life history component factors such as food preference, colony size and the duration of life cycle, do not overlap with those of other species. As a result, these five *Vespa* species can evade or reduce any overlap in the peaks of their reproductive periods, during which demand for food is greatest, despite the fact that the available period is restricted in temperate zones because of the seasons (Fig. 6.24).

The above also produces interspecific lags in the periods available for the collection of carbohydrate sources such as mushrooms and tree sap, which are common to all five species as provisions for the new queens. As has been described in this chapter, objects for provisioning by workers shift from larvae to adults, and their foraging activities also change in quality after the emergence of reproductives. In new queens, the rate of increase of fresh weight reaches about 20% from their emergence to nest-leaving, and accumulation of fat body in their gasters occurs in preparation for their dormancy (cf. Chap. 2). To workers, therefore, the foraging activity for these new adults of the next generation becomes the most important form of labor which completes their seasonal cycle. It can be considered that the differences in the life history types among these five species contribute to the reduction of interspecific competition for their foraging activities, with regard to their common food resources at the final step of the life cycle.

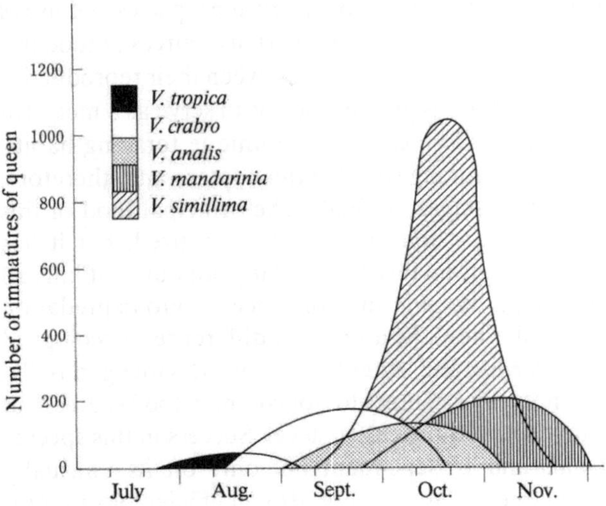

Fig. 6.24. Schematic representation of rearing time of new queens in five *Vespa* species in southwestern Honshu, Japan (After Matsuura 1984)

Thus, the five sympatric *Vespa* species achieve their coexistence in the southwestern region of Japan by securing ecologically different niches through the devleopment of various life history strategies. It seems that such a phenomenon acts as an important factor for the assortment of species which are able to coexist sympatrically, even in *Vespa* in the tropics and subtropics.

5 Survivorship of Colonies and Factors Influencing Mortality

5.1 Survivorship of Colonies

In Vespinae it is quite difficult to obtain reasonably accurate estimates of the survivorship of colonies, because the density of nests in the field is much lower than those of Polistinae, etc., and it is difficult, therefore, to find many nests and examine them continuously from the time of nest initiation. However, since nests themselves generally can withstand wind and rain, a colony which has lost its queen through an accident could retain its original form at the time of its termination for several months or, depending on the nest site, for several years. Such unsuccessful old nests, therefore, can be used as samples for estimating the time of nest termination and mortality factors.

With regard to *Vespa*, there is an example of trace observations on 59 queen nests of *V. analis* whose nests can be found relatively easily. These observations were performed for 12 years from 1964 to 1975 in Wakayama Prefecture (Matsuura 1984). According to this report, 47.5% (28 nests) lost their queens before worker emergence and 13.6% lost their queens during the cooperative period (about 1 month after worker emergence). Therefore, more than half of the observed nests lost their queens and were terminated during early stages of the annual cycle. Thereafter, 5.1% of the nests terminated their nesting activities without producing new queens, 3.3% were destroyed by *V. mandarinia* during new queen production, and only 30.5% survived to produce new queens. These figures show that, in *V. analis,* the possibility of colony termination is highest between the solitary and cooperative periods.

Table 6.6 shows the number of unsuccessful nests at each developmental period in five *Vespa* species, based on data obtained over 12 years of observations in Wakayama Prefecture. The ratio of unsuccessful nests was high before worker emergence in all five species, and the ratios during such period exceeded half of all unsuccessful nests in the case of *V. crabro* and *V. analis.* In *V. mandarinia* the discovery of queen nests is very difficult, especially at the earliest stage, because of its exclusive subterranean nidification, as in the case of *V. simillima.* Therefore, though the number of unsuccessful nests of *V. mandarinia* at the early stages was lowest among the five species, it is most probable that the mortality of colonies of this species in the early period is higher in reality than in later periods. In *V. crabro, V. analis* and *V. tropica,* the number of unsuccessful nests decreased with the development of colonies after worker emergence, while those of *V. simillima* and *V. mandarinia* were highest during the reproductive period with percentage ratios of over 40. These high mortalities in the colonies of *V. simillima* and *V. mandarinia*

Table 6.6. Number of unsuccessful nests at each developmental period in five *Vespa* species (Matsuura 1984)

	Period						
Species	Solitary[a]			Cooperative	Polyethic	Reproductive	Total
	I	II	III				
V. simillima	7 (14.2)	4 (8.2)	5 (10.2)	6 (12.2)	7 (14.3)	20 (40.8)	49
V. crabro	3 (10.0)	6 (20.0)	7 (23.3)	9 (30.0)	3 (10.0)	2 (6.7)	30
V. analis	7 (6.8)	12 (11.8)	34 (33.3)	22 (21.6)	14 (13.7)	13 (12.7)	102
V. mandarinia	0 (0.0)	2 (7.4)	5 (18.5)	6 (22.2)	3 (11.1)	11 (40.7)	27
V. tropica	3 (8.1)	5 (13.5)	7 (18.9)	11 (29.7)	7 (18.9)	4 (10.8)	37
Total	20 (8.2)	29 (11.8)	58 (23.7)	54 (22.0)	34 (13.9)	50 (20.4)	245

[a] I: before hatching of larvae; II: from hatching of first larvae to spinning of first cocoon; III: thereafter till emergence of first worker. Parenthesized figures show percent ratio in each period.

were mostly brought about by attacks from *V. mandarinia* and by artificial destruction by predators against apiculture.

Records on the survivorship of colonies of *Vespula* species are scarce, because the discovery of early nests is difficult, due to their generally underground nesting. In *Vl. flaviceps,* three of four early nests lost their queens before worker emergence (Shida 1959). Of seven queen nests of the same species discovered on the shore of Lake Nojiri in Nagano Prefecture in May, 1978, only two were found to have survived with scores of workers in mid-July. These two nests too, however, disintegrated afterwards without producing new queens (Okubo and Matsuura unpubl.). It seems probable that, in the species of *Vespula* as well, the mortality of colonies caused by the loss of queens during the preemergence period is highest throughout the whole annual cycle (Scott 1944).

Information on the survivorship of colonies is also scarce in *Dolichovespula.* Matsuura collected 81 abandoned nests of *D. saxonica* in and near Memuro-chô, Hokkaido, every August in 1979 and 1980. These nests were sorted according to their developmental stages as follows: 64 nests (79%) before worker emergence, 7 (10.9%) before new queen emergence, 5 (6.2%) after new queen emergence and 5 (6.2%) probably after the departure of new queens. These data show that the mortality of the colonies of this species is remarkably high during the solitary period. In *D. sylvestris* in England, only 1 of 12 nests completed the annual cycle, 6 nests failed before worker emergence and each of the remaining 5 nests produced less than 10 workers. The main cause of such a low survivorship is considered to be due to physiological or physical weakness in the foundress (Brian and Brian 1948, 1952).

5.2 *Mortality Factors*

Table 6.7 shows the mortality factors contributing to the termination of 245 colonies of five *Vespa* species, including disappearance of queen, natural enemies, human interference, climatic factors and other unknown causes. These factors affect *Vespa* colonies mainly at the nest level rather than at the individual level and operate on the colonies in an "all or nothing" manner.

Disappearance of the Queen. The main cause of colony demise during the early stage of nesting was attributed to the disappearance of the queen, and such "orphaned" nests represented 38.9% of the 245 unsuccessful nests of the five *Vespa* species. Factors causing the disappearance of founding queens include attacks by natural enemies and accidents during foraging activities, climatic factors such as low temperature and storms, lack of food, human interference, physiological death, etc. In *V. crabro* and *V. simillima,* a founding queen was killed by the queen and workers of a different colony when she had mistaken it for her own nest site in the course of nest relocation during the cooperative period (Matsuura 1970b), and some queens failed to find their new nest sites and perished afterwards. However, it is generally impossible to determine the cause of disappearance of the queens in their foraging trips by direct observation.

Table 6.7. Factors contributing to colony termination in five *Vespa* species (Matsuura 1984)

Species	*analis*			*simillima*			*crabro*			*mandarinia*			*tropica*			Total		
Period[a] / Factors	A	B	A+B (%)	A	B	A+B (%)	A	B	A+B (%)	A	B	A+B (%)	A	B	A+B (%)	A	B	A+B (%)
Loss of queen	50	0	50 (49.0)	3	5	8 (16.3)	9	0	9 (30.0)	8	0	8 (30.8)	18	2	20 (54.1)	88	7	95 (38.9)
Natural enemies																		
Attack by honey buzzards	0	11	11 (10.8)	0	0	0 (0.0)	0	0	0 (0.0)	0	0	0 (0.0)	0	0	0 (0.0)	0	11	11 (4.5)
Attack by *Vespa mandarinia*	0	6	6 (5.9)	0	17	17 (34.7)	0	3	3 (10.0)	0	0	0 (0.0)	0	1	1 (2.7)	0	27	27 (11.1)
Stylopization by *Xenos crabronis*	0	0	0 (0.0)	0	0	0 (0.0)	1	0	1 (3.3)	1	0	1 (3.8)	0	0	0 (0.0)	2	0	2 (0.8)
Parasitism by lepidopteran	0	0	0 (0.0)	0	0	0 (0.0)	0	0	0 (0.0)	0	0	0 (0.0)	0	1	1 (2.7)	0	1	1 (0.4)
Human interference																		
Destruction by man	3	3	6 (5.9)	1	3	4 (8.2)	2	1	3 (10.0)	0	10	10 (38.4)	0	0	0 (0.0)	6	17	23 (9.4)
Effect of insecticide spray	4	2	6 (5.9)	2	1	3 (6.1)	4	0	4 (13.3)	0	0	0 (0.0)	0	2	2 (5.4)	10	5	15 (6.1)
Effect of land reclamation work	0	0	0 (0.0)	0	0	0 (0.0)	0	0	0 (0.0)	0	1	1 (0.0)	0	0	0 (0.0)	0	1	1 (0.4)
Climatic factors																		
Strong wind with heavy precipitation	7	5	12 (11.8)	0	0	0 (0.0)	0	0	0 (0.0)	1	2	3 (11.5)	1	0	1 (2.7)	9	7	16 (6.5)
Unknown factors	11	0	11 (10.8)	16	1	17 (34.7)	9	1	10 (33.3)	3	1	4 (15.4)	7	5	12 (32.4)	46	8	54 (22.0)
Total	75	27	102	22	27	49	25	5	30	13	14	27	26	11	37	161	84	245

[a] A: Periods from solitary to cooperative period; B: periods from polyethic to reproductive period. Parenthesized figures show percent ratio throughout A + B.

The fate of these orphaned nests may be one of the followings: (1) end of colony caused by death of the larvae due to starvation, (2) disintegration of the colony caused by disappearance and death of newly emerged workers due to a shortage of provisions and (3) production of males by laying workers during and after the cooperative period. In the case of (3), the workers produce only male grandsons of the lost queen. Such males of a short-cycle species, *V. tropica,* have the opportunity to copulate with new queens of other nests, but those of the other species emerging during the cooperative period can hardly contribute to the production of the next generation, because they emerge before the period of new queen production. However, orphaned colonies which have lost their founding queens during and after the polyethic period sometimes produce a large number of males in the fall. In such a case, these males may be able to copulate with new queens which emerge from other nests during the same period.

Natural Enemies. In addition to several natural enemies which prey mainly upon larvae, pupae or adults in a nest at the individual level, causing less marked effects at the colony level, there are some natural enemies which are regarded as serious mortality factors operating at the colony level (cf. Chap. 9). Among the latter, the congeneric species, *V. mandarinia* and the honey buzzard, *Pernis apivorus*, can be regarded as "colony predators". As described in Chapter 5, *V. mandarinia* destroys all adults in the nests of other congeneric species or *Vespula* species by group predation, and brings larvae and pupae back to its own nest as provisions for its larvae. Such attacks continue from mid-August to early November, causing catastrophic damage to *Vespa* and *Vespula* colonies during and after the polyethic period. Among *Vespa* species, *V. simillima* suffers a particularly high level of damage from such attacks, because this species is the commonest among *Vespa* and its nests are large in size compared with the other congeneric species and constructed in conspicuous situations, such as under the eaves of buildings and on rock walls. Its large colony size compared with the other congeneric species also attracts the attention of *V. mandarinia,* since a large amount of food can be obtained by an attack.

The honey buzzard is known to be a predator of *Vespula* colonies in Europe (Schremmer 1962; Trap-Lind 1962; Edwards 1980). In Japan, *V. analis* is occasionally attacked by this predator, in addition to the subterranean nesting species of *Vespula*, such as *Vl. flaviceps* and *Vl. shidai*. The honey buzzard is a migratory bird and, in Japan, its brood leave the nest in or around mid-August, and they migrate to Southeast Asia in mid-October (Miyazaki 1981). Once the honey buzzard discovers a nest of vespine wasps, it takes out a comb by breaking the envelope and then takes the comb back to its nest. This process is repeated until all the combs are removed. Since attacks by this bird are limited to the pillaging of combs, wasps which have escaped from the nest repair the envelope immediately, but the preyed-upon nest can be easily distinguished by the considerably small size of its reconstructed combs compared with the appearance of its external structure (Fig. 6.25). Such specific attacks on *V. analis* among the *Vespa* species may be related to the weak counter-attacking ability of

Fig. 6.25. Nest of *Vespa analis* attacked by the honey buzzard, *Pernis apivorus*

the workers and/or the tendency to nest in trees, making them vulnerable to attack.

Stylopization by Xenos. A strepsipteran, *Xenos crabronis,* is occasionally found as a parasite, and such stylopized workers spend most of their time resting on the comb within the nest without performing any tasks. In the early stage of nesting just after worker emergence, such parasitism often reaches a high level. Two colonies each of *V. mandarinia* and *V. crabro* were destroyed because all the worker wasps emerging in the cooperative period were stylopized by *Xenos crabronis* (cf. Sect. 2 in Chap. 9).

Human Interference. Nests of vespine wasps once discovered by local inhabitants are generally regarded as a nuisance to be exterminated, since wasps are considered to be poisonous stinging insects. Apiarists regard *V. mandarinia* with hatred as the greatest natural enemy of honeybees, and destroy or burn its nests as soon as they are discovered. In addition, a large number of nests of *Vespa* and *Vespula* are collected every year and the immature wasps are eaten, mainly in Nagano and Gifu Prefectures.

On the other hand, application of various insecticides to crops and garden trees seems to act as a mortality factor on colonies of Vespinae. Many queens and workers die when poisoned directly by such insecticides in the field or after contact with residual chemicals. Not only adults but also larvae die when water and food contaminated with such insecticides is carried back to their nest by foraging individuals and given to each individual in the nest orally by trophallaxis.

Climatic Factors. To some tree- and underground-nesting species of vespine wasps, strong wind and heavy precipitation sometimes become factors of nest destruction and colony termination.

In the preemergence and cooperative periods, nests of *V. analis* which is a typical tree-nesting species among *Vespa,* are frequently blown off by strong winds and heavy rain. Such accidents occur because the nest of this species in the early stages is connected to the tree branch only by the upper part of the nest, which is not enough to stabilize it, in addition to the relatively poor waterproofing properties and elasticity of the material. In the later stage of nesting after September, however, the nest becomes strong enough to resist storm or typhoon. The stability of the nest improves as the colony develops, by the adherence of adjoining tree branches and leaves to the envelope and combs, or by their incorporation into the nest; this is coupled with simultaneous waterproofing of the surface of the envelope, especially the upper half, by closely painted nest materials. *V. simillima* also improves the resistance of its nests to strong wind and rain in the later stage of nesting in the same manner. *V. affinis,* as well as *V. analis,* is also a typical tree-nesting species. However, when Typhoon No. 5 passed Ishigaki Island on July 31, 1977, (the maximum instantaneous wind speed at the time being 70.2 m/s), all six nests of *V. affinis,* which had been observed by Kazumi Matsumura in an area near the Yaeyama Branch of Okinawa Agricultural Experiment Station, were blown off, leaving no remains. None of the workers returned to the site afterwards (Kazumi Matsumura, pers. comm.).

For *Vespa* species of subterranean nidification, rainfall occasionally causes destruction of their nests as a result of water erosion or flooding, although other species which construct their nests in closed terrestrial situations are hardly affected by wind and rain.

Since most of the species of *Vespula* construct their nests underground, their colonies, as well as the subterranean nidification species of *Vespa,* are affected by rain from the early to later stages of nesting. Especially in Japan, the precipitation during the rainy season from June to July and that accompanying typhoons in and after August seems to be the most important item among climatic mortality factors, although no actual data are available. Madden (1981) has studied annual fluctuations of *Vl. germanica,* a naturalized species in Tasmania, and reported that the heavy rains in spring and fall obviously contributed to colony termination. Also, along the northwest coast of North America, *Vespula* spp. occurred in large numbers only in years with warm, dry springs (Akre and Reed 1981). In Europe, the effects of rain and low temperature in springtime have also been discussed in relation to the population dynamics of *Vespula* spp. (Latter 1895; Beirne 1944; Fox-Wilson 1946; Döhring 1960; Archer 1981a).

Dolichovespula nests are usually aerial. The nest, however, seems to have a resistance to wind and rain superior to that of tree-nesting species of *Vespa,* even in the early stages of nesting, because the nest has high waterproofing properties and elasticity. Since the distribution of this genus is biased toward northern and highland areas, it seems probable that a long spell of low temperatures or rain in the early stages of nesting reduces the activities of queens and workers, or increases the mortality of foraging queens in the field, though almost no data are available to prove such an assumption in Japan, Europe or North America.

6 Annual Fluctuations in Number of Colonies

Vespine wasps generally have low distribution densities and broad foraging distances, compared with those of polistine wasps of social wasps. In addition, their life history consists of about half a year of solitary life by a single queen, including her dormancy period, during which she does not contribute to any nesting activity, while the remaining half year is one of collective life after the commencement of nest foundation until the production of new queens. Consequently, factors contributing to annual fluctuations in the numbers of vespine colonies are very complex. In this section, examples so far examined are introduced, mainly on the annual fluctuations in the number of colonies of vespine wasps.

Regarding *Vespa*, data have been collected on five species during a 12-year period in and near Kibi-chô, Wakayama Prefecture (Table 6.8). The figures can

Table 6.8. Fluctuations in number of mature nests in five *Vespa* species collected in and near Kibi, Wakayama (Matsuura 1984)

Year	*simillima*	*crabro*	*analis*	*mandarinia*	*tropica*	Total
1964	10	4	8	2	2	26
1965	10	6	12	3	2	33
1966	9	5	2	6	1	23
1967	9	5	19	8	2	43
1968	13	5	17	3	6	44
1969	9	0	3	2	2	16
1970	25	4	1	2	2	34
1971	23	3	7	7	6	46
1972	26	3	13	4	2	48
1973	16	7	15	2	5	45
1974	6	7	3	4	1	21
1975	10	8	19	2	4	43
Total	166	57	119	45	35	422

be used as a relative index for monitoring relative interspecific abundances, because these data show the numbers of nests collected or observed in situ, by selecting only colonies which reached the stage of new queen production. In this district, an almost stable relationship can be seen among these five species in terms of the abundance of mature nests as follows: *V. simillima* > *V. analis* > *V. crabro* > *V. mandarinia* > *V. tropica*. The annual fluctuations are broad to some degree in *V. analis*, but the abundances of nests of the other four species are stable, showing the fluctuations of only a few-fold.

In *V. simillima*, it is possible to estimate almost perfectly the abundance of its nests in a defined area, because the nests in the later stage of nesting are quite large (several tens of centimeters in diameter) and constructed in easily discoverable open situations, such as under the eaves of buildings and on rock walls. Table 6.9 shows fluctuations in the number of nests of *V. simillima* examined during a 12-year period in four adjacent areas covering 300 ha in Arida-gun, Wakayama Prefecture. In these areas, the number of nests per year varied only 3.3-fold from the minimum to the maximum, with an average of 6.2 nests per year, showing quite stable but low numbers of nests per year. However, the frequency of queens surviving until the later stage of nesting is considerably low, compared with not only solitary wasps but also insects in general, on the basis of the fact that these colonies of *V. simillima* are estimated, from their colony sizes, to have produced a total of 1800–6000 new queens every year.

Regarding annual fluctuations in the number of nests of *Vespula*, it is believed that periodic changes of high and low abundance with intervals of 2–3 years exist in Nagano and Gifu Prefectures by people who collect nests of *Vl. flaviceps* and *Vl. shidai* as a source of edible insects, although no actual data are available to prove this. In England, a report by Fox-Wilson (1946) on three *Vespula* species (*Vl. vulgaris, Vl. germanica, Vl. rufa*) and two *Dolichovespula* species (*D. norwegica, D. sylvestris*) collected in a 40.5-ha garden during a 25-year period (1921–1945) is famous. According to this report, annual fluctuations in these five species were 1–84 nests per year with periodic changes of high and low abundance at intervals of 2–4 years. The three *Vespula* species were the dominants in every year, thus occupying most of the data. The annual fluctuations in *D. sylvestris* were 1–5 nests per year, and those in *D. norwegica* were smaller still.

In England, there are records on the numbers of artificially exterminated nests of vespine wasps, mainly *Vespula*, during a 15-year period (1963–1977). According to these records, the number of nests per year varied about ten-fold, and the number also varied significantly in different localities, although it was difficult to discern any regularity in these records (Edwards 1979).

On the other hand, Archer (1980a,b) noticed that the number of nests of *Vespula* per year was relatively stable but showed repeated two-year cycles of abundance and scarcity, and thought that certain population-controlling mechanisms were active. That is, when a large number of new queens were produced in a certain area, severe competition would start the following spring among queens for their nest sites, which could result in a drop-off of a large number of founding queens. This would have a negative effect even on successful colonies

Table 6.9. Fluctuations in number of nests of *Vespa simillima* in four adjacent areas covering 295 ha near Kibi-chô, Wakayama (Matsuura 1984)

Subareas	Area (ha)					Year								Total
		1964	1965	1966	1967	1968	1969	1970	1971	1972	1973	1974	1975	
A. Mutsukawa in Kanaya-chô	75	1	1	3	3	2	1	2	3	1	0	1	1	19
B. Yoshikawa in Kanaya-chô	80	2	1	2	1	2	3	4	1	2	2	1	2	23
C. Yoshimi in Kibi-chô	50	2	0	2	4	1	0	2	2	1	1	0	1	16
D. Yamada in Yuasa-chô	95	0	1	3	1	1	0	2	3	1	1	2	2	17
Total No. nests observed		5	3	10	9	6	4	10	9	5	4	4	6	75
No. nests/ha	300	0.02	0.01	0.03	0.03	0.02	0.02	0.03	0.03	0.02	0.02	0.02	0.02	
No. of cells per colony[a]	Minimum	2381	3263	3053	3504	3418	2500	1885	2572	4802	3438	2949	2988	
	Maximum	10058	9323	8656	8433	7980	8656	9936	8465	6873	9602	8744	7444	
	Average	6313	6173	5844	5899	4325	5124	4666	5211	5106	6322	5450	6708	

[a] All nests were taken after cessation of nesting activities.

and, as a result, each colony might produce only a few new queens. Consequently, in the following spring after the year of severe competition, competition among founding queens would be reduced, colonies would develop smoothly and a large number of the next generation queens would be produced. Repetition of such cycles, therefore, could create regular annual changes in the numbers of individuals and repress overproduction. In the earlier reports, such regularity was not found because of the bunched-up data treatment, with no attempt being made to sort out the data at species level. MacDonald and Matthews (1981) have also indicated that intra- and interspecific relationships among founding queens could affect to some degree the number of individuals in a given year.

Chapter 7

Life of Wasps in the Tropics

In the tropical and subtropical regions of Southeast Asia two interesting genera are found, *Vespa*, allegedly retaining many ancestral traits, and *Provespa*, with the most specialized behavioral pattern. This contrasts with tropical Africa and South America, where almost no vespine wasps are distributed. A total of 16 species of *Vespa* and three of *Provespa* have been recorded from Southeast Asia and New Guinea, Southeast Asia here including Burma, Indochina, Malaya, the Philippines, Borneo, Sulawesi, Sumatra, Java, and some small islands in the Lesser Sunda and West Pacific. Species of *Vespa* often show strong geographical variation, and are sometimes dominant in a tropical ecosystem. Most species, however, are mountain inhabitants, and relatively few species penetrate lowland tropics (see Chap. 11). Eleven of the 19 species are found in continental Asia, and apart from *Provespa*, which are well adapted to the tropical climate, some species of *Vespa* (e.g., *V. analis, V. affinis, V. tropica*) extend their distribution range to the temperate regions, including Japan.

In this chapter the life cycle of wasps in the lowland tropics, characterized by high temperature and humidity, and high bioproductivity, will be compared with that of temperate zone species and populations. The following descriptions are mainly based upon the unpublished data amassed by Matsuura, as a member of the Entomology Team of the 1981–1983 Sumatra Nature Study (SNS). The chapter will focus on the life cycle peculiarities in *Vespa analis, V. affinis,* and *V. tropica,* and two species of nocturnal wasp, *Provespa anomala* and *Pv. nocturna* which have developed behavioral specialization.

1 Life Cycle Characteristics

For the wasps inhabiting the tropics, though there is a cyclic change in climate, i.e., dry and rainy seasons, there is no severe season with restrictive factors comparable to the winter of the temperate regions. In the temperate regions, owing to the coldness and the shortage of food in the winter season, it is hardly possible for vespine wasps to maintain social life all the year round. Wasps in temperate regions usually adopt an annual life cycle as in *Polistes* wasps; their colonies dissolve in fall, only inseminated new queens overwinter, and each queen founds her nest in the next spring. Three *Vespa* species (*V. analis, V. affinis, V. tropica*) distributed in plains and hilly regions of equatorial Sumatra have different subspecies in temperate and subtropical Japan. The main differences between the Sumatran and Japanese populations are described in the following Sections.

1.1 *Asynchronism*

Periods of nest building and dissolution in the temperate *Vespa* are generally settled by species. In the tropics it is possible for wasps to start and dissolve colonies at any time of the year. Of 35 *V. affinis indosinensis* nests collected in mid-November 1981 (rainy season) in the neighborhood of Padang, West Sumatra, 11 were in the preemergence, 15 in the polyethic, and 9 in the reproductive stage. In the case of *V. analis analis*, of 15 nests observed in August 1982 (dry season) in the same locality, 5 were in the preemergence, 7 in the polyethic, and 3 in the reproductive stage. Similar cases have been reported for *V. analis* and *V. velutina* of West Java, where all nesting stages occur throughout a year (van der Vecht 1957).

1.2 *Various Modes of Nest Foundation*

It has been thought by many vespidologists that all nests of the Vespinae, whether they inhabit temperate or tropical regions, are founded by individual queens, or haplometrosis. Pleometrotic nest foundation and swarming have reportedly occurred in some tropical *Vespa* (Rothney 1877; Bequaert 1936), but were denied by van der Vecht (1957), who had made intensive surveys in various parts of Indonesia over a long period. The present study, however, must conclude that *Vespa affinis* in Sumatra does found nests, not only by haplometrosis but also by pleometrosis including two to eight queens, and that pleometrotic nest foundation is generally more common (Table 7.1). Further, the fact that large-sized workers which are supposed to have emerged late in the nesting period participate in nest founding suggests the possible occurrence of swarming in this species (Matsuura 1983a). In the course of our study, clear dominance order and division of labor, except for oviposition, have not been observed among queens and

Table 7.1. Colony composition in some incipient nests of *Vespa affinis indosinensis* (Matsuura unpubl.)

Nest No.	Date	No. of adults Queen	Founding worker	New worker	Egg	No. of cells Larva	Cocoon	Total No. of cells
01	Nov. 24 1981	1	1	2	4	8	1	33
02	Nov. 24 1981	5	0	8	13	32	1	45
03	Nov. 26 1981	5	0	0	5	17	0	22
04	Nov. 26 1981	4	0	0	14	18	0	49
05	Nov. 28 1981	1	0	0	2	0	0	2
06	Nov. 28 1981	4	0	0	7	17	7	31
07	Nov. 28 1981	2	0	1	10	10	3	23
08	Dec. 1 1981	2	0	0	6	17	4	27
09	Dec. 2 1981	5	0	0	10	19	0	28
10	Dec. 3 1981	1	0	0	5	9	0	15
11	Dec. 3 1981	4	1	7	15	14	6	58

workers. Although all the founding queens have sperm in their spermathecae, only one (rarely two) individuals take part in oviposition. The coexistence of multiple queens continues even after the emergence of the first workers, and with the increase of the labor force within the colony every queen starts to lay eggs. The production of new queens and males follows rapid colony development. The coexistence of several queens following pleometrotic nest foundation poses the question as to the kin selection theory for the origin of insect sociality advocated by Hamilton (1964a,b) and debated by various authors. At present the formation process of multi-female nests and the origins of participating queens and workers are not known. The initial number of cells (30–50) does not differ between the pleometrotic and haplometrotic nests, and approximately equals that in the temperate *Vespa* species (cf. Matsuura 1971b). In *V. affinis* in Sumatra, there was no clear correlation between the number of cells finally produced and the number of co-foundresses (Table 7.1; Matsuura 1983a).

Nest founding by swarming is commonly performed by *Provespa* (see below). On the other hand, in *Vespa analis*, only haplometrotic nest foundation has been observed as in the temperate species (van der Vecht 1957; Matsuura unpubl.).

Table 7.2. Colony composition in some nests of *Provespa anomala* (Matsuura unpubl.)

Nest No.	Date	Queen	No. of adults New Queen	Worker	Male	No. of cells Egg	Larva	Cocoon	Total No. of cells
01	Oct. 16 1981	1	0	320	79	269	103	333	1579
02	Oct. 18 1981	1	0	573	17	317	468	539	2585
03	Oct. 24 1981	1	0	465	6	231	940	556	1600
04	Oct. 30 1981	1	0	170	5	126	252	330	913
05	Oct. 30 1981	1	3	529	12	265	525	509	1387
06	Nov. 9 1981	1	0	415	22	205	461	917	2276
07	Nov. 24 1981	1	0	69	0	77	176	121	384
08	Nov. 28 1981	1	0	149	0	104	161	118	419
09	Nov. 28 1981	1	0	59	0	26	28	0	59
10	Nov. 28 1981	1	0	47	0	48	98	45	197

1.3 Annual Life Cycle

No case of nesting exceeding 1 year has been observed in three species of *Vespa* and two of *Provespa* under favorable conditions in lowland tropics. Seven nests noticed before worker eclosion, four of *V. affinis* and three of *V. analis*, were checked for their development at intervals of 2 weeks to 1 month in West Sumatra. Only one nest of *V. affinis* produced males and new queens, and survived 6 months after the first worker eclosion (Jamalius and Matsuura unpubl.).

As in the temperate species, colonies of tropical *Vespa* species undergo 4 to 6 weeks of slow-development phase during the emergence of the first workers, then enter the rapid-development phase with a further increase in worker number. Finally, when worker numbers reach a peak, reproductives (males and new queens) begin to emerge. Two- to 3-month observations on *V. affinis* and *V. analis* nests at various nesting stages showed that after males and new queens began to emerge, worker emergence ceased, much as in the temperate *Vespa*. Workers thereafter devoted all their remaining energies to taking care of reproductives, and then died. As males and new queens leave the nest one after another, colonies in the late nesting period, like those in temperate regions, rapidly decline and disintegrate. The incubation period and the frequency in the use of one cell do not significantly differ between temperate and tropical species or populations. These facts suggest that the length of the nesting period in the tropics is 2 to 3 months longer than in temperate regions; thus perennial nesting seems unlikely to occur.

According to native reports from observations of *Vespa*, many colonies lasted 5 to 6 months after the worker activity had increased so far as to make the nest noticeable.

1.4 Large Colony Size

Wasps inhabiting the tropics, thanks to the perpetual summer and supported by the abundant food supply, make larger colonies than are encountered in the temperate or subtropical congeners and conspecific populations. The populations of *Vespa affinis affinis* in the Yayeyama Is. and *V. tropica pulchra* in western Japan (the northern limits of their distribution range) belong to the small to medium-sized colony types (Matsuura 1975a). However, these species in Sumatra (represented by different subspecies) produce more than ten times as many comb cells and new queens as the Japanese conspecifics. *V. tropica leefmansi* in particular produces 5000–6000 cells and 500–1000 new queens, while still depending on the larvae and pupae of other social wasps such as *Polistes* and stenogastrines for their food source. For such a huge colony to be maintained it is assumed that several thousand nests of *Polistes* and stenogastrines in the vicinity of one *V. tropica* nest are attacked and victimized. Four flourishing colonies of *V. tropica* have been found in an area of 1 km^2 (Matsuura 1983b). In this case the food source would have had to be immense, as the foraging areas of the four nests would have overlapped. This demonstrates the hardly imaginable scale of the tropical ecosystem.

In contrast to this, one colony of the Japanese *V. tropica pulchra* victimizes about 150–200 polistine colonies during its nesting period (Matsuura 1984), and produces on average only 200 cells and 50–60 new queens (Matsuura 1983b). The life cycle of *V. tropica* is restricted by that of prey species (polistines), so that the populations inhabiting temperate regions must complete the colony cycle within as little as 4 months. Because of this, as well as because of the low density of polistine nests, *V. tropica* is thought to survive the fluctuating surroundings with distinct seasons by drastic reduction of its colony size.

1.5 Nest Architecture

The initial (embryo) nests of tree-nesting *Vespa affinis* and *V. analis* in Sumatra do not differ in shape from those of temperate conspecifics; they are flask-shaped and consist of a single envelope sheet. However, with the increase in nest size following worker emergence, thick shell-like structures or scallops of envelope are added to the top of the nest. This addition finally becomes a layered and conical roof. No cells for rearing brood are constructed within this structure, which is not scraped off from inside with colony growth as in other parts of the envelope (Fig. 7.1C). The outer surface is hardened by applying material with oral secretions, and has a waterproof function. This conical roof is 50–80 cm long and 10–50 cm wide in large nests of *V. affinis*, and 20–30 cm long and 10–20 cm wide in *V. analis*. Its function was viewed by van der Vecht (1957) as an adaptation to climatic factors, particularly to protect nests from heavy rains which are an everyday occurrence in tropical rainforests. Van der Vecht assumes that the nest cannot weather a heavy rain if the nest roof is brittle and not resistant to the pressure from

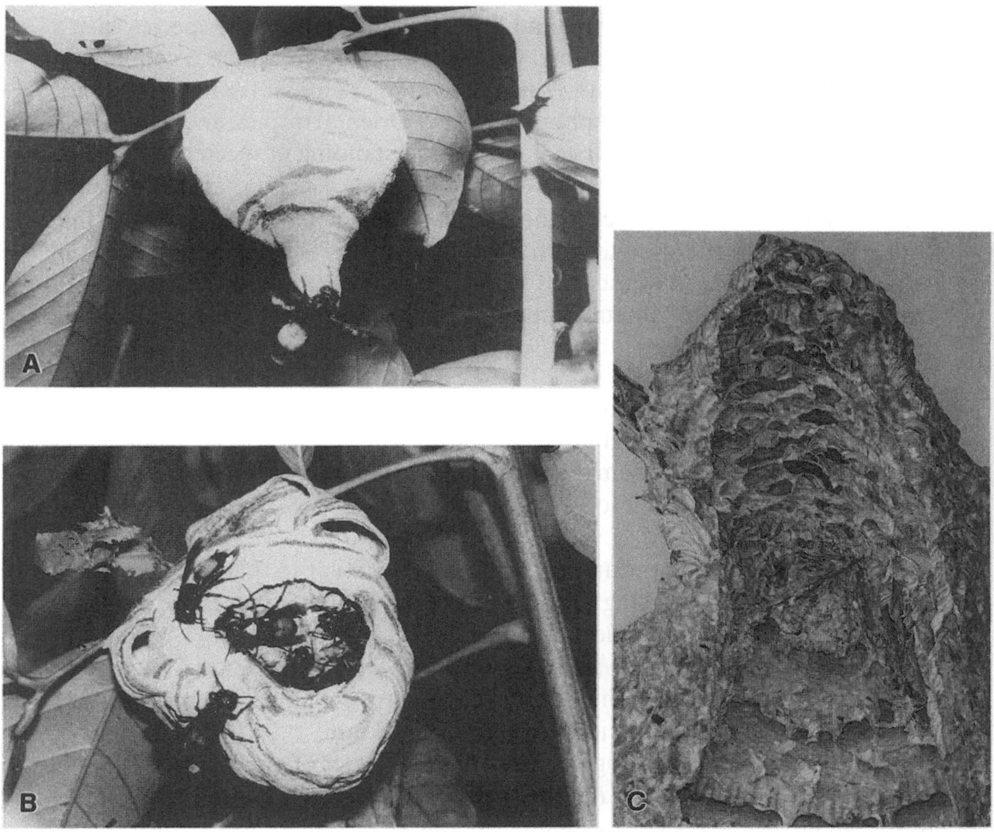

Fig. 7.1. Nests of *Vespa affinis indosinensis* in Sumatra. **A** Embryo nest; **B** pleometrotic nest foundation; **C** roof of mature nest (front part of envelope removed)

outside, as the lateral and lower parts of the envelope are. In the temperate or subtropical populations of these two species in the late nesting period, the upper parts of the envelope become somewhat thickened relative to other parts, and pillar-like structures sometimes develop in between the layers. If nests at the same stage of development are compared between temperate (or subtropical) and tropical populations, nests in the latter develop a more specialized "amejimai" structure and much thicker suspensoria between substrate and top comb; the thicker suspensoria strengthen the nest as a whole (Fig. 7.1C). The nest envelope of *V. analis analis*, except for the roof, consists of one to two layers of air pockets, and measures only 0.5–1.5 cm in thickness. On the other hand, in *V. analis insularis* in Japan, the envelope has four to eight layers of air pockets; the thick and multi-layered envelope may serve to keep warmth within the nest when rearing the brood of new queens in late summer and fall.

2 Ecological Characteristics of Nocturnal Wasps *Provespa*

Three species in the genus *Provespa* are known to be distributed from the eastern Himalayas in the west to Sumatra and Borneo in the east. All the species have a yellow-brown colored body and big, well developed ocelli, which is common in nocturnal Hymenoptera, such as *Apoica* (Polybiini: Vespidae) in Latin America, *Nyctomelitta* (*Xylocopa*; Anthophoridae) in the Oriental tropics, and like *Netelia* (Ichneumonidae) in Japan. Our ecological understanding of species of *Provespa* was quite poor and it has been reported that *Provespa* is nocturnal and attracted to light. Van der Vecht (1957) cited the observation made by C. J. H. Franssen that during World War II it used to catch adult flies in latrines at Japanese prisons in Palembang, Sumatra. As described below, *Pv. anomala* and *Pv. nocturna*, observed by Matsuura in Sumatra, appear to be the most specialized group of the subfamily Vespinae.

Nest Sites. Both species make nests in trees, in bushes or in the holes of trees where neither rain nor sunlight can reach directly (Fig. 7.2). *P. anomala* makes nests from as low as 10 cm from the ground to tree tops more than 20 or 30 m high. Usually nests are found at 7 to 15 cm.

Nest Architecture. The appearance of the nest is similar to that of the arboreal Vespinae of the temperate regions (Fig. 7.3). Leaves and branches are built into the nest to make it more stable and strong.

The initial nest of *P. anomala* is egg-shaped with a diameter of 4–8 cm. One or two combs are covered with two or three layers of envelope. There is only one small exit on the bottom and its diameter is only 8 mm.

Mature nests are also subglobular, with a diameter of approximately 20 cm in *Pv. anomala* and 30 cm in *Pv. nocturna*. Four to six combs are arranged one above the other and covered with an envelope consisting of three to five paper sheets. A single entrance is made laterally in the middle or slightly toward the bottom of the nest. The envelope in the upper portion consists of several layers of

Fig. 7.2. Nest of *Provespa anomala* (*top*), and workers of the same species sharing food pellet on the comb (*bottom*)

air pockets that are densely arranged; the outer surface is thickly coated with and strengthened by nest material mixed with saliva. Central and lower parts of the envelope consist of large air pockets enclosed by scallops, or a few paper sheets, as in the *Dolichovespula* nests; the outer surface is almost lacking in a coating of saliva. The nest envelopes of *Provespa* species have an excellent waterproof quality, and are stronger than those of any *Vespa* species. The nest combs are also stronger than those of any other vespine genera. Each cell is used several times for brood rearing, and the re-use of cells gradually strengthens the whole comb. Accumulated meconia excreted to the cell bottoms by mature larvae immediately after spinning cocoons no doubt contribute to this strengthening process.

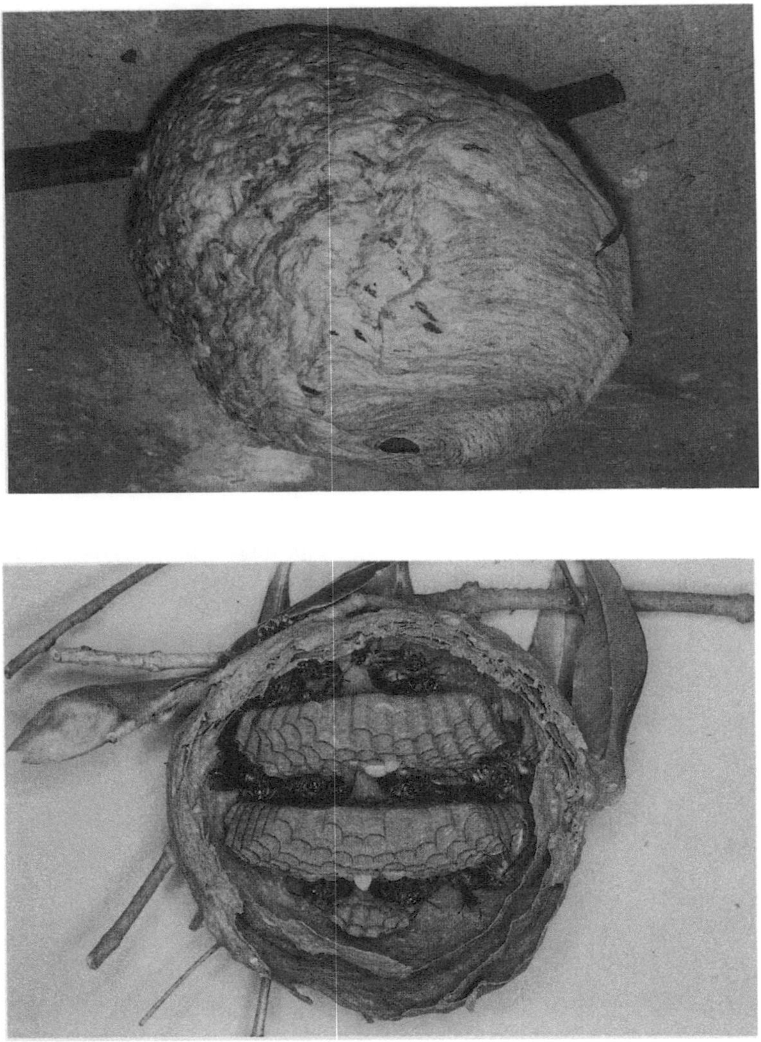

Fig. 7.3. Nests of *Provespa*. External view of a *Pv. nocturna* mature nest (*top*), and internal view of a *Pv. anomala* young nest with the cocoons of first brood (*bottom*)

The envelope and cell walls in *Provespa* are much thicker, stronger, more flexible and more fibered than in any species of the other three vespine genera. The nest is uniformly brown or greyish brown in color, without the distinct, multi-colored stripes as seen in *Vespa* and *Vespula*. Although only one example of dead banana leaves as nest material has been observed, judging from the mono-colored nest, *Provespa* may restrict their nest materials to specific plant species. Comb suspensoria consist of the thick main pillar and various numbers of smaller auxiliary pillars. In the late nesting period, suspensoria of the top combs often fuse into buttress-like structures (Fig. 7.4A).

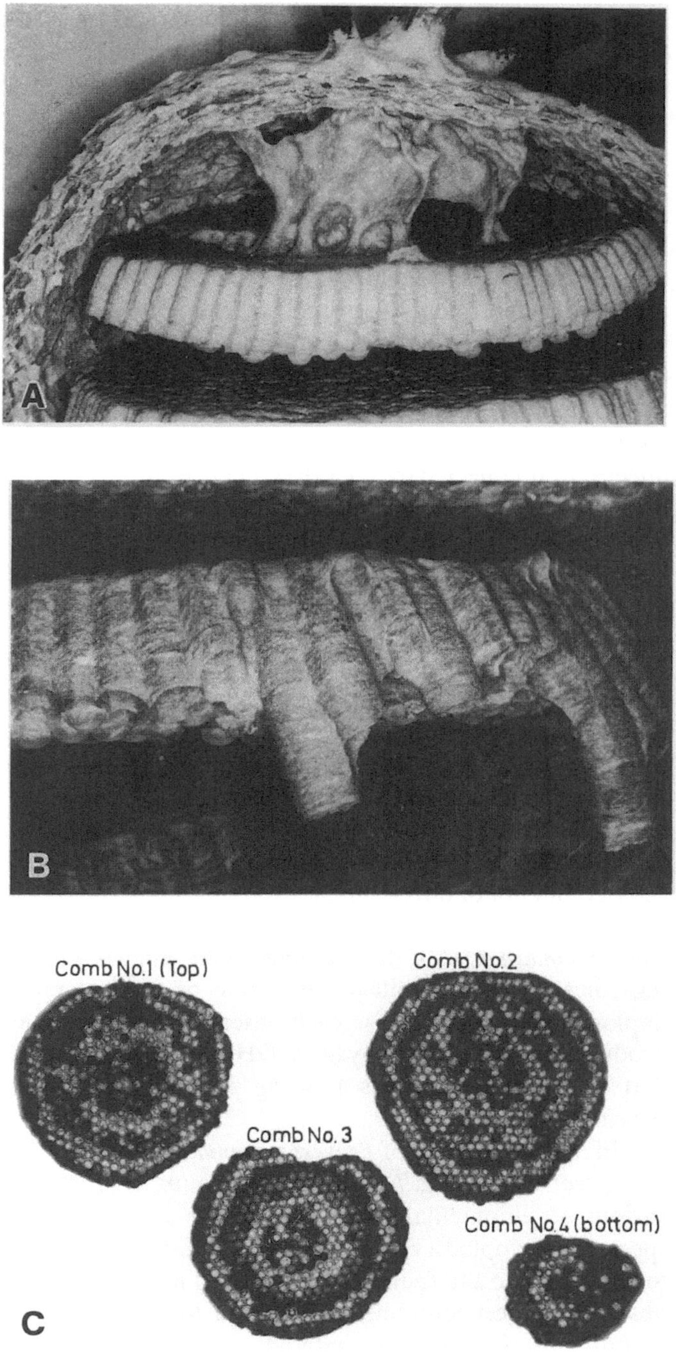

Fig. 7.4. Nests of *Provespa* in Sumatra. **A** Roof of a *Pv. nocturna* mature nest; **B** combs of the same species with large cells rearing new queens; **C** combs of a *Pv. anomala* mature nest

Comb cells are hexagonal at their opening and are used several times. Broods (eggs, larvae, and cocoons) are distributed irregularly in the combs, not necessarily in a series of rings, as seen in the other genera.

Colony Size. Colony composition of some nests of *Pv. anomala* is shown in Table 7.2. As mentioned earlier, the males and new queens of this species are produced after approximately 1000 cells have been constructed. The largest colony collected (Pa8102) had 317 eggs, 468 larvae and 539 cocoons (136 males, 401 workers, two new queens), and a total of 2585 cells, of which 1219 were empty, having been used two to three times for brood rearing; 591 adult wasps (1 foundress, 573 workers, 17 males) were collected from this nest. A smaller nest (Pa8105) had fewer cells (1387), though its brood and adult composition was comparable to that of the larger Pa8102; 265 eggs, 525 larvae, 509 cocoons (264 males, 240 workers, 5 new queens), and 545 adult wasps (1 foundress, 529 workers, 12 males, 3 new queens) were collected from Pa8105. There were only 21 empty cells that had been used more than twice for brood rearing. These figures suggest that the smaller Pa8105 was at the peak of activity, just rearing and producing new queens. The largest nest Pa8102, on the other hand, may already have passed its peak of development. This may indicate that the ultimate size of a *Pv. anomala* colony size reaches around five combs of approximately 2500 cells.

Pv. nocturna is the largest of the *Provespa* species in body size, and may form a much larger colony than does *Pv. anomala*. The single nest (Pn8101) collected during the course of this survey had 617 eggs, 2014 larvae, 1521 cocoons (1308 workers, 209 males, 4 new queens), and 1695 adult wasps (1 foundress, 1559 workers, 135 males) and a total of 4946 cells, of which 746 (15.1%) were empty, having been used two to three times for rearing brood. This nest (with 6 combs and ca. 5000 cells) may represent a normal colony size of *Pv. nocturna* at the peak of activity, but not of development.

Nest Foundation. All the vespines with known mode of nest foundation regardless of their habitats (temperate or tropical regions) initiate the nest haplometrotically; that is, each inseminated new queen finds a nest and produces the first workers by herself. However, the present survey must conclude that in this genus there is a strong possibility of founding new colonies by swarming.

In mature colonies of *Provespa anomala* (cf. Table 7.2) large cells for rearing new queen brood (Fig. 7.4B) were few in number (7–29), and such brood were in various developmental stages even in the same comb. This implied that this species had adopted a very efficient way of initiating new colonies: one nest is able to contribute to the foundation of several new nests. Much effort was made to observe the nest founding process and to locate initial nests. Although nest founding itself was not encountered, some initial (without the first cocoons) and preemergence nests were collected. Judging from the brood and adult composition of these nests (cf. Table 7.2), the following nest founding process is likely to occur in this species:

1. At a stage when cells reach 800–900 and workers 150 in number, a few adult males are produced. Large cells for new queens are then initiated, the eggs of new queens are laid, and their larvae reared.
2. Laying and rearing of worker and male eggs continues, with intermittent laying of new queen eggs and their rearing, until cells reach 2500–3000 in number.
3. There is a heterogeneous eclosion of new queens and coexistence with the foundress queen.
4. New queen-male copulation takes place (copulation sites unknown).
5. Swarming by a single, inseminated new queen and approximately 50 workers occurs, followed by new nest foundation (males are not included). Oviposition by the new queen (now foundress) takes place.
6. Intermittent swarmings continue from the mother nest.

In the above process, stages 4 and 6 have not been directly evidenced, and the very moment of nest founding, stage 5, has also not been observed. It is necessary to increase available samples of these various nesting stages, especially 4, 5 and 6, in order to determine conclusively the occurrence of swarming in this species, this strategy being unknown in the other vespine genera.

In *Pv. nocturna*, which constructs a colony more than three times as large as that of *Pv. anomala*, one nest with 6 combs and 4935 cells had only 11 new queen cells. This may be a clear indication of the presence of the efficient colony multiplication mode, swarming, also in this species.

Royal Court. There was no dominance display between the queen and the workers in *Provespa*. But as in *Vespa*, when a queen was observed on the comb, ten to 20 workers were around her with their heads toward her. This action (the "royal court") is known in honeybees and *Vespa*, and such behavior can be explained on the assumption that the queen secretes a pheromone to which workers are attracted (Ishay 1964; Matsuura 1968a). It is known that this action is related to the construction of queen cells. However, the function of this action is not clear in *Provespa*.

Chapter 8

Social Parasitism

1 Vespine Social Parasites

The Hymenoptera are known to include a great many social parasites, among which are several species of vespines (Table 8.1). An understanding of their present behavior can provide insights into the evolution of the social-parasitic habit. In the best studied species, *Vespula austriaca, Dolichovespula adulterina* and *D. arctica*, the queen invades a nest of a closely related host species in the initial or preemergence stage, kills the resident queen, and forces the resident workers to rear her own brood of males and new queens. The parasite queen cannot found a nest by herself, does not produce workers, and exclusively depends upon the host colony's work force to support her and her offspring. Such obligate social parasites are often called "inquilines".

In some other species, such as *Vespula squamosa* and *Vespa dybowskii*, it is common for a queen to usurp a nest of congeneric species and kill the resident queen. She can, however, found her own nest, and in any case she produces workers, so that even initially parasitic colonies become self-supporting as the host workers die off. There is thus a transition stage in the colony cycle, in which both host and parasite workers coexist, apparently without conflict. In recent years there has been increasing evidence that intraspecific usurpation occasionally occurs even in many typically non-parasitic species (see also Sect. 2 of Chap. 3). Facultative social parasitism in vespines is therefore not as uncommon as was once thought.

In Japan only three social-parasitic species have been recorded: *V. dybowskii* (Fig. 8.1; Sakagami and Fukushima 1957a), *Vl. austriaca* (S. Yamane and Kubo 1970) and *D. adulterina* (Sk. Yamane 1982c). Of the papers cited, only that of Sakagami and Fukushima (1957a) contains substantial new information.

Vespidologists have consistently seen intraspecific usurpation as relevant to the evolution of social parasitism. What exactly is the evolutionary relationship between the various forms of parasitism known in living vespines? The earliest attempt to answer this question may have been L. H. Taylor's (1939) hypothesis of a series of steps in the behavior of queens, leading away from the purely free-living, non-parasitic habit: (1) facultative temporary parasitism within the species, in which queens usurp nests if conditions allow and produce their own workers, (2) facultative temporary parasitism between species, (3) obligatory temporary parasitism between species, in which queens never found their own nests, but nonetheless produce their own workers within the host nest, and (4) obligatory permanent parasitism between species (inquilinism), in which the worker caste has been lost in the parasite species.

Table 8.1. Social parasites and their hosts

Parasites	Type of parasitism[a]	Hosts	Locality	References
Dolichovespula arctica	4	*D. arenaria*	N. America	Wheeler and Taylor (1921); Jeanne (1977b); Greene et al. (1978)
		D. spec. close to *saxonica*	N. America	Wagner (1978)
D. adulterina adulterina	4	*D. saxonica saxonica*	Europe	De Beaumont (1944); Haeseler (1981)
		D. norwegica norwegica	Europe	Bischoff (1931)
D. adulterina montivaga	4	*D. saxonica nipponica*	Japan	Sk. Yamane (1982c)
D. omissa (= *D. ingrica*)	4	*D. sylvestris*	Europe	Weyrauch (1937); De Beaumont (1944)
Vespula austriaca	4	*Vl. acadica*	N. America	Reed et al. (1979)
		Vl. rufa rufa	Europe	Robson (1898); Archer (1977)
		Vl. schrenckii	Japan	S. Yamane and Kubo (1970)
Vl. squamosa	3[b]	*Vl. maculifrons*	N. America	MacDonald and Matthews (1975); Matthews (1982)
		Vl. flavopilosa	N. America	MacDonald et al. (1980)
		Vl. vidua	N. America	L.H. Taylor (1939)
Vl. pensylvanica[c]	2	*Vl. vulgaris*	N. America	Akre et al. (1977); Roush and Akre (1978)
Vespa dybowskii	2	*V. crabro flavofasciata*	Japan	Sakagami and Fukushima (1957a)
		V. simillima xanthoptera	Japan	Sakagami and Fukushima (1957a)

[a] See text, pp. 196–198.

[b] This species has been considered a facultative parasite, though Matthews' (1982) recent work suggests it to be an obligate parasite.

[c] Parasitization of this species on *Vespula vulgaris* is not uncommon in regions such as Oregon, where both species inhabit (Akre and MacDonald 1986). *Vl. pensylvanica* remains a facultative parasite.

This pathway is illustrated in Fig. 8.2, with examples of each step given in Table 8.1. It connects the two extremes through a series of intermediates in a way which is widely accepted today by entomologists, including ourselves. At the same time, Taylor (1939) seems to have regarded the intermediate stages as only intermediates, which is to say that the species presently at these stages are necessarily evolving toward inquilinism. This assumption seems to be shared by some authors today (e.g., MacDonald and Matthews 1975; Matthews 1982; Akre 1982), yet there is no good reason to suppose that a shift from partial to full dependence on the host is inevitable or that inquilinism need be the evolutionary endpoint (Sk. Yamane 1978).

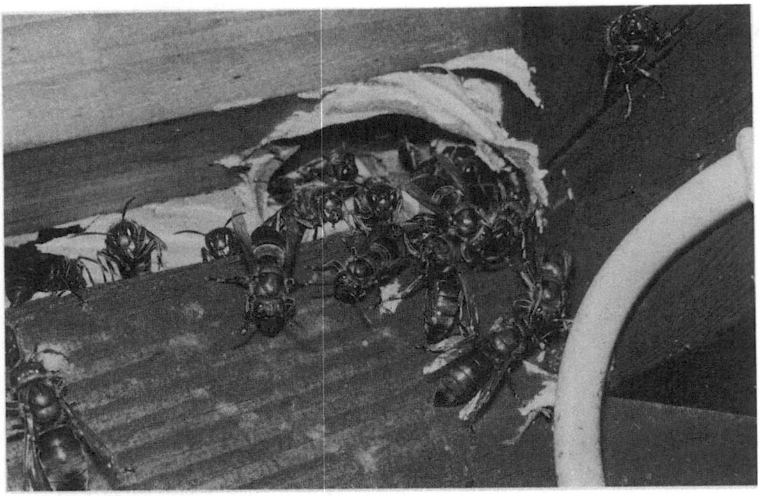

Fig. 8.1. Entrance of a *Vespa dybowskii* nest. Interspaces between wall boards are sealed with carton

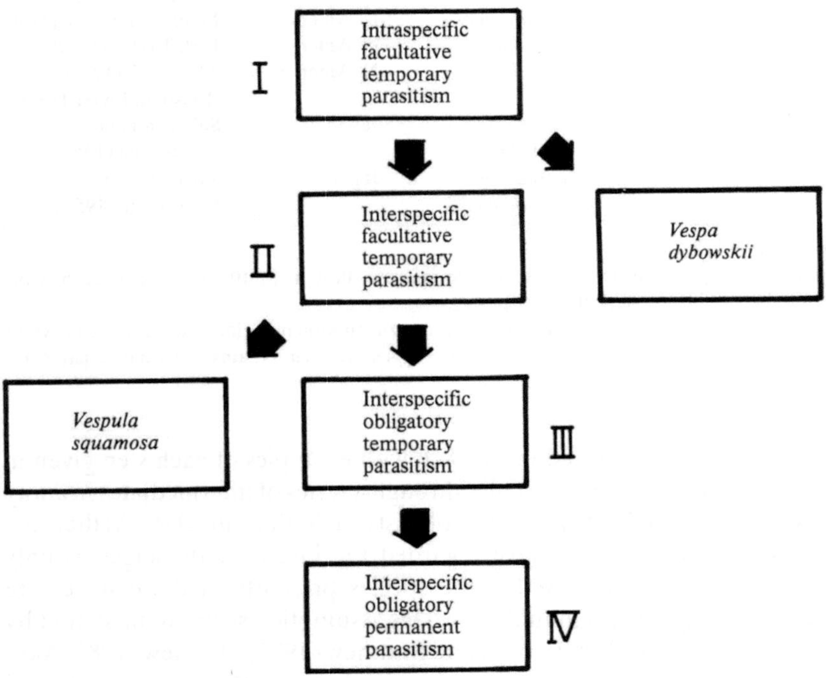

Fig. 8.2. L.H. Taylor's (1939) scheme explaining evolutionary route to social parasitism among vespines, with some extant examples

2 Social Parasitism as an Adaptive Strategy

The various forms of social parasitism in Hymenoptera are not treated in detail here, as they have already been the subject of several reviews (Richards 1927; L.H. Taylor 1939; G.E. Bohart 1970; E.O. Wilson 1971, 1975; Sk. Yamane 1978; Starr 1979; Akre 1982). Instead, we will discuss some key issues relating to the evolution of social parasitism.

2.1 Why Usurp?

In discussing the evolution of social parasitism in bumble-bees (Apidae: Bombini), Richards (1927) presented an allopatric model for the origin of interspecific parasitism. Assuming the occasional intraspecific usurpation in a free-living species, be then allowed for an extension of its range into that of a closely related species. If the extension is from south to north, physiological constraints may cause queens of the immigrating species to emerge from diapause later in the spring than queens of the native species. This will put the immigrants at a reproductive disadvantage, compared with the native species, as they will get a later start in developing their colonies. Under these circumstances the immigrants may be better off attempting to usurp already growing colonies of the native species than founding on their own. This plausible scenario has remained unchallenged during the last half century. If it accounts for the origin of the inquiline bumble-bees, however, it is uncertain how well it also applies to vespine social parasitism. For example, the parasitic *Vespula squamosa* is apparently not especially closely related to its host, *Vl. maculifrons* (MacDonald and Matthews 1975).

Richards (1927) assumed intraspecific (facultative) usurpation prior to the origin of interspecific social parasitism. Under what conditions is such usurpation an adaptive option? Any situation in which nest-founding is delayed or the nest is destroyed can increase the relative advantage of attempting to usurp. This is especially so where the growing season or colony cycle is short. As a queen's expected reproductive success (in males and new queens) from founding on her own approaches zero, usurpation should become the preferred option even with low probability of success.

Some queens within a well adapted population may have poor prospects of reproductive success because of a miscalculated emergence from winter diapause (due to subtle differences in the conditions within different hibernacula), unsuccessful early colony development, or losses to natural enemies. If such reverses are sufficiently common, the usurpation tendency may increase in the population, especially if the risk involved in a usurpation attempt is not very high. In some cases a queen searching for a nest to usurp may succeed without risk, by finding by chance a recently abandoned nest with live brood.

The evolution of facultative parasitism may also be facilitated by other factors. Although a delayed emergence from diapause decreases potential reproductive success, it may also confer protection from early-season predators.

Birds presumably learn through painful experience to avoid social wasps, and such learning is quite likely fatal to the individual wasps involved. There should thus be an advantage in remaining inactive until such time as other queens have already paid the price of "teaching" naive birds. The significance of bird predation to the survival of spring queen wasps has not been well known. But it is striking that sympatric vespines often have very similar color patterns, so that a bird which learns to avoid one species almost certainly avoids others with which it has had no direct contact. For example, queens of the Nearctic *Dolichovespula maculata* and *Vespula consobrina* closely resemble each other, and those of *Dolichovespula media media* and *Vespa simillima simillima* are often confused even by hymenopterists. Such cases of Müllerian mimicry should prove more widespread as local faunae are more closely examined. It is interesting to note in this regard that, although *D. maculata* is the Nearctic vicariant of *D. media*, the two are unmistakably different in color pattern.

Competition for nest sites may often be unavoidable among vespines which nest in enclosed places (*Vespula* and some *Vespa*), and this is a plausible factor in promoting usurpation. In excavating underground nests of *Vespula* spp., one often finds dead bodies of conspecific queens within the nest cavity (Shida 1959; Akre et al. 1976; see also Sect. 1 of Chap. 2). It seems clear that these are the losers in fights, but some care is needed in deciding whether the issue of the fight is the cavity itself or a developing nest and colony, i.e., whether it has to do with competition or usurpation. *Vespa analis* and *Dolichovespula media* nest in open situations in which there is no apparent basis for nest site competition, yet usurpation fights are not rare (Makino 1980; see also Sect. 2 of Chap. 3). This suggests that fights are often made over developing nests rather than nest sites, even in underground nesters. Thus, there is reason to doubt that such nest site competition has been an important contributor in the evolution of social parasitism in vespines.

2.2 Fights over Attempted Usurpation

Little is known of success rates in usurpation attempts. When a queen first invades a conspecific nest in the preemergence stage, she typically has an uncertain manner and flees upon becoming aware of the resident queen. She returns, though, and becomes less and less inclined to flee, until the two queens finally grapple in unrestrained combat, which may result in the death of one or the other (cf. Sect. 2 of Chap. 3). The invading queen's aggressiveness seems largely determined by her reproductive prospects and the value of the nest and brood.

Matthews (1982) studied the outcomes of staged intra- and interspecific contests involving *Vespula squamosa* and *Vl. maculifrons*, in which one queen (the invader) was introduced into a nest-box containing an early nest with a queen (the resident) and sometimes a few workers. Resident queens won 13 of 16 contests, both intra- and interspecific. Losers were not always killed, but sometimes flew off. When *Vl. squamosa* attempted to usurp *Vl. maculifrons* they were successful in two of six contests, while in the reverse arrangement the resident always won.

Matthews enumerated four factors affecting the outcome of contests: (1) home advantage, (2) relative size of the combatants (*Vl. squamosa* queens are significantly larger), (3) duration of the conflict, and (4) the presence or absence of workers. He concluded that home advantage may be the decisive factor. The reason for this is that a resident stands to lose everything she has established, while an invader may have other options. Accordingly, invaders take fewer risks and are more ready to flee.

2.3 Behavior of Host and Parasite Queens

Little is known of the fate of usurped queens. In cases of intraspecific usurpation there are no confirmed instances of peaceful coexistence between the two queens. This is not surprising, as pleometrosis is unknown in most vespine species, so that behavioral mechanisms for mutual tolerance must be almost absent. This is in contrast to the coexistence of foundresses in a great many polistines.

Following the observations of Weyrauch (1937) and de Beaumont (1958), it has been generally believed that in cases of (interspecific) inquilinism the host queen is killed by the parasite. However, L.H. Taylor (1939) reported the coexistence of a parasitic *Dolichovespula arctica* queen with the *D. arenaria* host queen, and Jeanne (1977b) has added supporting observations. In 12 of 14 nests of *D. arenaria* examined up to about mid-summer and which were parasitized by *D. arctica*, he found both queens living together. The *D. arenaria* queen presumably remained in the nest in a subservient position in each case. The process by which she was dominated by the *D. arctica* queen was observed in a lab colony. On the day of invasion, the resident attacked the invader seven times between 18:05 and 20:00, in which she struck and pulled at the invader's legs, wings and antennae. On at least one occasion the resident mounted and licked the invader's abdomen. The invader made no attempt to escape or retaliate during this period, but kept her wings tucked under as in the pupal stage. She was largely ignored by the three resident workers. By the following morning the attacks had ceased. The *D. arctica* queen became increasingly active over the next few days and the *D. arenaria* less so. When the nest was later invaded by another *D. arctica* queen, the host queen did not resist.

Evans (1975) and Greene et al. (1978) have made similar observations, in which resident queens initially resisted an invader, but then became less and less active while the invader took over the queen functions of the colony. On the other hand, among those obligately parasitic species which usually invade colonies in which there are already workers (e.g. *Vespula austriaca*) it is usual to immediately kill the resident queen (Reed and Akre 1982).

The rule of strict monogyny in almost all vespines must have posed a serious barrier to the coexistence of more than one queen, whether of one or two species. On the other hand, pleometrotic nest founding is common in *Polistes*, and the three known social parasitic species (= *Sulcopolistes*) show a pattern of dominating and coexisting with their hosts (de Beaumont and Matthey 1945). It has also been noted that labor parasites have evolved in a number of lineages of bees in which associations of foundresses are common (Starr 1979). The occasional appearance of a similar pattern of coexistence among vespines is therefore something of a puzzle. If the host queen submits to the parasite, rather than resisting to the end, she may still enjoy some very limited reproductive success. Note also that if domination of her workers by the parasite is not complete, these may lay a few male-producing eggs (Greene et al. 1978; Sk. Yamane 1982c). If the colony is still in the preemergence stage, the parasite queen should also prefer to delay a final struggle if she can dominate the host queen into continuing to go out foraging. Outside activities are generally considered much more hazardous than within-nest activities, so the parasite queen thereby minimizes risks to herself.

In the early phase in the evolution of interspecific parasitism, fierce fightings among queens may have been common. Jeanne's (1976b) observation and that by Greene et al. (1978) on parasite-host interactions suggest that as some species became increasingly specialized for interspecific parasitism, they evolved physical features which made the combat less and less equal. Once the outcome of their physical struggle was settled in favor of the parasite, the host's submission, possibly also promoted by parasite's pheromonal weapons (Sk. Yamane 1978), should have been both in her interests and those of the parasite queen if the host is allowed to live for a time. It is because the host queen may have been driven into situations where she could not reproduce at all without submitting to the parasite, and because the parasite could draw labor force from the host or her offspring by not killing her immediately.

2.4 Adaptive Significance of Permanent and Temporary Parasitism

As discussed above, it is the rule for intraspecific usurpation conflicts to end in the death or flight of one of the queens, while in obligate permanent parasitism the defeated host queen remains in the nest in a subordinate position, at least to the end of the preemergence stage. The invasion process by a temporary parasite (*Vespula squamosa*) has been directly observed by Matthews and Matthews (1979) and Matthews (1982). *Vl. squamosa* queens usually invade nests that have produced some workers and kill the host queens in the course of the struggle. The fact that workers are already present may be closely connected with the timing of killing the host queen.

It has often been considered that the possession of worker caste indicates a stage less advanced than "true inquilinism". We believe this view is mistaken. Data on the colony span (i.e., the time from founding to dissolution) and distribution of parasitic species suggests an interesting relationship between climate and the type of parasitic habit. All known inquiline vespines (four species in two genera) are mainly restricted to the north temperate and north frigid zones and have a short colony span, with the brood (of reproductives only) emerging by mid- or late summer. Their hosts have similar distributions and colony spans. With such a restricted growing season, the new generation of parasites can only be reared by host workers. That is to say, it would do the parasite queen no good to rear a brood of her own workers, as there would remain too little time for them to then rear a second (reproductive) brood. She can maximize her reproductive success if all fertilized eggs become queens (Starr 1979).

The final productivity of parasite colonies is usually much less than that of unparasitized colonies of their hosts (Archer 1977; Greene et al. 1978). One way for the parasite queen to increase her reproductive success may be to secure more host workers by allowing the host queen to live and oviposit instead of producing her own workers. This may in turn lead to another difficult situation, as the parasite queen is threatened by the host's increasing worker force. Taking over a host colony later in the season may also lead to serious risks, as the workers may join their mother in attacking the invader. Their genetic interests are certainly much more closely allied with hers. Colonies parasitized by *Dolichovespula arctica* or *D. adulterina* are often found to be orphaned, with both queens lost (Greene et al. 1978; Haeseler 1981; Sk. Yamane 1982c), and it is reasonable to think that in some of these the workers were an important factor in the invader's death.

The conflict of interests between the resident workers and invading queen may take on another aspect some time later, in a conflict over the laying of unfertilized (male) eggs. Although *D. arctica* queens appear to use a pheromone to pacify host workers, they cannot entirely suppress development of the workers' ovaries. This leads the queen to fight with and kill reproductively developing workers. On the other hand, if the parasite queen can take over a nest very early in the preemergence stage it will be to her advantage to produce some workers of her own. There are thus some particular situations in which the loss of the worker caste remains in part unexplained (Sk. Yamane 1978).

The three known temporary social parasites (*Vl. squamosa, Vl. pensylvanica* and *V. dybowskii*) have more southerly ranges and longer colony spans than the permanent parasites. In such a situation, a complete dependence on the host workers after the death of the host queen might restrict the parasite's reproductive potential. Conversely, if the parasite spares the host queen, such that the latter produces additional workers, these latter may later turn against the parasite. It stands to reason, then, that the parasite will do its best to kill the host queen, utilize the small available work force as long as it lasts, and produce her own workers to rear the reproductive brood. The abundance and large colony size of the two Nearctic temporary social parasites, *Vl. squamosa* and *Vl. pensylvanica* (Akre and Davis 1978), goes against the idea that they are evolving toward inquilinism. Rather, it is consistent with the view that temporary and permanent parasitism are each adaptations to particular ecological/climatic conditions. From this it is expected that inquilinism will be more common in cold temperate and arctic regions than warm temperate and tropical regions, as in fact it is, at least in vespines. At the same time, this does not contradict Taylor's (1939) hypothesis that temporary parasitism is a necessary evolutionary precursor to permanent social parasitism (Sk. Yamane 1978).

Chapter 9

Natural Enemies

There are many known natural enemies of the vespine wasps, and it is hardly possible to deal with all of them. Further, numerous insects or other small arthropods, which may be commensals, live in and around vespine nests. In this Chapter, we explain habits of representative natural enemies mainly observed in Japan. Readers can consult Spradbery (1973a) for a more comprehensive list of predators, parasites and commensals of vespine wasps. See also Section 3 of Chapter 6 for natural enemies as factors contributing to failure of colony.

1 Main Natural Enemies

The natural enemies of vespine wasps are classified into four categories: (1) predators feeding on adult wasps outside the nest, (2) endoparasites of adult wasps, (3) parasitoids of immatures within nests, and (4) predators attacking nests and capturing immature and/or adult wasps. Fungi that are parasitic on overwintering female wasps (Fig. 9.1) or infecting nest combs and broods are also natural enemies, but they are not included here because of scarcity of information.

1.1 Predators of Adult Wasps

Birds may be the most important vertebrates predaceous on foraging wasps. Numerous examples of Batesian and Müllerian mimicry, in which vespines play the role of model, have been reported throughout the world. This strongly but indirectly suggests that bird predation on wasps occurs frequently, rather than that vespine wasps have been excluded from the menu of birds. However, there are virtually no data that quantitatively show the significance of birds in controlling wasp populations. Although hornets and bumblebees are occasionally observed to be pinned by shrikes to thorns as a larder (mozu-no-hayanie, Fig. 9.2), they seem not to be a major prey of shrikes.

Robber flies (Asilidae) and spiders also feed on adult vespines. Tagawa (1981) reported that a robber fly (*Promachus yesonica*) captured a male *Vespula flaviceps*. We also observed robber flies that feed on workers of *Vl. flaviceps* in Japan. According to Kumata (personal observation) the Taiwanese robber fly *Microstylum* sp. is predaceous on foraging workers of *Vespa*. Web-constructing spiders can be both victims and predators of *Vespula*. As mentioned in Chapter 5, some species of *Vespa* often hunt smaller congeneric species and *Vespula*.

Fig. 9.1. Queen (*top*) and worker (*bottom*) of *Dolichovespula media* infected by a fungus, *Cordyceps* sp. near *oxycephala,* found in decayed wood, probably during hibernation. Not to the same scale

Fig. 9.2. Larder of a killed *Vespa simillima* worker by the bullshrike *Lanius bucephalus*

1.2 Endoparasites of Adults

The major endoparasites of adult wasps are Stylopidae, Conopidae and Nematoda.

Stylopids (Strepsiptera) are unfamiliar insects with curious morphology and habits. The genus *Xenos* (= *Vespaxenos*), parasites of Vespinae, is known from the Holarctic region, but reports of stylopization mainly come from eastern Asia (Salt and Bequaert 1929). In Japan, two species have been recorded. Bionomic studies were made by Matsuura on *X. crabronis* parasitic on *Vespa* (see Sect. 2 of this Chap.), and by Kifune and Maeta (1975) on *X. vespularum* parasitic on *Vespula* (*Paravespula*; see also Kifune 1971). Indirect evidence suggests that first instar larvae of *Xenos* are scattered from females protruding from hosts' intersegments on flowers and other suitable places, then transported to nests by foraging wasps, to which they are firmly attached. They then enter the body cavity of host larvae, feed on body fluids and pupate when the hosts have grown into adults. *Xenos* males are winged insects capable of flying to seek females, but females are wingless, completing their life within the host. As host wasps do not always die of parasitization, strepsipterans are usually distinguished from parasitoids such as ichneumon wasps, which kills their host. The behavior of stylopized wasps drastically changes (see Sect. 2).

Three species of conopid flies (Diptera) are reported to be parasitic on *Vespula* (Spradbery 1973a). Female flies congregate at the nest entrance of a wasp colony and dive toward returning foragers, on which eggs are deposited during the brief moment of contact. When the fly pupates, the host, filled with the fly's puparium, dies. No conopid fly has been found in *Vespula* in Japan.

A nematode, *Pheromermis pachysoma*, lives in the gaster of adult vespines. Eggs of this nematode are deposited in water-side soils, where they are eaten by the larvae of aquatic or semi-aquatic insects. These insects, in turn, are captured by foraging wasps that transport the eggs of the nematode to their own nests (Welch 1958; Edwards 1980). In Japan, a species of Nematoda is often found within adult Vespinae, but nothing is known of its biology.

1.3 Parasitoids and Predators Within Colonies

Nests of Vespinae are usually covered with a thick envelope which has in general only one entrance, and are guarded by aggressive workers. This condition has made it difficult for parasitoids and predators to successfully enter nests. About 40 species of enemies (chiefly insect parasitoids) have been recorded from the nests of *Polistes* (Makino 1983a), but only 20 species are known for Vespinae (Spradbery 1973a; Edwards 1980). Harmful species, i.e., those severely restricting a host's colony productivity, such as the ichneumon wasp *Latibulus argiolus* in *Polistes* colonies (Makino 1983b), are not known for vespine species. Two species of *Sphecophaga* (Hymenoptera, Ichneumonidae), four of *Bareogonalos* (Hymenoptera, Trigonalidae), and three of *Metoecus* (Coleoptera, Rhipiphoridae) are major parasitoid insects found in vespine nests. *Sphecophaga* females enter host's nests by themselves and lay eggs there, but *Bareogonalos* and *Metoecus* are transported to hosts' nests via complicated routes when they are first instar larvae. These three insect groups together are distributed throughout most of range of the Vespinae, but their ranges only partly overlap with each other. *Sphecophaga* is known from Europe and North America, *Bareogonalos* from Eastern Asia and North America, and *Metoecus* from Europe and eastern Asia; yet there is no region inhabited by all the three groups.

Sphecophaga. At least two species, *S. vesparum* and *S. thuringiana,* occur in Europe (Spradbery 1973a), though owing to marked intraspecific variation there is doubt regarding the number of species involved (Edwards 1980). *Sphecophaga* is found in nests of *Vespula* (Spradbery 1973a), and occasionally of *Dolichovespula* and *Vespa.* Females are assumed to enter the nests of host species, and deposit eggs on or in the larvae. *Sphecophaga* larvae that develop within hosts' abdomens kill the hosts when they have spun their cocoons, then spin a cocoon for themselves. In North America, *S. vesparum burra* parasitizes *Vespula* and *Dolichovespula. Sphecophaga* appears to adversely affect incipient colonies (Akre 1982), and occasionally mature colonies as well (Gauss 1968).

Bareogonalos. Five species of this trigonalid genus, including an undescribed species, occur in North America and eastern Asia (Fig. 9.3A,B). Harrington (1896) was the first to confirm the parasitism of this group on vespines. He ascertained the association of the Nearctic *B. canadensis* with *Vespula pensylvanica.* Later, van der Vecht (1933), Sk. Yamane (1973), and Sk. Yamane and S. Yamane (1975) reported hosts (various vespines) of Asiatic species (Table 9.1). No information is available for the biology of the Mexican *B. scutellaris.*

Not much is known of the life cycle of these interesting trigonalids. S. Yamane and Sk. Yamane (1975) infer the life cycle of *B. jezoensis* based upon results by van der Vecht (1933) and their own observations (Fig. 9.4A-E). The female wasp lays eggs on tree leaves which must be later eaten by caterpillars of Lepidoptera or sawflies (intermediate hosts). Swallowed eggs that receive scratches on their shells hatch in the bodies of caterpillars that may be hunted by vespine wasps. The wasps feed their larvae with body fluids and pellets of caterpillars which contain surviving microtype first instar larvae of *Bareogonalos.* The trigonalid larva molts twice within the immature wasp, growing into a mandibulate third instar. The larva in this stage is very peculiar, having large mandibles with which it fights other larvae, leaving only one individual, when more than one mandibulate larva is present in a single host. The surviving third instar larva molts to the fourth instar of a normal type and emerges from the host's prepupa. Thus, the *Bareogonalos* larva is endoparasitic during the first to third instars and ectoparasitic during the fourth and fifth instars. When mature (fifth instar), it pushes the host remnants toward the cocoon cap spun by the host, then spins two silken partitions to isolate itself from the host remnants by arranging a narrow empty space between their rooms (Fig. 9.4C). Even if a worker wasp notices the occurrence of something unusual with an infested cell, it only breaks open the cocoon cap and pulls out the

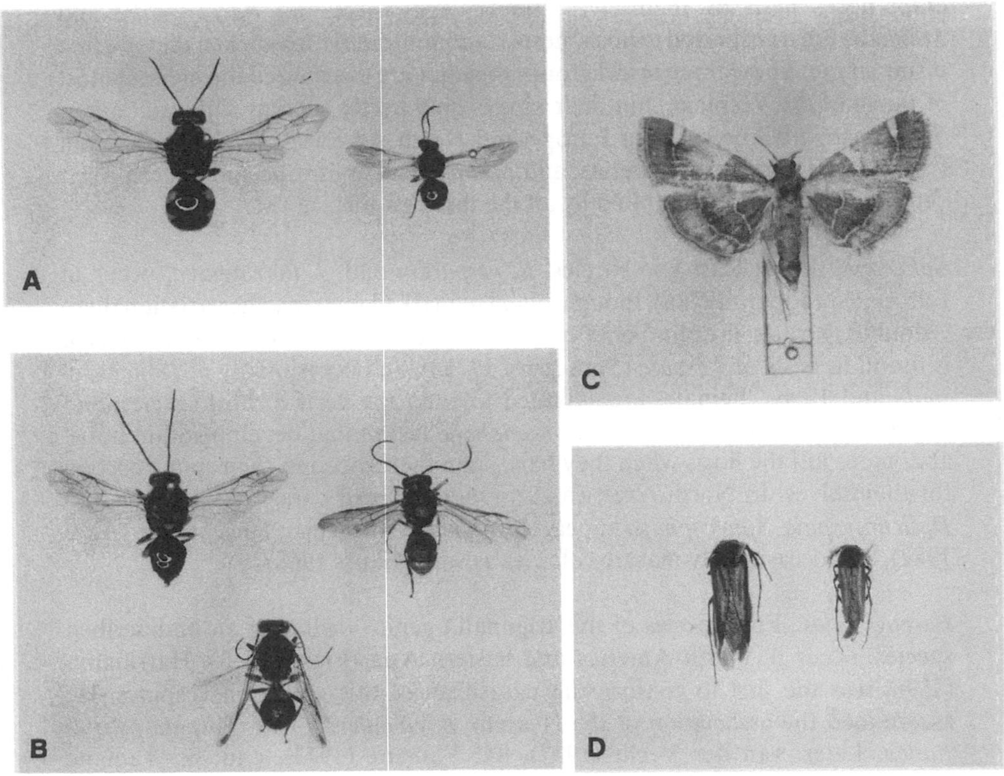

Fig. 9.3. Parasitic or pradatory insects attacking wasp colonies. **A** Female of *Bareogonalos jazoensis* (Trigonalidae) emerged from a *Vespa* colony (*left*) and a *Vespula* colony (*right*); **B** males of three *Bareogonalos* species found in Asia, *top left*: *B. jezoensis* from Japan, *top right*: *B. huisuni* from Taiwan, *bottom*: *B.* sp. from Sumatra); **C** *Pyralis regalis*; **D** *Metoecus paradoxus* (Rhipiphoridae) (Photos: K. Tomiyama)

victim; it appears to be indifferent to the parasite pupa situated in the innermost part separated by double silken wall.

In Hokkaido, parasite adults emerge from early July to late September, and possibly overwinter (Sk. Yamane 1973). In Fukui Prefecture, Honshu, however, females are observed to lay eggs upon foliage of *Albizzia julibrissin* and *Salix* in late fall (Tsuneki 1973; Nozaka 1976). Thus, at present it is not confirmed at which stage the parasite overwinters. According to van der Vecht (1933) and Nozaka (1976) eggs are placed singly in minute incisions in the leaf tissue near the margin. In *Albizzia*, approximately 2500 eggs are laid per leaf. The number of eggs deposited by a single female is not known, but judging by the very risky nature of the life cycle, and based upon dissection of adult females (Tsuneki 1973), more than 10 000 eggs can be laid. *Bareogonalos* species are relatively rare, and are collected almost exclusively from host colonies and at oviposition sites. Ratio of parasitism in host colonies (percentage of infested cocoons to total cocoons) is

Table 9.1. Trigonalid wasps parasitic on Vespinae

Trigonalid species	Hosts	Reference
Bareogonalos jezoensis Uchida	*Vespa velutina velutina*	van der Vecht (1933)
	V. analis analis	van der Vecht (1933)
	V. simillima simillima	Sk. Yamane (1973)
	V. crabro flavofasciata	Sk. Yamane (1973)
	Vespula vulgaris	Sk. Yamane (1973)
	Vl. shidai	Sk. Yamane (1973)
	Vl. schrenckii	Sk. Yamane (1973)
	Dolichovespula saxonica nipponica	Sk. Yamane (1973); Makino (1982)
B. canadensis Harrington	*Vl. pensylvanica*	Harrington (1896); G.W. Taylor (1898)
	Vl. vulgaris	Carmean et al. (1981)
	Vl. acadica	Carmean et al. (1981)
	D. arenaria	Stage and Slobodchikoff (1962)
B. huisuni Sk. et S. Yamane	*Vl. flaviceps karenkona*	Sk. Yamane and S. Yamane (1975)
Bareogonalos sp.	*Provespa nocturna*	Matsuura (unpubl.)

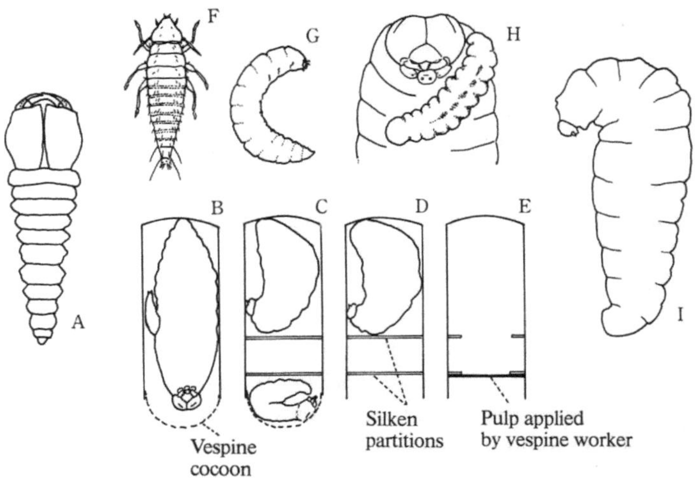

Fig. 9.4. Immature stages of *Bareogonalos jezoensis* (**A-E**), and *Metoecus paradoxus* (**F-I**). **A** 3rd instar larva; **B** 4th instar larva just emerging from host prepupa; **C** grown-up larva in uppermost room, host remnants in lowermost room (cell top). Silken partitions were spun by parasite larva; **D** infested cell with cell top cleaned by vespoid worker; **E** infested cell after emergence of adult trigonalid. Exit hole is sealed with pulp by vespid worker; **F** migratory 1st instar larva; **G** endoparasitic 1st instar larva; **H** ectoparasitic 2nd instar larva; **I** full-grown larva (4th instar?)

Table 9.2. Records of parasitism of *Bareogonalos jezoensis* in Hokkaido (After Sk. Yamane 1973)

Host	Date & locality	Cocoons infested to Total No. cocoons	Host sex or caste	Combs infested[a]
Vespa simillima simillima	Aug. 21, 1972 Sapporo	11/50 (22.0%)	♀	I/3
V. crabro flavofasciata	Aug. 16, 1972 Otoineppu	16/97 (16.5)	☿	I,II/2
V. crabro flavofasciata	Sept. 13, 1972 Sapporo	–	♀	IV
Vespula vulgaris	Aug. 28, 1970 Oketo	2/328 (0.6)	?	I,II/3
Vl. shidai	Aug. 11, 1972 Sapporo	1/453 (0.2)	☿	IV/5
Vl. shidai	Sept. 4, 1972 Sapporo	1/3516 (0.03)	☿ (?)	V/9
Vl. schrenckii	Aug. 7, 1969 Sapporo	10/380 (2.6)	☿♂♀	I,II/2
Dolichovespula saxonica nipponica	July 31, 1972 Sapporo	2/101 (2.0)	?	I,II/3
D. saxonica nipponica	July 31, 1972 Sapporo	1/67 (1.5)	☿ (?)	I/2
D. saxonica nipponica	July 30, 1972 Sapporo	1/24 (4.2)	☿ (?)	I/1

[a] Roman numerals: combs infested; Arabic numerals: total No. combs in nest.

generally low, but occasionally exceeds 15% (Table 9.2). The body size of the *Bareogonalos* adult is dependent on the size of the host species and type of cell from which the parasite emerges (Sk. Yamane and S. Yamane 1975; Carmean et al. 1981). Indeed, adults that have emerged from *Vespula* look specifically different from those from *Vespa*.

Metoecus. Three species of this peculiar coleopteran genus occur in Japan, i.e., *M. vespae* (host: *Vespula schrenckii*), *M. satanus* (*Vl. flaviceps*) and *M. paradoxus* (*Vl. flaviceps*; Fig. 9.3D). A species of this genus was found in a colony of *Dolichovespula* only once in Japan (Makino 1982). Though there is no reliable record of this genus from *Vespa*, van der Vecht (1957) got a rhipiphorid (genus not identified) from a nest of the Javanese *V. velutina velutina*. Only *Metoecus paradoxus* occurs in Europe, where Chapman (1870, 1891) studied its life cycle. In Japan two species (*M. vespae* and *M. paradoxus*) have been bionomically studied (Okutani 1952; Hattori and Sk. Yamane 1975). The outline of the life cycle of *M. paradoxus* is inferred as follows:

The female of this beetle lays eggs in cracks of decaying wood. First instar larvae (triungulinid larvae) wait for vespine workers that will come to collect wood as nest material. Triungulinid larvae (free-living phase) that have successfully jumped onto the wasps are transported to the colony, where they seek fourth or fifth instar wasp larvae and enter their bodies. The parasite at this stage possesses a well sclerotized cuticle and three pairs of functional thoracic legs. Within the host body, it grows into a fat sausage-like insect (endoparasitic phase), but does not soon molt into the second instar. Just after the host spins the cocoon, the parasite emerges through a puncture in its cuticle and molts into a second instar without legs. The parasite larva thereafter feeds on the host externally and molts again (now third instar). During second and third instar the larva attaches to the host, and the mature (fourth instar) larva sucks the host dry (Fig. 9.4F-I), while the leaked host body fluid smears the cocoon cap spun by the host. This smear and the loose cocoon spun by the infested host are good indications of parasitism by *Metoecus*. After pupating in the host cell, and adult beetle appears in summer and fall.

Metoecus usually does only trivial damage to the host colony, though we have observed a *Vl. flaviceps* colony in which 42 percent of the cocooned cells were infested by this beetle (Hattori and Sk. Yamane 1975).

Finally, brief mentions are made of syrphid flies, *Volucella* (Diptera), and a pyralid moth, *Pyralis regalis* (Fig. 9.3C). The female *Volucella* directly enters vespine nests to lay eggs on the nest envelope (Fig. 9.5). During larval stage, *Volucella* species are known as scavengers, feeding on wasp corpses and other colony-produced debris dumped on the soil below *Vespula* nests (Spradbery 1973a). They may also be parasites or predators of wasp larvae and pupae within

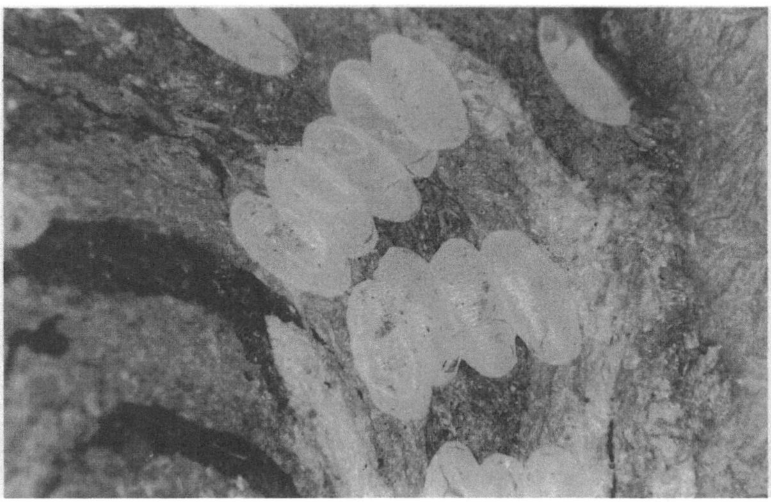

Fig. 9.5 Eggs of a syrphid fly, *Volucella* sp., deposited on the envelope surface of a *Vespula* nest (Photo: S. Aoki)

nest cells. In Japan, Katô (1950) claims that *Volucella tabanoides* is parasitic on *Vl. flaviceps*. Since *Volucella* mature larvae are occassionally found in the brood cells of *Vl. flaviceps* and show a carnivorous tendency under rearing conditions, Akima (1979) suspects that *Volucella* larvae are predators rather than scavengers. They were also found in comb cells of a *Vl. vulgaris* nest (Sk. Yamane et al. 1980b), and recently many young larvae of *V. nitobei* (determined by Dr. K. Ohara) attached to the larvae of *Vl. flaviceps* were collected by Sk. Yamane in Sendai, Miyagi Prefecture, Honshu. These observations strongly suggest that *Volucella* larvae are at least on occasion parasitoids or predators of vespine larvae. Although eggs and larvae of these flies have not been found within nests of *Vespa*, Matsuura (1974c) collected adult flies of *V. nitobei* from a nest of *Vespa simillima xanthoptera*. *Volucella* species associated with vespine nests in Japan are listed in Table 9.3.

Table 9.3. Hover-flies of genus *Volucella* found from vespine nests in Japan[a] (After Ohara unpubl.)

Species of *Volucella*	Vespine species	References
V. tabanoides Motsch.	*Vespula flaviceps lewisii*	Katô (1950); Akima (1979); Okuno (1970)
	Vl. schrenckii	Takamizawa (1981a)
V. nitobei Matsum.	*Vespa simillima xanthoptera*	Matsuura (1974c)
	Vl. flaviceps lewisii	Ikezaki and Ariga (1975); Sk. Yamane et al. (1980b)
	Vl. schrenckii	Takamizawa (1981a)
V. nigricans Coq.	*Vl. flaviceps lewisii*	Okuno (1970)

[a] In most cases larval flies were collected or observed. Only Matsuura (1974c) recorded adult flies.

Pyralis regalis is well known because its larvae eat vespine nests. Our observations suggest that the moth larvae may wound or even feed on vespine larvae and pupae when they bore tunnels throughout the nest combs. This is generally a rare species, but in one nest of *V. simillima*, collected on September 28, 1975 on Haguro-san, Yamagata Prefecture, Honshu, many cocoons had been destroyed by *Pyralis* larvae, the colony being almost completely ruined (Sk. Yamane, pers. obser.).

1.4 Predators on Colonies

In Japan, wild mammals are rarely observed to attack the nests of social wasps. Aoi (1979, personal communication) reported that body cuticles of adult *Vl. vulgaris*, *Vl. schrenckii* and *V. simillima xanthoptera* were found in feces of Japanese bears (*Ursus arctos* and *U. thibetanus*). Since many corpses of wasps including males were recognized in these samples, the bears had probably attacked colonies rather than fed on individual wasps. Among some Mongoloid and Nigroid peoples, insect-eating prevails, and in many districts in Asia people use wasp immatures as protein food. This suggests that humans have long been predators of vespine colonies. One informant in Taiwan said that the Taiwanese monkey occasionally attacks nests of aerial *Vespa* nests. It may be interesting to survey the relationships between the predation by primates, including humans, and the evolution of the extreme aggressiveness of some vespine species.

Among Japanese birds, the honey buzzard *Pernis apivorus* is known to attack nests of *Vespula* and *Vespa analis*, carrying nest combs away to feed its young with immature wasps (Miyazaki 1981; Matsuura pers. obser.). In a survey carried out in Wakayama Prefecture, Honshu, 11 of the 27 nests of *V. analis* that failed after the polyethic period were those destroyed by this bird.

In Europe and North America, badgers, bears, racoons, skunks, moles and honey buzzards are known to attack vespine nests (Fox-Wilson 1946; Trap-Lind 1962; Spradbery 1973a; MacDonald 1977; Edwards 1980). Among these vertebrates, damage by racoons may be the most severe. In Athens, Georgia, USA in 1975, 13 of the 27 mature *Vl. maculifrons* colonies left for ecological sampling were destroyed by racoons during early September into December (MacDonald 1977).

Ants are generally thought to be one of the most serious natural enemies of social wasps (Jeanne 1975). The relative scarcity of Vespinae in lowland tropics may be explained by the fact that their haplometrotically founded nests are vulnerable to ant predation. Vespine nests have structural aspects that seem to protect them from ant predation, though there are no quantitative data indicating that ant predation operates as a factor detrimental to colony development. At present it is not clear whether vespine species have already escaped attacks by ants or still suffer from ant predation even in the temperate region.

2 Effect of Stylopization by *Xenos* spp.

Only occasionally are *Xenos* spp. found as parasitic natural enemies, but once *Vespa* adults are stylopized by them various interesting effects can be observed. Since very little information on the ecology of *Xenos* spp. is available so far, the behavior of stylopized wasps (Matsuura unpubl.) is described in outline here.

2.1 Host and Parasitism

Stylopization by *Xenos* spp. was found in all six species of *Vespa* in Honshu (Table 9.4). The hosts included workers, new queens and males, but no stylopization could be observed in founding queens performing nesting activities. Males show no copulatory reaction to stylopized new queens who emerge during the fall, as will be described later in detail, and so it seems that stylopized queens do not found nests.

Rate of parasitism by *Xenos* spp. was 0.9%, according to examinations of all castes of five *Vespa* species observed in and near Kibi-chô, Wakayama Prefec-

Table 9.4. Parasitism by *Xenos* spp. on Japanese *Vespa* (Matsuura unpubl.)

Species	No. of individuals surveyed	Caste of host					Para-sitism (%)
		Foundress	New queen	Worker	Male	Total	
V. tropica	273	0	1	3	2	6	2.2
V. dybowskii	480	–	0	0	0	0	0.0
V. analis	3653	0	3	29	22	54	1.1
V. crabro	4964	0	6	26	24	56	1.1
V. mandarinia	7559	0	2	43	12	67	0.8
V. simillima	57200	0	62	259	149	470	0.8
Total	74129	0	74	360	209	643	0.9

ture, during a 10-year period from 1966 to 1975, in addition to data on a sixth species, *V. dybowskii,* obtained in Nagano Prefecture. The degree of parasitism, however, varies according to region, nest and stage of nest development. Such parasitism often reaches especially high levels in *Vespa* nests in a given region, regardless of species. The extent of parasitism also varies according to the stage of nest development: the ratio of nests containing stylopized individuals increased with the emergence of workers, and more than half of the total nests of five *Vespa* species contained stylopized individuals during the reproductive period (Table 9.5).

Table 9.5. Number and developmental period of *Vespa* nests attacked by *Xenos* spp. (Matsuura unpubl.)

Stage \ Species	*V. tropica*	*V. analis*	*V. crabro*	*V. mandarinia*	*V. simillima*	No. of nests stylopized	No. of nests surveyed	Parasitism (%)
Solitary p.	0	0	0	0	0	0	28	0
Cooperative p.	3	5	2	0	3	13	64	20.3
Polyethic p.	1	3	3	8	22	37	100	37.0
Reproductive p.	4	16	5	8	38	71	139	58.3
No. of nests stylopized	8	24	10	16	63	131	–	–
No. of nests surveyed	34	88	33	41	135	–	331	–
Parasitism (%)	23.5	27.3	30.3	39.0	46.7	–	–	39.6

2.2 Life History of *Xenos* spp.

Hibernating *Vespa* individuals are generally new queens, but workers stylopized by *Xenos* spp. may also hibernate. All individuals of *Xenos* spp. found in stylopized *Vespa* during the winter and after hibernation were females (Matsuura unpubl.). Since every *Xenos* male produced during October and November emerges during the nesting period of *Vespa* and dies within a single day, it seems that only female *Xenos* adults which have completed copulation outside wasp's nest hibernate in the body of their host. Stylopized *Vespa* individuals found during winter, or just after hibernation, are workers or queens, but not males. The overwintered stylopized wasps start their activities in late April to early May, searching for tree sap from *Quercus* spp., etc., and they survive until around early July. Since a large number of the first instar larvae (triungulin) of *Xenos* spp. can be seen attached to the bodies of stylopized *Vespa* taking tree sap during late May and early June, it seems that the parasites start to hatch inside their hosts during this period, waiting an opportunity to escape and become transferred to *Vespa* nests.

On the basis of its unique life history, it is known that *Xenos* is a prolific insect in general (Maeta 1963). Examination has revealed that the number of first instar larvae of *Xenos moutoni* appearing per *V. mandarinia* worker was 29 843–36 844 (average 34 581.3, n = 6). The following is an example of a study on the period of appearance of the first instar larvae (Matsuura 1975a): One tree sap-visiting worker of *V. mandarinia* (fresh weight, 1.51 g) was collected on May 17, 1966; when the worker was reared in a cylindrical bottle, the first larva started to creep out of the worker's body on June 9; thereafter, appearance of larvae was observed every day for a month until the host died on July 8; the number of larvae emerging per day was 1000–2000 from June 9 to June 30, and the maximum number of 2362 was observed on June 2; on July 8, however, only 13 larvae appeared, and the host died.

It seems that these larvae become attached to the bodies of founding queens and workers of various vespine wasps visiting tree sap, in order to be transferred to vespine nests. They spend their parasitic life in the nest by creeping into their hosts, although no information is available concerning their behavior inside the nest. It seems probable that *Xenos* spp. repeat at least three generations in 1 year, because the ratio of nests containing stylopized wasps has a tendency to increase along with the development of nesting activity of *Vespa*. The first parasite adult appears in mid- to late June, a period during which *Vespa* workers start to emerge. The appearance of adults continues without break until late November.

2.3 Activities of Stylopized Wasps in the Field

Generally, it is difficult to find *Vespa* wasps stylopized by *Xenos* spp. in the field, but such wasps can be found occasionally among hibernating *Vespa* wasps or those visiting tree sap in spring. A stylopized *V. mandarinia* worker has been observed in winter, hibernating in an old nest of *V. analis*, and a stylopized queen of *V. analis* was also discovered in rotting wood being used as her hibernaculum (Matsuura unpubl.).

Overwintered stylopized *Vespa* wasps visit sap-exuding trees, such as *Quercus* spp., together with their non-stylopized queens. Next to the queen of the same species, stylopized workers of *V. mandarinia* predominate over individuals of various *Vespa* species who gather around any tree's sap-exuding aperture (Matsuura 1969). They spend time around such apertures until early July. However, they seem to die soon after the appearance and creeping-out of *Xenos* larvae.

2.4 *Intranidal Behavior of Stylopized Hornets*

Remarkable behavioral differences can be seen between wasps stylopized and non-stylopized by *Xenos* spp. (Matsuura 1967b, 1976a). Table 9.6 shows the percentage of time spent in each activity by stylopized (x12) and non-stylopized (No. 105) workers of *V. mandarinia*, which emerged during the polyethic period,

Table 9.6. Percent of time spent in each activity of stylopized and nonstylopized *Vespa mandarinia* workers (Matsuura unpubl.)

	Activity		Nonstylopized worker (No. 105)	Stylopized worker (X-12)
Extra-nidal act.	Foraging for	Pulp	6.0	0
		Prey	6.2	0
		Sap	4.0	0
		Water	1.0	0
	Orientation		0.1	0
Intra-nidal act.	Spitting water		2.1	0
	Sucking water		2.4	0
	Fanning		1.0	0.1
	Patrol		0.2	0.1
	Nest cleaning		0.1	0
	Prey malaxation		4.9	0.1
	Feeding larvae		4.6	0
	Food exchange	larvae	4.9	0.9
		adults	1.5	0.1
	Pulp malaxation		3.0	0
	Enlarging envelope		1.0	0
	Making cell		1.3	0
	Removing cocoon caps		0.2	0
	Resting		29.8	95.8
	Walking		20.0	2.8
	Royal court		3.5	0
	Body cleaning		2.2	0.1

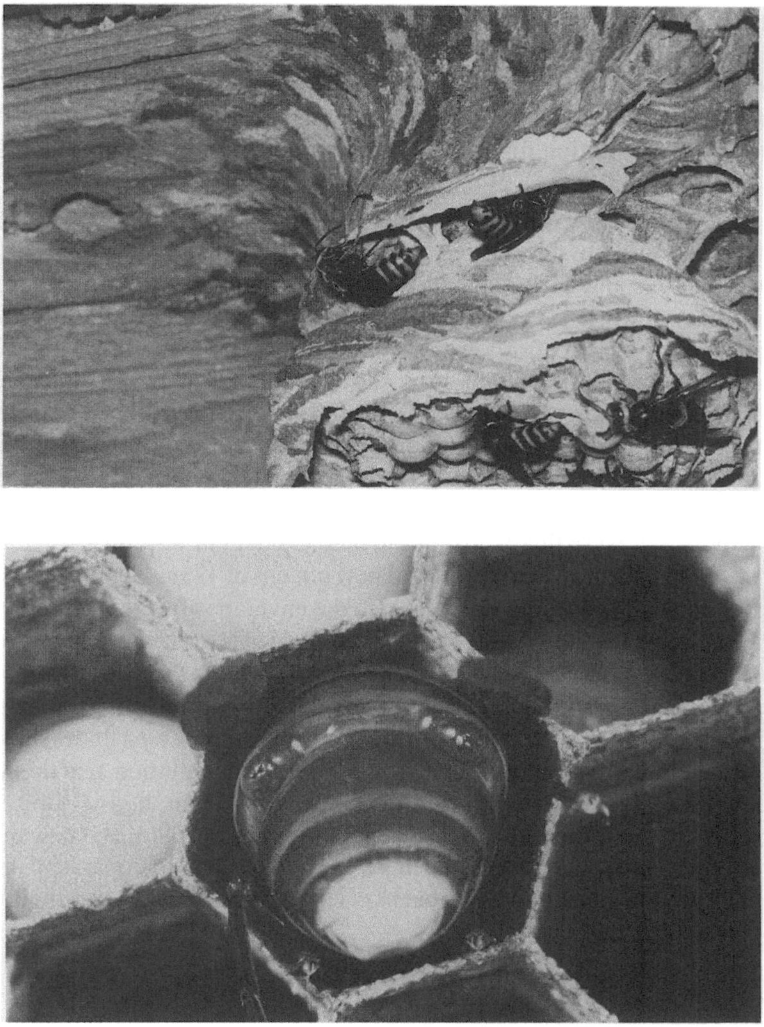

Fig. 9.6. Stylopized workers of *Vespa crabro* (*top*), and gaster of a *V. analis* male parasitized by *Xenos crabronis* (*bottom*)

during 13 days from emergence to death (total observation time, 48 h). Based on the results of this examination and some other fragmentary observations on four other *Vespa* species (*V. analis, V. crabro, V. tropica* and *V. simillima*), the behavioral features of stylopized workers, as compared with those of non-stylopized individuals, can be summarized as follows: (1) lack of orientation in flight 2–3 days after emergence, (2) non-participation in extranidal activities for the collection of nest materials, prey, tree sap, water, etc., (3) non-participation or only scant participation in intranidal activities, such as nest preservation, brood care and nest construction (occasionally prey malaxation, provisioning of larvae and fanning can be observed, but no nest-construction activities occur).

(4) non-participaton in counterattacking enemies (threatening inside the nest and on the envelope can be observed), (5) non-participation in royal court, (6) dependence on the secretion of fifth instar larvae for most of their food, (7) resting on the top comb or the envelope when they are at home, and (8) non-orientation flight and non-return to the nest, finally leaving 9–12 days after emergence.

As a result, stylopized workers become non-laboring individuals, and their presence disturbs nesting activity and colony development, especially during the cooperative and early stage of the polyethic periods, because of the insufficient number of workers at these times (see upper photo in Fig. 9.6). A *V. crabro* nest in which the ratio of stylopized workers reached 45.5% (5/11) at the time of the cooperative period in mid-July, showed a great delay in colony development compared with normal nests (Matsuura 1973a). In addition, it has been observed that the proportion of stylopized workers reached 100% in a nest of *V. mandarinia* and a nest of *V. crabro* during the cooperative period and that these nests were disorganized thereafter; their founding queens could not retire from foraging activity and were finally lost in the field (Matsuura unpubl.).

When the hosts are males and new queens of *Vespa*, stylopized wasps remain still on the upper comb or in a gap in the envelope after their emergence, similar to the case of workers, except for occasional visits to larvae in order to ingest their secretions. After 7–12 days of such a resting period, they leave their nest without an orientation flight. Non-stylopized males and new queens gain their copulation abilities as soon as they leave the nest and, therefore, *Vespa* males show an interest in new queens. In a rearing box, however, no copulation reaction can be seen between any pair of stylopized male and new queen, non-stylopized male and stylopized new queen, or stylopized male and non-stylopized new queen having left their nests. Therefore, it seems that stylopization by *Xenos* spp. has the effect of depriving males and new queens of copulatory ability during their later nesting period.

Chapter 10

Phylogeny and Classification

Morphologically and in their behavior, the Vespinae are relatively homogeneous. In this chapter we (1) enumerate characters which are important in describing the subfamily and in dividing it into genera and subgenera, (2) infer the polarities within these characters (ancestral vs. derived states), and (3) infer the phylogenetic position of the Vespinae within the Vespoidea and the relationships among vespine genera.

Although adult morphological characters are generally considered most useful in such an analysis, we will also make use of larval, behavioral and ecological characters as needed (Figs. 10.1, 10.2). Many of these are known only in a few species, but we have tried in each case to include at least one representative from each subgenus or species-group. This cannot be done for the subgenus *Rugovespula*, however, as nothing is yet known of its nesting biology. For reasons of space, we will list and discuss only a selection of the entire array of characters used in vespid systematics.

In deriving the phylogenetic polarity of vespine character states, we use the standard approach of out-group comparison, taking various other vespoids as out-groups. On the assumption that the Vespinae originated in the Old World, we have largely left aside polistine genera endemic to the New World. There is reason to believe that each of the vespine genera, subgenera and species groups mentioned here is monophyletic.

The cladistic method of Hennig (1966) is used here in analyzing phylogenetic relationships. The core of this approach is the linking of related groups through shared derived character states, or "synapomorphies". Its power is exemplified in Carpenter's (1982) treatment of the Vespoidea and Brothers's (1975) of the Aculeata as a whole. We draw a distinction, though, between the analytical method of cladistics and the doctrine of cladistic classification that only strictly monophyletic taxa are acceptable, i.e., each must include all and only the descendants of a single ancestor species. Insofar as we follow Hennig's method in constructing phylogenetic trees, or cladograms, we are cladists. Insofar as we do not convert these trees directly into classification, we are not complete cladists.

To date, only de Beaumont (1958) and Sk. Yamane (1976) have presented relationships among the genera, subgenera and species of vespines in the form of cladograms. Yamane's treatment has been especially criticized with regard to inferred character states in the common ancestor of *Vespula* and *Dolichovespula* (Greene 1979; Akre 1982; Akre and MacDonald 1986). In Section 3 of this Chapter we examine these criticisms and try to arrive at a deeper understanding of phylogeny within the subfamily. At the end of the chapter is a list of the known species and subspecies which we consider valid, together with their approximate ranges.

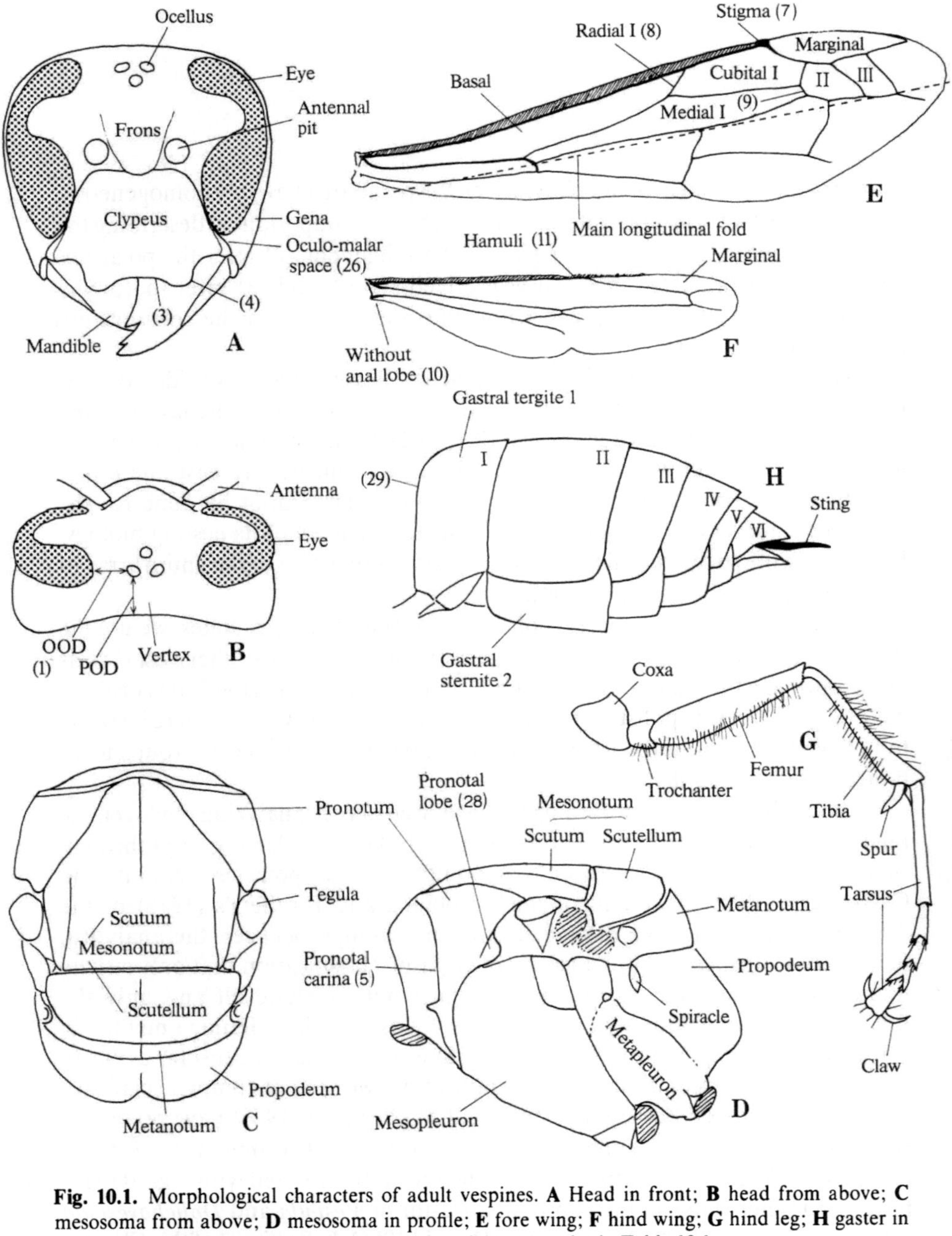

Fig. 10.1. Morphological characters of adult vespines. **A** Head in front; **B** head from above; **C** mesosoma from above; **D** mesosoma in profile; **E** fore wing; **F** hind wing; **G** hind leg; **H** gaster in profile. *Numerals in parentheses* correspond to character codes in Table 10.1

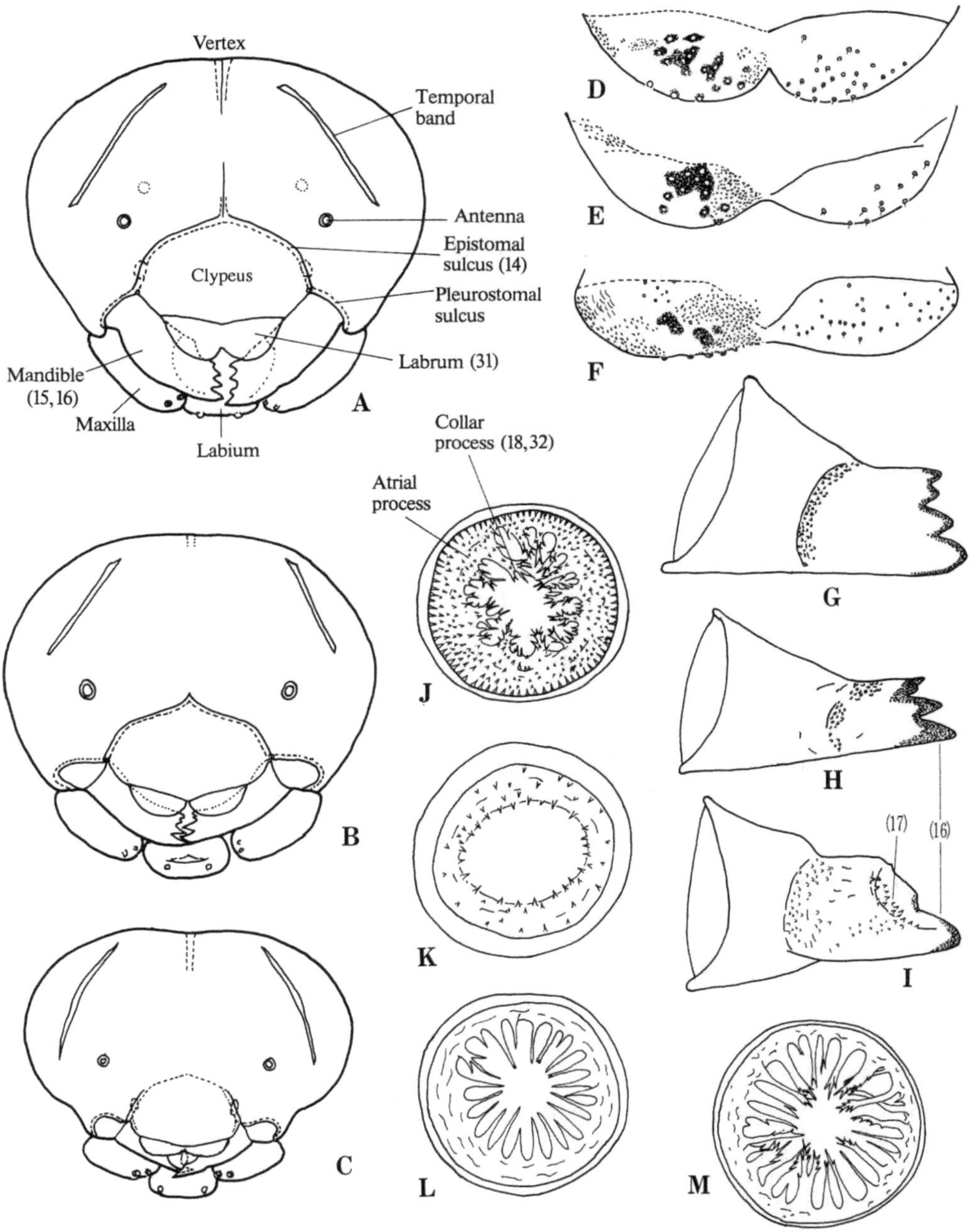

Fig. 10.2. Morphological characters of larval vespines. **A-C** Head in front in *Vespa* (**A**), *Vespula* (**B**) and *Dolichovespula* (**C**); **D-F** labrum in *Vespa* (**D**), *Vespula* (**E**) and *Dolichovespula* (**F**); **G-I** inner side of mandible in *Vespa* (**G**), *Vespula* (**H**) and *Dolichovespula* (**I**); **J-M** spiracle in *Vespa* (**J**), subgenus *Paravespula* of *Vespula* (**K**), subgenus *Vespula* (**L**) and *Dolichovespula* (**M**)

1 Characters Used

In any cladistic analysis, each character must be differentiated, or polarized, into an ancestral (plesiomorphic) and one or more derived (apomorphic) states. For present purposes this requires an extensive investigation, not only within the Vespinae (in-groups) but among an array of vespoid relatives (out-groups). The most important out-group is the Polistinae, the nearest relatives of the Vespinae. It should be noted that no character state can be absolutely ancestral or derived. Rather, these terms indicate the relative positions of states within the transformation series of a given character. For example, the first gastral tergum is clearly separated into an anterior vertical and a posterior horizontal region in all vespines except *Provespa*. In the polistines, as in *Provespa*, there is no such separation, a state which we interpret as ancestral within the Vespidae. If we consider, though, that the state in *Provespa* represents a secondary modification, or reversal, then this would be derived relative to the state in other vespines, even though a very similar condition remains ancestral in the Polistinae and for the family as a whole. Some ecological and behavioral characters (e.g., nest paper type: character 38) vary considerably within groups whose monophyly is strongly supported on morphological grounds, while others (e.g., embryo nest architecture: characters 22, 24) seems at least as informative as many adult morphological characters.

The polarization of characters is often so difficult that different investigators arrive at sharply differing conclusions. Here we list 19 adult, 7 larval and 16 behavioral characters, 42 in all, of which 24 are relatively easily polarized (Tables 10.1a, 10.2a), the remaining 18 being rather difficult (Tables 10.1b, 10.2b). In some cases we have improved upon Sk. Yamane's (1976) conclusions in this regard. MacDonald (1977) gave a useful comparison of the Nearctic yellowjackets, chiefly discussing the nest-structural and behavioral characters.

Table 10.1a. Ancestral and derived states of 24 characters of Vespinae

Characters	Ancestral states (–)	Derived states (+)
1. Vertex length (♀, ☿)	Ocello-occipital distance (POD) >oculo-ocellar distance (OOD)	OOD >2POD (rarely OOD = POD)
2. Ocellar size	Ocellar diameter < OOD	Ocellar diameter >OOD
3. Apical emargination of clypeus	Narrow	Wide
4. Lateral teeth of clypeal apex (♀, ☿)	Round or slightly angulate	Sharply pointed
5. Pronotal carina	Complete	Absent dorsally
6. Carina on hind coxa	Absent	Present
7. Pterostigma length	Pterostigma longer than prestigma	Pterostigma shorter than prestigma
8. Prestigma length	Prestigma < 2 × radial sector	Prestigma >2 × radial sector

Table 10.1a. *(Cont.)*

Characters	Ancestral states (–)	Derived states (+)
9. Base of 1RS cell	Obliquely truncate	Vertically truncate or nearly so
10. Anal lobe of hind wing	Present	Absent
11. Hamuli of hind wing	Beginning before the tip of marginal cell	Beginning at or around the tip of marginal cell
12. Ovariole number (A)[a]	3 pairs	More than 6 pairs
13. Ovariole number (B)[a]	6 pairs	7–10 pairs
14. Epistomal sulcus of larva	Complete	Reduced dorsally
15. Shape of larval mandible	Wide and stout	Slender and feeble
16. Larval mandible	Tridentate	Monodentate
17. Inner processes of larval mandible	Absent	Present
18. Collar processes of larval spiracle	Absent	Present
19. Prey range	Live insects, etc.	Vertebrate carrion in addition
20. Colony founding	Independent	Swarm
21. Cocoon cap	Consisting of 1 or 2 layers, simple in structure; no sex or caste differences	Consisting of 2 lavers in ♀ and ♂, and 3 in ♀; complex in structure
22. Embryo nest pedicel	Simple, stalk-shaped	Thin, with a twist
23. Coating of pedicel	Dry	Adhesive (a) or lacking (b)
24. Embryo nest disk	Absent	Present

[a] A and B indicate evaluations of ovariole number at different levels.

Table 10.1b. Eighteen Vespinae characters, in which it is difficult to determine ancestral or derived conditions

Characters	State a [a]	State b [a]
25. Occipital carina	Present close to mandibular base	Lacking in the lower one-third
26. Oculo-malar space (♀, ☿)	Shorter than half the length of antennal segment 11	Longer than antennal segment 11
27. Labial palps (♀, ☿)	Segment 3 with 1–2 strong setae	Segment 3 with weak hairs alone
28. Pretegular carina	Well developed	Obscure or absent
29. Shape of gastral tergite 1 (TI)	T1 sharply angled between anterior and dorsal faces	This angle indistinct
30. Aedeagal tip (♂)	Round	Bifid
31. Larval labrum	Widest at base, with somewhat deep apical emargination	Widest more apically, with shallower emargination
32. Collar processes of larval spiracle	Short, simple	Long and simple (1) or long and branched (2)
33. Nest site	Open spaces	Covered spaces
34. Colony duration	Early decline	Decline in late fall
35. Larval hunger signal	Cell wall scraping consistent	Frequency of scraping extremely low
36. 1st envelope sheet of embryo nest	Attached directly to substratum	Attached to nest pedicel
37. Subsequent sheets	Sheets after 2nd or 3rd each starting from the preceding sheet	All sheets starting independently from pedicel
38. Nest paper type	Brown, brittle	Grey, pliable
39. Suspensoria between 1st comb and envelope	Pillarlike, rarely platelike	Ribbonlike
40. Other suspensoria	Pillarlike	Ribbonlike
41. Envelope of mature nest	Scalloped, sometimes with long vertical tubes	Laminar
42. Hibernaculum	Pre-existing cavities used	At least partly dug by queen herself

[a] "a" and "b" do not indicate nor even suggest ancestral and derived states respectively.

Table 10.2a. States of 24 characters (in Table 10.1a) in the four vespine genera[a]

Characters	*Vespa*	*Provespa*	*Vespula*				*Dolicho-vespula*
			Pl.	*Vl.*	*Sq.*	*Rl.*	
1. Vertex length (♀, ☿)	−	+	+	+	+	+	+
2. Ocellar size	±	+	−	−	−	−	−
3. Apical emargination of clypeus	+	+	+	+	+	+	+
4. Clypeal teeth (♀, ☿)	−	−	−	±	−	−	±
5. Pronotal carina	−	+	+	+	+	+	−
6. Carina on hind coxa	+	(−)	+	+	+	+	+
7. Pterostigma length	+	+	+	+	+	+	+
8. Prestigma length	+	−	−	−	−	−	−
9. Base of 1RS cell	−	+	+	+	+	+	+
10. Anal lobe of hind wing	+	+	+	+	+	+	+
11. Humeli placement	−	(+)	+	+	+	+	+
12. Ovariole number (A)[b]	+	?	+	+	+	(+)	+
13. Ovariole number (B)	+	?	−	−	(+)	(−)	−
14. Epistomal sulcus of larva	−	−	−	−	−	(−)	+
15. Larval mandibular shape	−	−	−	−	−	(−)	−
16. Denticulation of larval mandible	−	−	−	−	−	(−)	+
17. Inner processes of larval mandible	−	−	−	−	−	(−)	+
18. Collar processes of larval spiracle	+	+	+	+	+	(+)	+
19. Prey range	−	−	+	−	−	?	−
20. Colony founding	−	+	−	−	−	?	−
21. Cocoon cap	−	?	+	−	−	?	−
22. Embryo nest pedicel	−		+	+	+	?	+
23. Coating of pedicel	−		+a	+a	?	?	+b
24. Embryo nest disk	−		+	+	?	?	−

[a] +, − Derived and ancestral character states respectively.
(+, −) Evaluation of character state is not very reliable, or character has not been observed but its state is supposed.
+a, +b Different conditions in a derived state.
[b] A and B indicate evaluations of ovariole number at different levels.

Table 10.2b. States of 18 characters (in Table 10.1b) in the four vespine genera[a]

Characters	*Vespa*	*Provespa*	*Vespula*				*Dolicho-vespula*
			Pl.	*Vl.*	*Sq.*	*Rl.*	
25. Occipital carina	a	b	a,b	b	b	a	b
26. Oculo-malar space (♀, ☿)	a, b	a	a	a	a	a	b
27. Labial palp (♀, ☿)	a	a	b	b	b	b	b
28. Pretegular carina	a	b	b	b	b	b	b
29. Shape of T1	a	b	a	a	a	a	a
30. Aedeagal tip	b	b	a	a	a	a	b
31. Larval labrum	a	a	a	a	a	a	b
32. Collar processes of larval spiracle	b2	b1	a	b1	b1	?	b2
33. Nest site	a,b	a,b	(a),b	b	b	?	a,b
34. Colony duration	(a),b		b	a	b	?	a
35. Larval hunger signal	a	?	b	b	?	?	a
36. 1st envelope sheet of embryo nest	a		b	b	?	?	b
37. Subsequent sheets	a		a	a	?	?	b
38. Nest paper type	a		a,b	b	b	?	b
39. Suspensoria between 1st comb and envelope	a	a	a	b	a	?	b
40. Other suspensoria	a	a	a	a	a	?	b
41. Envelope of mature nest	a	a+b	a	b	b	?	b
42. Hibernaculum	b		a	?	?	?	b

[a] a, b Different states of character.
(a, b) Character state not commonly seen in the taxon.
b1, b2 Different conditions in a character state.

2 Phylogenetic Position of the Vespinae

Several investigators have proposed classifications of the vespoid wasps (de Saussure 1852–58; Bequaert 1918; Richards 1962; Charnley 1973). In his discussion of phylogeny within the superfamily, Richards (1962) gave special attention to traits restricted to particular groups. This implies that he was attempting to establish taxa on the basis of synapomorphy. He stated that the Polistinae shared the most recent ancestor with the Vespinae, though he did not explicitly support this conclusion. In his treatment of phylogeny within the Polistinae, on the other hand, Richards (1978) virtually abandoned the use of synapomorphy.

In 1982 the young systematist (and strict cladist) J.M. Carpenter reanalyzed the relationships of vespoid families and subfamilies. He set aside Richards's (1962) classification within the Masaridae and Eumenidae, on the grounds that some groups were not monophyletic, but agreed with Richards in recognizing the social subfamilies Stenogastrinae, Polistinae and Vespinae. He upheld the monophyly of the Vespinae on the basis of eight derived characters: (1) Absence of an anal lobe in the hind wing (character 10 in our Tables 10.1, 10.2), (2) broad mandibles (ratio of mandibular length to clypeal width = 0.50–1.13) (Richards 1962, gave the opposite polarization), (3) hind coxae with dorsal carinae (character 6), (4) very short gonocardo of the male genitalia reduced laterally, (5) volsella with a digitus, and short cuspis, (6) six ovarioles/ovary (character 12), (7) larval spiracular atrium with processes, and (8) advanced eusociality. In our view, the anteriorly broadly truncated clypeus (character 3) and reduced stigma of the forewing (character 7) are also important in this regard. Carpenter further enumerated six synapomorphies of the Vespinae and Polistinae, such as simple claws (reversal) and an externally visible van der Vecht's gland. For the most part, our observations are in agreement with his.

In order to make classification isomorphic with the cladogram, Carpenter (1982) reduced the superfamily Vespoidea to a single family, Vespidae, comprising six subfamilies. Our own approach to classification is more conservative. Given the extreme difficulty in inferring phylogeny, we prefer not to be hasty in setting aside a well accepted classification while the basis for it is still under discussion. Figure 10.3 shows Richards's (1962) scheme along with Carpenter's (1982).

On the assumption that the Polistinae originated in Southeast Asia, Richards (1971) explained their occurrence in the New World as dispersal across the Bering land bridge from Asia. Van der Vecht (1957) implied an East Asian origin of the

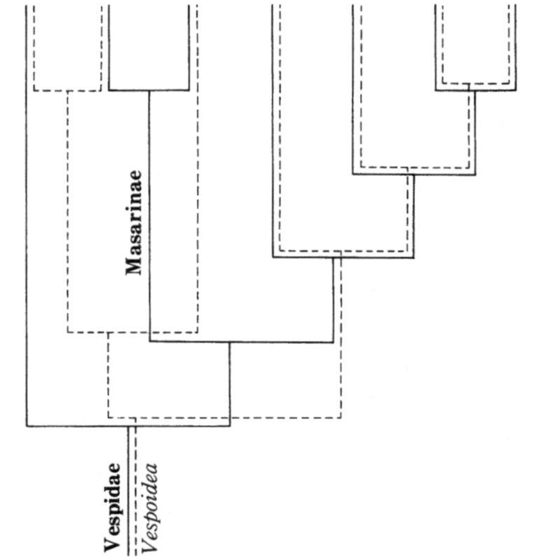

Fig. 10.3. Phylogenies of the Vespoidea proposed by Richards (1962) (*dotted lines* and *italic letters*), and Carpenter (1982) (*solid* and *block*)

Vespinae. Carpenter (1982), on the other hand, drew attention to the fact that the Vespinae are mainly a north temperate group and suggested a vicariant event giving rise to the (laurasian) Vespinae and (gondwanian) Polistinae, a scenario which does not require their dispersal from Asia into the Americas. Such an approach may well be fruitful in the phylogeny of social wasps, but Carpenter's criticism of Richards (1971) and Sk. Yamane et al. (1980b) in this question does not strike us as very powerful; at best, a contending viewpoint. The distribution of vespine species is discussed in Chapter 11.

Yoshikawa (1973, p. 114–15; Fig. 4.26) proposed a multilinear evolution of sociality among vespoid wasps, culminating independently in the Vespinae in Europe, Asia and North America. This hypothesis is unsupported by cladistic analyses.

3 Phylogeny of the Vespinae

3.1 Phylogenies Proposed up to Now

In a discussion of the origin of social parasitism in the Vespinae and bumblebees, de Beaumont (1958) presented a phylogenetic tree of the European species as a way of showing the supposed close relationships between parasites and their respective hosts. Sk. Yamane (1976) derived a cladogram of the genera and subgenera of Vespinae, using larval and nest architectural characters, as well as adult morphological characters. Yamane had not seen de Beaumont's paper at the time, but their conclusions are mostly in agreement, showing *Vespula* and *Dolichovespula* most closely related to each other and *Vespa* as their common sister-group. Sk. Yamane enumerated four synapomorphies for *Vespula* + *Dolichovespula*: (1) Absence of strong setae on the labial palps (our character 27), (2) hamuli of the hindwing beginning at the tip of the subcosta (character 11), (3) humeral calli not separated from the sides of the pronotum (character 28) and (4) embryo nest with a twisted pedicel (character 22). This relationship is now generally accepted for these three genera (Richards 1971; Greene 1979; Akre 1982). We extend the discussion below to include *Provespa*.

Sk. Yamane (1976) also presented an evolutionary scenario of the Vespinae, drawing attention to the remarkable similarity between *Vespa* and *Vespula* in some larval characters. Greene et al. (1976) and Greene (1979), on the other hand, emphasized the striking behavioral similarities between *Vespa* and *Dolichovespula* and regarded the latter as much more primitive than *Vespula*, a view shared by Akre (1982) and Akre and MacDonald (1986). Given the widespread tendency of American entomologists (e.g., Snelling 1981) to lump together *Dolichovespula* and *Vespula* as "yellowjackets" on the basis of superficial similarities, Greene's attitude can have a positive effect. Nonetheless, Greene's and Akre's arguments are not entirely sound. While they claimed to arrive at a phylogenetic sequence opposite to that of Sk. Yamane (1976), they did not draw a clear distinction between branching pattern and polarity with regard to the common ancestor of *Dolichovespula* and *Vespula*. The conclusions of the two parties concerning the branching sequence are in fact the same (see below).

3.2 Phylogenetic Relationships at Genus Level

Let us first show to what extent each of the four vespine genera can be regarded as monophyletic. *Vespa* retains ancestral states in a great many characters, while reliable derived states are scarce. The very small pterostigma of the forewing (character 8) may be the most important unique derived character, or autapomorphy, of this genus, and the 7+ ovarioles/ovary may represent another. Several peculiar characters firmly establish the monophyly of *Provespa*, e.g., the extraordinarily large ocelli and nocturnal habits (character 2), absence of an occipital carina (character 25), and colony-founding by swarming (character 20).

Sk. Yamane (1976) treated the long oculo-malar space (character 26) of *Dolichovespula* as a good autapomorphy of the genus, but this character now seems rather uninformative. More reliable are the monodentate larval mandible (character 16) and cusps on the inner surface of the larval mandible (character 17). In the few species of *Dolichovespula* in which it has been examined, the embryo nest pedicel is coated with a rubbery oral secretion; this may also be autapomorphic for the genus as a whole.

The monophyly of *Vespula* (s.l.) has been problematic and uncertain, as it is a heterogeneous assemblage of subgroups with no very strong adult or larval characters to unite them. On the other hand, within this genus *Paravespula*, *Vespula* (s.str.), *Rugovespula* and the *Vl. squamosa* species-group are very probably each monophyletic. Only the presence of a disk just above the comb in the embryo nest (character 24) is at present a good candidate for an autapomorphy of the entire genus, and this has yet to be studied in *Rugovespula* and the *Vl. squamosa* group. The absence of an oral secretion on the embryo nest pedicel (character 23) also merits attention in this regard; it has been examined in only a few species. The subgenus *Vespula* resembles *Dolichovespula* in a number of characters, such as the range of prey taken (character 19), colony life-span (34), type of nest paper (38), and the suspensoria connecting adjacent combs in the mature nest (39). However, no common character of the two groups has yet given strong evidence of synapomorphy. We tentatively treat the genus *Vespula* as monophyletic on the basis of the embryo nest disk and its overall phenetic impression.

Four characters were previously adduced as support for a sister group relationship of *Dolichovespula* + *Vespula* (Sk. Yamane 1976). Our present analysis calls two of these (characters 27 and 28) into question, owing to difficulty in determining their polarity, but we here introduce two new ones: the very short ocellar-occipital distance (character 1) and the vertically truncate apex of the first discoidal cell in the forewing (9). This leaves us with four apparently reliable characters (1, 9, 11, 12), of which the twisted embryo nest pedicel is unique among the Vespidae. Opposing this general conclusion are some characters suggestive of a close relationship between *Dolichovespula* and *Vespa*, such as the apically bifid penis (character 30), branched collar processes of the mature larval spiracles (32) and the hibernacula dug by new queens (42). However, none of these is as reliable as the characters linking *Dolichovespula* with *Vespula*. There is no strong indication of any synapomorphy of *Vespa* + *Vespula*.

The genus *Provespa*, often regarded as the most primitive group of vespines, in fact shows many derived and specialized traits in both its physical makeup and behavior (Matsuura pers. obser.; see also Chap. 7). Three of the four derived character states mentioned as shared by *Vespula* and *Dolichovespula* are also found in *Provespa*, which strongly suggests that the three together form a monophyletic group. The situation is complicated, though, by some derived character states common to *Provespa* and *Vespa*, sucn as the apically bifid penis and the formation of a royal court around the queen.

Even given that *Provespa* + *Vespula* + *Dolichovespula* appears monophyletic, the phylogenetic relationship between the three genera is far from certain. *Provespa* and *Vespula* share a derived condition, the absence of the pronotal carina above (character 5), which also appears independently in the Stenogastrinae and the polistine genus *Polybioides*. As mentioned above, though, *Vespula* and *Dolichovespula* share a unique derived state, the twisted embryo nest pedicel, which seems very informative. One solution to this incongruency is the hypothesis that the common ancestor of all three genera had a twisted pedicel, which is secondarily lost in *Provespa* (Fig. 10.4). Almost uniquely among vespines, *Provespa* swarm-found, so that the nest is founded by a group of workers, rather than a queen. This means that the true embryo nest itself has been lost in this genus. The phylogenetic position of *Provespa* within the subfamily calls for further analysis, for which new and better characters are needed.

In conclusion, *Vespa* is monophyletic and apparently the sister group of the other vespines together. Among the latter, the monophyly of *Vespula* is open to question, and there are two plausible hypotheses regarding the relationships among *Provespa*, *Dolichovespula* and *Vespula* (Fig. 10.5). The problem of the

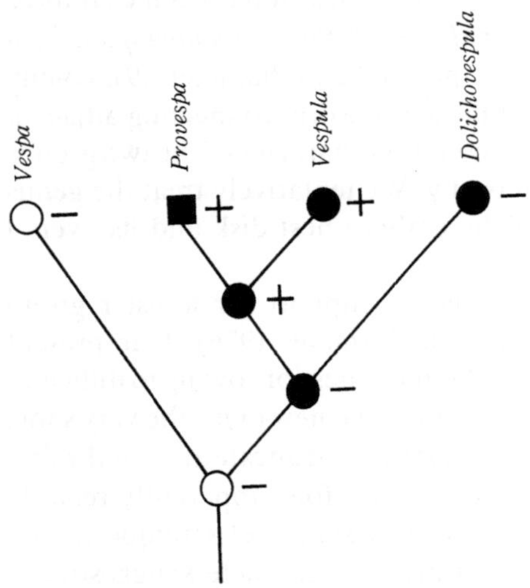

Fig. 10.4. Character state distribution for embryo nest pedicel and pronotal carina within the Vespinae. ○ Pedicel without twist (ancestral); ● with twist (derived); ■: condition without embryo nest (secondarily derived). – Pronotal carina developed (ancestral); + carina absent in dorsal part (derived). This cladogram seems consistent for these two characters, but see text

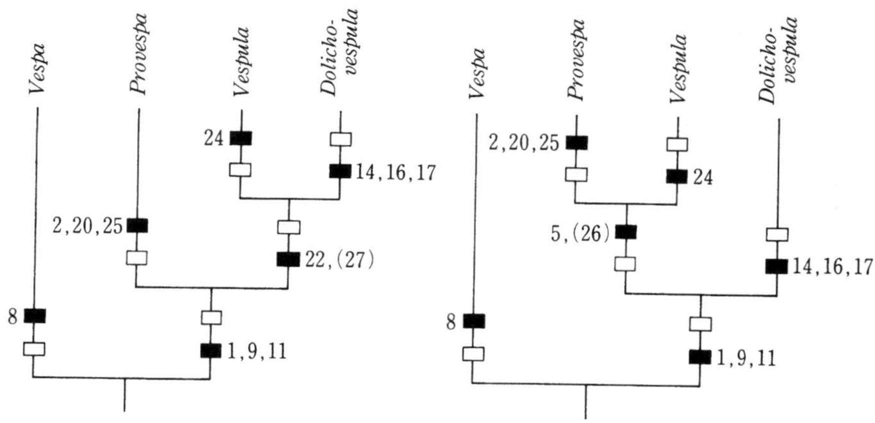

Fig. 10.5. Phylogenetic relationship among the four vespine genera, reconstructed on the basis of shared, derived character states. Two possibilities are suggested, according to different evaluations of some characters for their polarity. ■ Derived state; □ ancestral state. *Numerals* correspond to character codes in Table 10.1

genera of Vespinae illustrates that even a four-taxon problem can be difficult. The classification in Table 10.3 reflects the hypothesized phylogeny of vespines but is not a one-to-one translation of it, as it preserves most of the infrageneric taxa proposed to date.

3.3 The Common Ancestor of Vespula and Dolichovespula

According to Greene (1979), *Dolichovespula* is more primitive than *Vespula* and closely related to *Vespa*. His arguments rest on the conception of social complexity as a principal indicator of phylogeny and the idea that social evolution is generally a one-way process. He apparently views the extant taxa as standing approximately in ancestor-descendant relationships to each other, rather than each comprising a mosaic of ancestral and derived traits. This view is not in accord with that of phylogenetic systematists. The characters enumerated by Greene in support of close phylogenetic relationship between *Dolichovespula* and *Vespa* are as follows:

Morphological. (a) Apically bifid penis, (b) tyloides on the male antennae, (c) laterally developed pronotal carina, (d) developed collar processes, with many branches, of mature larval spiracles, (e) large size (applicable only to the subgenus *Dolichovespula*), and (f) posterior pair of abdominal ganglia.

Behavioral. (g) Caste differences less marked than in *Vespula*, (h) the mature nest with a thick, “cellular” roof, (i) the embryo nest with an extended entrance tube below, (j) nest envelope incorporating elongate shells or tunnels (at least in *Vespa crabro* and the subgenus *Dolichovespula*), and (k) conspicuous scraping of cell walls by the larvae.

Table 10.3. A classification of Vespinae

Group (genus, subgenus, species group)	Species
Genus *Vespa* (23 species)	
Subgenus *Vespa*	*crabro* etc., 21 spp. in total
Subgenus *Nyctovespa*	*binghami, suprunenkoi*
Genus *Provespa* (3 species)	*anomala, nocturna, barthelemyi*
Genus *Vespula* (22 species)	
Subgenus *Paravespula*	
Group 1	*flaviceps, shidai*
Group 2	*vulgaris, structor, maculifrons flavopilosa*
Group 3[a]	*germanica, pensylvanica*
Subgenus *Vespula*	
Group 1	*rufa. schrenckii, kingdonwardi, nursei, intermedia, atropilosa, acadica, consobrina, vidua*
Group 2	*austriaca*
Vl. squamosa group	*squamosa, sulphrea*
Genus *Dolichovespula* (13 species)	
Subgenus *Dolichovespula*	*media, maculata*
Subgenus *Boreovespula*	
Group 1	*norwegica, saxonica, norvegicoides lama, panda, arenaria*
Group 2	*adulterina, arctica*
Subgenus *Metavespula*	
Group 1	*sylvestris, asiatica*
Group 2	*omissa* (= *ingrica*)[b]

[a] According to Hoffman and MacDonald (1982), who made an immunological comparison of venoms among four *Vespula* species (*vulgaris, maculifrons, flavopilosa* and *pensylvanica*), *vulgaris* and *pensylvanica* have the most recent common ancestor among them.
[b] The position of this parasitic species (host: *D. sylvestris*) should be reexamined in the light of informative characters.

Nonetheless, two of these traits (c and g) are undoubtedly ancestral, and the others are either difficult to polarize or found only in some members of each of the two genera. They therefore do not support the monophyly of *Vespa* + *Dolichovespula*, while Greene's (1979) and others' results indicate a number of traits found both in the common ancestor of *Vespula* and *Dolichovespula* and in extant *Vespa*. The larvae of *Dolichovespula* are for the most part highly specialized, while those of *Vespula* retain many primitive traits (e.g., characters 14, 16 and 31).

The twisted, tape-like pedicel of embryo nests can be regarded as an adaptation to a subterranean nesting habit, in which the pedicel has the sole function of supporting the weight of the comb. Among the Old World vespids, this peculiar type of pedicel is known only from *Vespula* and *Dolichovespula*. The implication is that it was present in their subterranean-nesting common ancestor and was retained in *Dolichovespula* when they switched to nesting above ground. These two genera also differ from *Vespa* in the attachment of the first envelope sheet to the pedicel (character 36). Direct attachment to the pedicel may be correlated with subterranean nesting. Most *Dolichovespula* species (*D. media* and *maculata* are exceptions) nest in closed spaces if these are available, but they usually occupy open sites in areas where *Vespula* species dominate. This suggests that the ancestral *Dolichovespula*, which probably differentiated at the northern periphery, nested both below ground and in the open but was increasingly forced out of underground cavities as it came into sympatry with the ancestral *Vespula* (Sk. Yamane 1976). It follows from this line of reasoning that the ancestral *Dolichovespula* secondarily evolved a secreted rubbery coating for the embryo nest pedicel as its nesting habit became mainly aerial (Sk. Yamane et al. 1981). If this is so, then the uncoated pedicel of *Vespula* is apomorphic in relation to the coated pedicel of *Vespa*, but plesiomorphic to the coated pedicel of *Dolichovespula*.

4 List of the World's Vespinae

The arrangement of taxa mainly follows Edwards's (1980) checklist. We have made changes in line with newer information and have indicated the approximate range of each form, insofar as this is known. Lee's (1982) treatment adds greatly to our distributional information on the Chinese species and subspecies. Only valid names are given here; for synonyms see Edwards (1980). Species are numbered consecutively, with subspecies enumerated below their respective species.

1. Genus *Provespa* Ashmead, 1903

1. *Provespa anomala* de Saussure 1854 Borneo, Sumatra, Bangka, Malaya, Tenasserim, Thailand
2. *P. barthelemyi* du Buysson 1905 Malaya, Indochina, Thailand, Burma, Sikkim, China (Yunnan, Kwangsi)
3. *P. nocturna* van der Vecht 1935 Borneo, Sumatra, Bangka

2. Genus *Vespa* Linnaeus, 1758

2a. Subgenus *Nyctovespa* van der Vecht, 1959

4. *Vespa binghami* du Buysson 1905 Tenasserim, Yunnan, Burma, N.E. India
5. *V. suprunenkoi* Birula 1925 Sakhalin

2b. Subgenus *Vespa* Linnaeus, 1758

6. *Vespa affinis* Linnaeus 1764
   affinis (Linnaeus) Burma, S.E. China, Taiwan, Yaeyama Is., C. & N. India
   alduini Guerin 1831 Molucca Is. (Buru)
   alticincta van der Vecht 1957 New Britain
   archbordi van der Vecht 1957 N. New Guinea
   continentalis Bequaert 1936 S. India
   indosinensis Pérez 1910 Java (?), Krakatau Is., Sumatra, Bangka, Malaya, Tenasserim, Indochina

moluccana van der Vecht 1957 W. New Guinea, S. Moluccas
nigriventris van der Vecht 1957 Palawan, Labuan, Samar (?), Mindanao (?)
picea du Buysson 1905 S. New Guinea
rufonigrans van der Vecht 1957 N.W. New Guinea, Waigeo, Salawati, Misool, N. Moluccas, Sulawesi, Borneo

7. *V. analis* Fabricius 1775
analis Fabricius W. Java, Panaitan, Krakatau Is., Sumatra, Bangka
barbouri Bequaert 1939 Sikkim
insularis Dalla Torre 1894 Japan
kuangsiana Bequaert 1939 China (Kwangsi)
nigrans Buysson 1903 Malaya, Tenasserim, Laos, Burma, S. China, E. Himalaya, (Taiwan)
paralleia André 1884 C. & N. China, S.E. Siberia, Korea
tenebrosa du Buysson 1905 Bali, Bawean, C. & E. Java
tyrannica Smith 1857 Singapore

8. *V. basalis* Smith 1852 Sumatra, Tenasserim, Indochina, Burma, S. China, Taiwan, E. Himalaya, Ceylon (?)

9. *V. bellicosa* de Saussure 1854 Sumatra, Borneo, Bengkalis

10. *V. bicolor* Fabricius 1787
bicolor Fabricius Indochina, S. China, Assam, Bhutan
citriventris du Buysson 1905 Burma, Yunnan, N. India

11. *V. crabro* Linnaeus 1758
altaica Pérez 1910 Altai
birulai Bequaert 1931 N.E. China
caspica Pérez 1910 Transcaucasia
crabro Linnaeus Scandinavia, N.E. USSR incl. Sakhalin, northernmost part of China
crabroniformis Smith 1852 Thailand, S. China
flavofasciata Cameron 1903 Japan, Korea (?)
germana Christ 1791 Europe, North America (introduced)
oberthuri du Buysson 1902 China (Szechwan)
vexator du Buysson 1902 England

12. *V. dybowskii* André 1884 Thailand, Burma, China, E. Siberia, Korea, Japan (Hokkaido & Honshu)
? *mutata* Ma 1937 China (Hangchow)

13. *V. fervida* Smith 1858 Sulawesi, Salajar, Butung

14. *V. luctuosa* de Saussure 1854
luctuosa de Saussure Luzon, Mindoro, Panay
luzonensis Bequaert 1934 Panaon, Leyte, Biliran
semperi du Buysson 1905 Mindanao

15. *V. mandarinia* Smith 1852
japonica Radoszkowski 1857 Japan
mandarinia Smith Thailand, Laos, China, Korea
magnifica Smith 1852 Tenasserim, Indochina, N.E. Burma, S. China, Himalaya
nobilis Sonan 1929 Taiwan

? sonani Matsumura 1930
soror du Buysson 1905 Indochina, Burma, Yunnan

16. *V. mocsaryana* du Buysson 1905 Sumatra, Malaya, Tenasserim, Indochina, S. China
17. *V. multimaculata* Pérez 1910
multimaculata Pérez Borneo
pendleburyi van der Vecht 1957 Sumatra, Malaya, Thailand
18. *V. orientalis* Linnaeus 1771
aegyptica Vallot 1802 Egypt
arabica Giordani Soika 1957 Oman, Yemen, Aden
orientalis Linnaeus China (Xinjiang), N. India, Middle East, Balkan, Madagascar (introduced ?)
somalica Giordani Soika 1934 Somalia
zavattarii Guiglia and Capra 1933 Sahara
19. *V. philippinensis* Saussure 1854 Luzon, Leyte, Samar, Negros, Mindanao
20. *V. simillima* Smith 1868
simillima Smith E. Siberia, Hokkaido, Kuril Is., Sakhalin, Korea (?)
xanthoptera Cameron 1903 Japan excl. Hokkaido, Korea
21. *V. tropica* (Linnaeus 1758)
anthracina Bequaert 1936 Samar, Leyte, Panay, Cebu, Mindanao, Sibuyan, Negros, Palawan
deusta Lepeletier 1836 Luzon, Lubang, Mindoro
ducalis Smith 1852 China, Korea
esakii Sonan 1935 Tsushima
eulemoides du Buysson 1905 Andaman Is.
haematodes Bequaert 1936 Tenasserim, Burma, S. China, Nepal, India, Ceylon
leefmansi van der Vecht 1957 Borneo, Sumatra, Malaya, Tenasserim, Indochina, Thailand, Burma, Assam, N.E. India
loochooensis Bequaert 1936 Yaeyama Is.
pseudosoror van der Vecht 1959 Indochina, Taiwan
pulchra du Buysson 1905 Japan excl. Hokkaido
trimeres van der Vecht 1957 New Guinea, Aru, Molucca Is., Sulawesi, Lesser Sunda Is.
trisignata Pérez 1910 S. Moluccas, Timor, Banda Is.
tropica (Linnaeus) Bali, Java, Sebesi, Krakatau Is., Panaitan, Bawean, Madura, Kangean, Sumbawa
unicolor Smith 1863 Buru
22. *V. variabilis* du Buysson 1905
fumida van der Vecht 1959 Bhutan, E. Himalaya
variabilis du Buysson S. China
23. *V. velutina* Lepeletier 1836
ardens du Buysson 1905 Lombok, Sumbawa
auraria Smith 1852 S. China, Burma, N. India, Nepal
celebensis Pérez 1910 Sulawesi
divergence Pérez 1910 Malaya

flavitarsus Sonan 1929 Taiwan
floresiana van der Vecht 1957 Flores, Komodo
karnyi van der Vecht 1957 Sumatra
nigrithorax du Buysson 1905 C. & N. China
pruthii van der Vecht 1959 Kashmir, Pakistan
sumbana van der Vecht 1957 Sumba
timorensis van der Vecht 1957 Timor
variana van der Vecht 1957 N. Thailand, Indochina, Burma
velutina Lepeletier Bali, Java

24. *V. vivax* Smith 1870 China (Szechwan), Tibet, Sikkim, Nepal
25. *V. walkeri* du Buysson 1905 China (Hai-ning)
26. *V. wilemani* Mead Waldo 1911 Taiwan

3. Genus *Vespula* Thompson 1869

3a. Subgenus *Paravespula* Blüthgen 1938

27. *Vespula flaviceps* (Smith 1870)
    flaviceps (Smith) N. India, China, Korea, Siberia
    karenkona (Sonan 1929) Taiwan
    lewisii (Cameron 1903) Japan
28. *Vl. flavopilosa* Jacobson 1978 North America
29. *Vl. germanica* (Fabricius 1793) Eurasia, England, N. Africa; introduced into many countries in the world.
30. *Vl. maculifrons* (du Buysson 1905) E. North America
31. *Vl. pensylvanica* (de Saussure 1857) W. North America
32. *Vl. shidai* Ishikawa, Sk. Yamane and Wagner 1980 Siberia, Japan, Kuril Is.
33. *Vl. structor* (Smith 1870) S. China, Himalaya
34. *Vl. vulgaris* (Linnaeus 1758) Europe, C. & N. Asia, Japan, North America

3b. Subgenus *Rugovespula* Archer 1982

35. *Vespula koreensis* (Radoszkowski 1887)
    koreensis (Radoszkowski) N. China, Korea, Siberia
    salebrosa Archer 1982 S. China
    stizoides (du Buysson 1905) Laos, China (Yunnan), N. India
36. *Vl. orbata* (Buysson 1903)
    orbata (du Buysson) Burma, Himalaya
    arisana (Sonan 1929) Taiwan

3c. Subgenus *Vespula* Thompson 1869

37. *Vespula acadica* (Sladen 1918) North America
38. *Vl. atropilosa* (Sladen 1918) W. North America
39. *Vl. austriaca* (Panzer 1999) Europe, N. Asia, North America
40. *Vl. consobrina* (de Saussure 1854) North America
41. *Vl. intermedia* (du Buysson 1905) N. North America
42. *Vl. kingdonwardi* Archer 1981 Burma, Tibet
43. *Vl. nursei* Archer 1981 S. China, Himalaya, Kashmir

44. *Vl. rufa* (Linnaeus 1758)
grahami Archer 1981 S. China
rufa (Linnaeus) Europe, USSR (west of River Lena)
45. *Vl. schrenckii* (Radoszkowski 1861) E. Siberia, Kamchatka, Mongolia, Japan, Kuril Is.
46. *Vl. vidua* (de Saussure 1854) S.E. North America

3d. *Vespula squamosa* Group
47. *Vespula squamosa* (Drury 1770) S.E. North America, Central America
48. *Vl. sulphurea* (de Saussure 1854) S.W. North America

4. Genus *Dolichovespula* Rohwer, 1916

4a. Subgenus *Boreovespula* Blüthgen, 1943
49. *Dolichovespula adulterina* (du Buysson 1905)
adulterina (du Buysson) Europe excl. England
colchica (Birula 1930) Caucasus
montivaga Sk. Yamane 1982 Japan (Hokkaido, Honshu)
50. *D. arctica* (Rohwer 1916) North America
51. *D. arenaria* (Fabricius 1775) North America
52. *D. lama* (du Buysson 1903) Sikkim
53. *D. norvegicoides* (Sladen 1918)
loekenae Eck 1980 N. Europe
norvegicoides (Sladen) North America
pacifica (Birula 1930) N.E. Asia, Japan
xanthicincta Archer 1980 N. Burma, Tibet
54. *D. norwegica* (Fabricius 1781)
albida (Sladen 1918) N. North America
norwegica (Fabricius) Europe, N. Asia excl. Japan
55. *D. panda* Archer 1980 China (Szechwan)
56. *D. saxonica* (Fabricius 1793)
kamtchatkensis Eck 1982 Kamchatka, N. Kurils
? *monticola* (Birula 1930) Caucasus
nigrescens Eck 1982 Sakhalin
nipponica Sk. Yamane 1975 Japan (Hokkaido, Honshu)
saxonica (Fabricius) Europe excl. England, N. Asia

4b. Subgenus *Dolichovespula* Rohwer, 1916
57. *Dolichovespula maculata* (Linnaeus 1763) North America
58. *D. media* (Retzius 1783)
geerii (Lepeletier 1836) C. & S. Europe
media (Retzius) N. Europe, N. Asia incl. Hokkaido
sugare Ishikawa 1969 Japan (Honshu)

4c. Subgenus *Metavespula* Blüthgen, 1943
59. *Dolichovespula asiatica* Archer, 1981 USSR (Turkmen)
60. *D. omissa* (Bischoff 1931) Europe, Middle East

61. *D. sylvestris* (Scopoli 1763)
sumptuosa (du Buysson 1905) Middle East
sylvestris (Scopoli) Europe, Asia, Morocco

Species and Subspecies Described Since the Publication of the Japanese Edition of this Book

Vespa tropica cebuana Kojima & Reyes 1984 Cebu I. (Philippines)
Vespa luctuosa negrosensis Kojima & Tano 1985 Negros I. (Philippines)
Vespa analis eisa Sk. Yamane 1987 C. Ryukyus (Japan)
Vespa analis nagatomii Sk. Yamane 1987 S. Ryukyus (Japan)
Vespula hainanensis Lee 1986 Hainan I. (China)
Vespula gracilia Lee 1986 Zhejiang Prov. (China)
Vespula yichunensis Lee 1986 Heilongjiang Prov. (China)
Vespula hirsuta Lee 1986 Tibet
Vespula obscura Lee 1986 Tibet
Vespula shidai amamiana Sk. Yamane 1987 Amami I. (Japan)
Dolichovespula alpicola Eck 1984 N. America
Dolichovespula xinjiangensis Lee 1986 Xinjiang (China)
Dolichovespula borealis Lee 1986 Heilongjiang Prov. (China)
Dolichovespula nyalamensis Lee 1986 Tibet
Dolichovespula stigma Lee 1986 Sichuan Prov. (China)
Dolichovespula flora Archer 1987 Adung Valley (Burma), Yunnan (China)
Dolichovespula sinensis Archer 1987 Szechwan (China)
Dolichovespula baileyi Archer 1987 Szechwan (China)

Chapter 11

Distribution and Phylogeny

The taxonomy of the Vespinae has almost been completed except in a few Asiatic groups. However, the distributions of species and subspecies have been only poorly understood in the Old World, except in Europe. In this chapter we present the outline of the distribution pattern of this subfamily from a worldwide view (Sect. 1), and discuss the vespine faunal similarity between Eurasia and North America (Sect. 2).

1 Distribution Pattern

1.1 Distribution of the Four Genera

The ranges of the four vespine genera are shown in Fig. 11.1. *Provespa,* consisting of only three species, is confined to Southeast Asia, ranging from the eastern Himalaya in the west through Burma, Indochina and the Malay Peninsula, to Sumatra and Borneo in the east. It is not found in Java, an island very close to Sumatra. Most species of *Vespa* are inhabitants of eastern Asia, and a few species extend to the Lesser Sunda islands, New Guinea and the Philippines. *V. crabro* and *V. orientalis,* on the other hand, extend their ranges into the western parts of Eurasia. The southern border of the range of this genus lies on the island of Roti situated at 10°S Lat., south to Timor I. (Indonesia) (van der Vecht 1957). The northern border has not been well defined. In Scandinavia there are records of *V. crabro* collected at 60°N Lat. in Norway (Løken 1964), and S.W. Batra (pers. comm.) found its nest on the island of Oland (57°N) in Sweden. In eastern Asia *V. simillima* and *V. crabro* are known to be distributed up to ca. 55°N Lat. The Ethiopian region, Australia, New Zealand and the New World lack this genus (however, *V. crabro* was introduced into North America between 1840 and 1860). The range of *Vespula* is more northerly than that of *Vespa,* and includes North America as well. In the subtropics *Vespula* occurs only in mountainous regions, indicating that its distributional center lies in the temperate region. *Dolichovespula* is also Holarctic in distribution, as in *Vespula,* and is dominant in the cold temperate to frigid zone. These two genera are not found in the Southern Hemisphere, but two species of *Vespula* have been introduced and have established themselves in South Africa (*germanica*), southern Australia (*germanica* and *vulgaris*), Tasmania (*vulgaris*), New Zealand (*germanica*), Chile (*germanica*), etc. (Edwards 1980; Sk. Yamane et al. 1980b). In New Zealand *Vl. germanica* has increased in numbers, becoming a significant pest, and often constructs huge overwintered colonies (Thomas 1960).

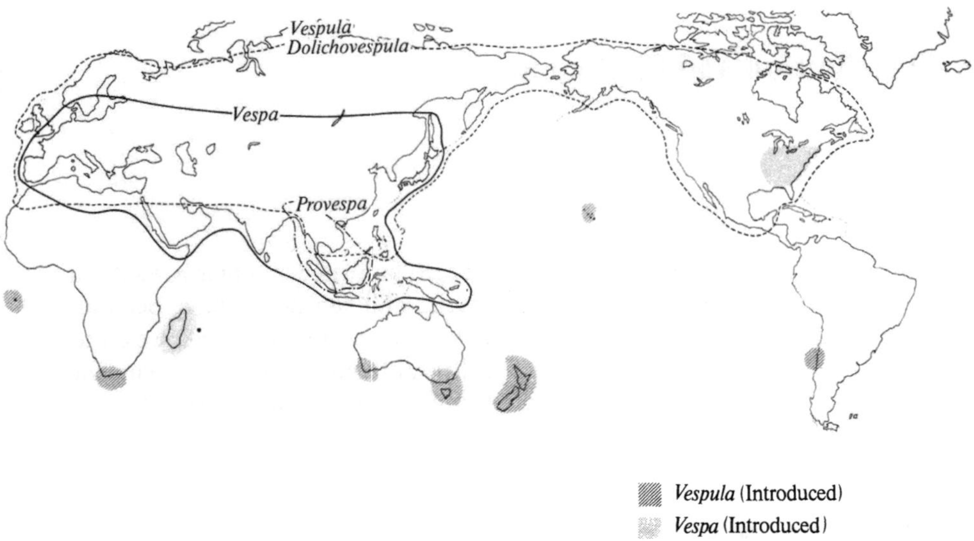

Fig. 11.1. World distribution pattern of the four vespine genera. Only one record is available for the occurrence of *Vespula* in the Philippines (*Vl. nursei*; Archer 1981b)

1.2 Center of Diversification

The first important aspect of the present distribution pattern is that the Southern Hemisphere almost completely lacks the wasps of this subfamily, except for introduced species. If we assume the Gondwanian origin of this subfamily, we must consent to the unlikely idea that vespine species have become extinct in all the Gondwanian regions other than India, i.e., in Africa south of the Sahara, Australia, New Zealand and South America, and that only those that have colonized Laurasia have survived. The hypothesis that vespine stock originated in Laurasia is, therefore, powerful, unless vespine fossils are not discovered from the Southern Hemisphere.

The absence of *Vespa* (hornets) in the New World is another interesting phenomenon. This is important because *Vespa* is thought to retain many ancestral traits, which are supposed to be conditions seen in the common ancestor of the four genera (Richards 1971; Sk. Yamane 1976; Greene 1979). According to this notion, the place of origin of *Vespa* is possibly that of the Vespinae. As clearly understood from Fig. 11.2, the region from the eastern Himalaya to southern China harbors the largest number of hornet species. This led van der Vecht (1957) to presume that this zone had been the center of diversification of this genus. It is generally a difficult task to determine the place of origin and the center of diversification (these two may often not be the same) of a particular animal group, since later decline or extinction in these places may be a frequent occurrence. However, at present there is no disadvantage to van der Vecht's hypothesis. If we accept his view, then the center of diversification of *Vespa* is better said to lie in

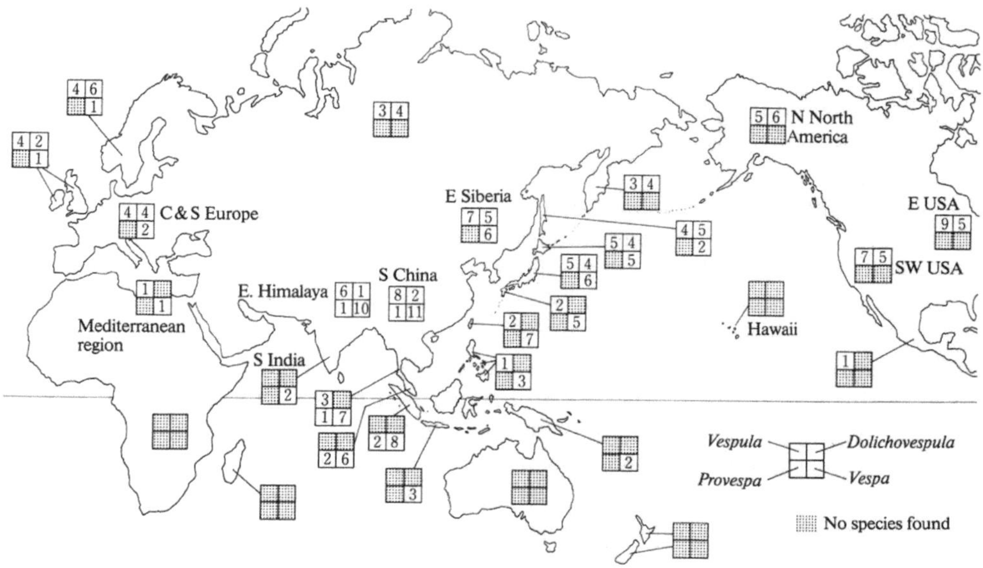

Fig. 11.2. Number of species in the four vespine genera in various localities in the world. Introduced species are not counted. Also see p. 238

the mountainous regions in the subtropics and warm temperate zone of eastern Asia, rather than in tropical Asia, as is often claimed.

A third important aspect is the remarkable similarity of the yellowjacket (*Vespula* + *Dolichovespula*) fauna in eastern Asia and North America. The reason for this will be discussed in the next section.

Virtually nothing is known of the time when the vespine stock branched off from the common ancestor shared with the Polistinae. That the Vespinae are a compact group without noticeable diversity in morphology and nest architecture (van der Vecht 1968) suggests the relative recency of this group. On the other hand, the subfamily has a history during which four genera have differentiated. These tell us nothing about the time of its origin. No fossil that belongs with certainty to the vespine stock has been discovered. The only candidate is the species *Palaeovespa baltica,* described by Cockerell from a Miocene amber in North America, though the possibility that it is a species of Polistinae cannot be eliminated (Bequaert 1930). An Eocene vespid species recorded from Blakenburn Mine (British Columbia) is said to resemble extant *Vespa* and *Vespula* in wing venation (M.V.H. Wilson 1977), but the short parastigma suggests that it may belong to the Polistinae or Eumenidae. Unfortunately, because the clypeus and hindwing (the region of anal lobe) are not illustrated, it is difficult to determine the precise position of this fossil.

By examining the distribution pattern of extant species we might get some clues concerning the time of evolution: First, there are species and species groups as well as subspecies of hornets endemic to the Greater and Lesser Sunda islands, the Philippines and Sulawesi (van der Vecht 1957; Kojima 1982). This means that successive invasions and isolations of populations occurred during more than one glaciation. The history of *Vespa*, not surprisingly, may have covered some glaciations during the Quaternary; second, because the New World lacks *Vespa*, the possibility exists that this genus had not yet been differentiated during late Mesozoic and Tertiary, when Beringia was still mild. This reasoning excludes the possibility that the colonization of the New World by vespines was made via a route other than the Bering land bridge, or that a species of *Vespa* that migrated to the New World via the Bering land bridge when the climate was mild became the ancestral stock of *Vespula*/*Dolichovespula*, which later returned to the Old World. There is an alternative explanation for the lack of hornets in the Nearctic region (see Sect. 2); third, there are species groups of *Vespula* endemic to Eurasia or North America (the northern limits of their ranges lying between 40° and 45°N. lat.) (Sk. Yamane et al. 1980b). This may be because these wasps might not have been able to migrate between Eurasia and North America during the four Quarternary glaciations via the Bering land bridge; that is, the ancestor of these groups migrated during milder ages before the ice epoch (this can be said whether *Vespula* originated in the Old World or New World).

1.3 Expansion of Vespa Range

As mentioned earlier, the region covering the eastern Himalaya and southern China is the richest area for species of *Vespa*. The mountainous regions of Burma, the Malay Peninsula and Sumatra also have rich *Vespa*-faunas. The hornet species number decreases, departing from these regions in any direction. The only exceptions are northeastern China, Taiwan, and central to southwestern Japan. Only two species, *V. crabro* and *V. simillima*, prosper in the northern parts of the temperate region.

Two species, *V. tropica* and *V. affinis*, are typically adapted to the lowland tropics, and extend onto numerous islands of varying size, including the Greater and Lesser Sundas, Borneo, Sulawesi, the Philippines and New Guinea in the extreme east, differentiating into many geographical races. A close examination of the distribution of hornets in these regions was made by van der Vecht (1957). The wasps of the *V. luctuosa* species-group also occur at lower altitudes in the tropics, with two species in Sumatra and Borneo, one in the Philippines, and one in Sulawesi. The West Javanese population of *V. analis* (nominate subspecies), and the Sumba population of *V. velutina* (subsp. *sumbana*) are also adapted to the lowland tropics (van der Vecht 1957). *V. velutina* over most of its range is confined to the mountains. It has been argued that most *Vespa* species have not been able to colonize the lowland tropics because they insist upon haplometrotic nest foundation, which is vulnerable to ant predation (Jeanne 1975). Recently, evidence was given by Matsuura (1983a; our Chap. 7) that the typically tropical *V. affinis* adopts pleometrotic nest foundation in lowland Sumatra. All the species of *Provespa*, prospering in lowland tropics, may found their new colonies by swarming not subject to attacks by ants (Chap. 7). Although this group has been believed to be archaic, in fact it is quite specialized in both morphology and behavior.

The southern belt of the Eurasian continent west of India forms a vast dry zone that is not suitable for the life cycle of forest-inhabiting *Vespa.* Only one species, *V. orientalis,* which is resistant to a dry climate, is distributed from western India to the Middle East and Mediterranean districts. This may be the only *Vespa* species that is found in Africa (Guiglia 1972), though du Buysson (1905) recorded *V. crabro* from Algeria. There is no reliable record of *Vespa* from the Ethiopian region (the record of *V. orientalis* from Madagascar may have derived from the introduced population; Bequaert 1918, p. 349). The Sahara became a steppe or savanna during the last glaciation, but did not form forests. Although this vast area is known to have repeatedly been relatively wet before this glaciation, these events probably did not permit *Vespa* to invade the Ethiopian region. *V. crabro* has the widest range among the hornets, being found widely over the northern parts of Eurasia. The time and route of expansion by this species is not clear. That only slight differences in color pattern are found between the east-Asian and European populations suggests that range expansion into Europe might have occurred in relatively recent time, through the forest zone of Siberia (this species principally nests in tree hollows under natural conditions).

Only two hornet species, *V. tropica* and *V. affinis,* are known from southern India. Although there are some mountain chains with peaks of more than 2500 m in altitude, such as the Nilgiris, neither mountain-inhabiting *Vespa* nor *Vespula* are found. This may be because the Ganges Plain in the east and the Thar in the west form barriers, preventing mountain-inhabitants from dispersing from the Himalayas to southern India. The fact that some cyprinid fish are distributed in the mountainous regions of southern India suggests that part of the Ganges Plain had once risen to connect the Himalayas with the mountains in southern India. However, little is known of the exact migration route of these fish.

1.4 Examples of Interspecific Competition

Vespine species often form huge colonies which require a large quantity of food. Some species nest almost exclusively in protected spaces such as underground cavities and tree hollows that are also limited resources. These circumstances possibly bring about interspecific competition for these resources, leading to competitive exclusion. This exclusion, if any, should occur more frequently on small islands. However, no reliable examples of this phenomenon have been reported. A possible competition over the same niche is suggested by the absence of *V. affinis* on the island of Java (van der Vecht 1957). It is difficult to explain why this tropical species, adapted to lowlands and almost dominant over its range, is found on Sumatra but not on Java. Since it is evident that these two islands were connected during the last glaciation (Darlington 1957), and since even now they are only 25 km apart from each other, the absence of *V. affinis* on Java is not explained in terms of historical geology or the dispersal ability of hornets. One plausible explanation is competitive exclusion. The candidate competitor is the

Javan population of *V. analis* that has invaded the lowlands (van der Vecht 1957) and is commonly found even along the coasts (Sk. Yamane 1983). This may present to *V. affinis* a different circumstance from that of the Malay Peninsula, where *V. analis* mainly inhabits mountain regions, permitting *V. affinis* to occupy lowlands. However, both the species coexist on some very small islands (Sebesi, Sebuk, Legundi) near Sumatra; this may contradict the competition argument. In Sumatra *V. analis* was confined to the southern extremity (cf. van der Vecht 1957 for the range before the 1960s), but recent investigation (Matsuura unpubl.) confirmed the occurrence of this species around Padang, central Sumatra, suggesting that it may be distributed widely in Sumatra. The populations of *V. analis* occurring on Sumatra and the islets between Java and Sumatra, such as Legundi, belong to the subspecies *analis*, primarily known from west Java. It is quite possible that *V. analis* recently migrated from west Java to these islands, where *V. affinis* is being oppressed by *V. analis*. Continued study of the change in range and abundance of these species will solve the question of whether interspecific competition is actually occurring.

Between Java and Sumatra lies another group of islets, the Krakataus, whose biota is said to have been exterminated by the great eruption in 1883. Fifty years after the eruption, Dammerman (1948) noted colonization by *V. analis* (and probably also *V. affinis*), and in 1982 *V. affinis* was collected on Sertung, *V. analis* on Rakata Kecil, and both on Rakata Besar and Anak Krakatau (Sk. Yamane 1983). The records of these two species on these islets from 1921 to 1982 are given in Table 11.1 (in addition to these *V. tropica* was collected in 1933 on Sertung, and in 1955 on Rakata Besar: see Dammerman 1948; van der Vecht 1957). These islands are much smaller than Java as targets of colonization, and are situated 15 km away from the nearest island (Sebesi), one of the stepping stones connecting the Krakataus with Sumatra where *V. affinis* dominates. Vacant niches on these islets may have been responsible for the rapid colonization of the Krakataus by *V. affinis*. In this regard it would be interesting to note whether the coexistence of the two species on some Krakatau islands (Fig. 11.3) continues or whether one of them becomes evicted. Although some workers and a male of *V. affinis* were captured from West Java in 1957 (van der Vecht 1959), the establishment of this

Table 11.1. *Vespa* species found on the Krakatau Islands in different years[a]

Name of island	1921	1933	1955	1982
Sertung	*affinis*	*affinis*	*analis*	*affinis*
Rakata Besar	–	–	*analis*	*analis* *affinis*
Rakata Kecil		–	–	*analis*
Anak Krakatau		–	–	*analis* *affinis*
Sebesi, Sebuk	*analis*		*analis* *affinis*	

[a] – Surveyed but *Vespa* were not found; blank, not surveyed.

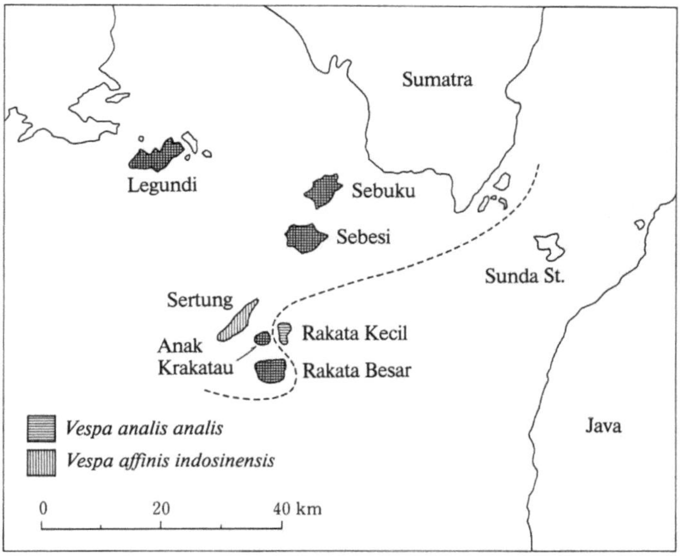

Fig. 11.3. Distribution of *Vespa analis* and *V. affinis* in the Sunda strait region. *Dotted line* denotes the eastern border of the range of *V. affinis indosinensis*

species in Java has not been confirmed. Since occasional migrations between Sumatra and Java may be a natural occurrence, the capture of adult wasps does not falsify van der Vecht's hypothesis.

Another interesting case comes from North America, which was invaded by the German yellowjacket *Vespula germanica*, a species widely spread over Eurasia. Though the precise year of its establishment is not known, the first capture was made in 1891, some were collected between 1920 and 1950, and a definite increase was confirmed by the end of the 1960s (Morse et al. 1977). The first worker was collected from Ithaca, New York; the species has subsequently spread into Maryland, Delaware, New Jersey and Pennsylvania, and is now rapidly expanding westward (Akre et al. 1981). The most notable behavioral change following its establishment in North America is the tendency for a queen to initiate her colony inside structures. Subterranean colonies, common in Europe, are very rare in United States (MacDonald 1980; Parrish and Roberts 1982). *Vl. germanica* seems to be replacing *Vl. maculifrons* as the most abundant pest yellowjacket in some areas of the eastern United States (Morse et al. 1977; Akre et al. 1981). Although the extinction of one of the interacting species is not likely to occur, the current balance of relative abundance may be lost soon. It is also quite interesting to know what happens when *Vl. germanica* continues to spread westward and contacts with *Vl. pensylvanica* that is native to the West and is thought to be most closely related to *Vl. germanica* (Sk. Yamane et al. 1980b).

2 Similarity of Fauna in Eurasia and North America

Bequaert (1930, 1932) extensively surveyed the world species of *Vespula* and *Dolichovespula* when he wrote the revision of Nearctic Vespinae. Thereafter, taxonomic studies of these groups proceeded independently in Europe, Asia and North America. Relationships of the vespine faunas between these realms was not given attention until recently (Sk. Yamane et al. 1980b). In this section we compare the vespine faunas of Eurasia, especially of northeastern Asia and North America, and discuss the formation process of the present distribution pattern, paying attention to the important role of the Bering land bridge.

2.1 Climate of Beringia

In many animal groups we can see a similarity in species composition between Eurasia and North America. This is because Europe had been connected with North America until the early Tertiary (they formed a single continent, Laurasia), and because after Laurasia began to split along the Atlantic Rift to separate and become Eurasia and North America both the continents were intermittently connected to each other by the Bering land bridge in the late Tertiary and during the great Ice Age (Darlington 1957; Cox et al. 1976). Whatever the exact time of appearance of the first Vespinae, the current distribution pattern might have been formed during the Cenozoic. It is known that during the Mesozoic the higher latitudes of the Northern Hemisphere were much warmer than they are now and that tropical plants were extending northerly up to the regions that are now north-temperate in climate. During the Cenozoic the climate of these regions gradually got colder, culminating in the glacial age in the Quaternary. However, except during the last two glaciations, forests persisted in Beringia, and there is no evidence of the predominance of tundra vegetation (Hopkins 1967). Regardless of the reason for the appearance of a land bridge (in the Quaternary the land bridges resulted from sea level reduction due to the accumulation of water as ice), the climate of Beringia is believed to have gradually deteriorated during 65 million years from the Tertiary to the present (Fig. 11.4). During the last two glaciations, Riss (Illinoian) and Würm (Wisconsin), the land bridges are thought to have been covered with tundra.

2.2 Vespine Distribution Patterns

Assuming that the wasps of this subfamily migrated between Eurasia and North America mainly via the Bering land bridge, the following predictions are derived (Sk. Yamane 1982b): (1) The species that can now inhabit the Arctic zone or tundra ought to be common to both Eurasia and North America, or to be a closely related vicariant of each other (owing to interchange in the last glaciation), (2) in this case the phenetic similarity of the populations is generally greater between North America and eastern Asia than between North America and Europe

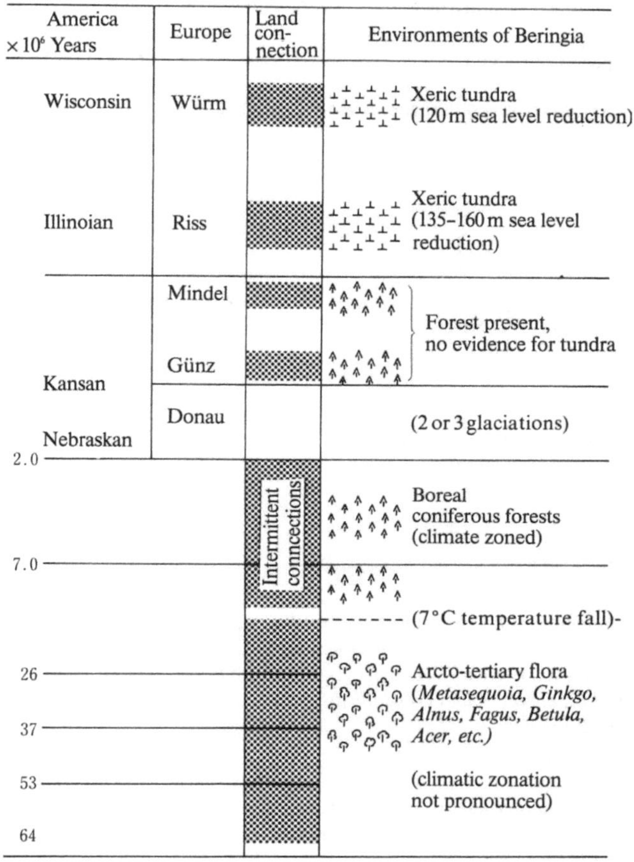

Fig. 11.4. Intermittent occurrence of Bering land bridge during the Cenozoic. Supposed change in the environment of Beringia is indicated. (Based upon Hopkins 1967 and others)

(owing to the location of the land bridge), and (3) the more southerly the northern limit of a given species' range, the more it has diverged from the corresponding species (vicariant) of the opposite continent. On some occasions a species may not have an apparent vicariant.

In order for these predictions to be met, one must assume that (1) wasps always migrated if the climatic and ecological conditions permitted, and (2) the degree of divergence resulted from geographical isolation parallels the time spent (for example, the color pattern first changed, then structural modifications took place). It is clear that these assumptions are occasionally not correct and are tendencies rather than laws.

Testing the above predictions with the vespines occurring in Eurasia and North America provided the following positive results (Table 11.2): Distributions of the wasps in North America are mainly cited from Miller (1961), Wagner

Table 11.2. Comparison of E. Asiatic and N. American vespine faunae[a]

Species of E. Asia		N. American counterparts
Vespula (Paravespula) germanica	———	*Vl. (Pl.) pensylvanica*
Vl. (Pl.) vulgaris	———	*Vl. (Pl.) vulgaris*+
Vl. (Pl.) structor		?
?		*Vl. (Pl.) flavopilosa*
–		*Vl. (Pl.) maculifrons*
Vl. (Pl.) flaviceps *Vl. (Pl.) shidai*		–
Vespula (Vespula) rufa *Vl. (Vl.) schrenckii*	———	*Vl. (Vl.) intermedia*+ *Vl. (Vl.) acadica*+
?		*Vl. (Vl.) consobrina* *Vl. (Vl.) vidua* *Vl. (Vl.) atropilosa*
Vl. (Vl.) nursei *Vl. (Vl.) kingdonwardi*		?
Vl. (Vl.) austriaca[b]	———	*Vl. (Vl.) austriaca*[b]+
–		*Vespula squamosa* *Vl. sulphrea*
Vespula (Rugovespula) koreensis *Vl. (Rl.) orbata*		–
Dolichovespula (Dolichovespula) media	———	*D. (D.) maculata*
Dolichovespula (Boreovespula) saxonica	———	*D. (B.)* sp.+
D. (B.) norwegica	———	*D. (B.) norwegica (= albida)*+
D. (B.) norvegicoides (= pacifica)	———	*D. (B.) norvegicoides*+
D. (B.) panda		?
D. (B.) lama		?
–		*D. (B.) arenaria*+
D. (B.) adulterina[b]	———	*D. (B.) arctica*[b]
Dolichovespula (Metavespula) sylvestris *D. (M.) asiatica*		–
D. (M.) omissa (= ingrica)[b]		–

[a] +, Species extending their range into the Arctic Zone; ———, connecting vicariants; –, vacariant absent; ?, at present no apparent vicariant specified.
[b] Permanent social parasites.

(1978), and Akre et al. (1981); those in Eurasia from Løken (1964), Guiglia (1972) and Sk. Yamane et al. (1980b).

1. Of the eight Nearctic species (with + in our Table 11.2) that are collected in the Arctic (the northern limit of species' range lies above 66° N) and seem to sustain their lives in tundra, six are also found or have a close vicariant in Eurasia. Among the remaining two species, at present the Eurasian counterpart is not strictly determined for *Vl. acadica.* The other (*Dolichovespula arenaria*) is endemic to North America, presenting an apparent exception to our prediction 1. The distribution maps in the above literature are mostly based on collection records of adult wasps, so it is often not known whether these eight species actually complete their life cycles in tundra. The northernmost record of nesting so far known is at 69°26′ N (Tuktoyaktuk, Canada)

for *D. norwegica albida*, and this may be the only nest record from typical tundra vegetation (Sk. Yamane et al. 1980a).
2. There are several exceptions to our prediction 2 (see below).
3. Of the five Nearctic species that are confined in distribution between 54° –65° N or seem to be unable to inhabit tundra, three have a Eurasian vicariant, but in each pair the degree of divergence is more pronounced than in tundra-inhabiting pairs or even differences in male genitalia are observed (*Vl. pensylvanica* vs. *Vl. germanica*; *D. maculata* vs. *D. media*). Among the others, the social parasite *D. arctica* resembles the Eurasian *D. adulterina* in structure, but its host is *D. arenaria*, which is endemic to North America (Greene et al. 1978). *Vl. consobrina* and *Vl. atropilosa* might have differentiated in North America where they have the close relative *Vl. acadica*, which may in turn be a vicariant of the Eurasian *Vl. rufa* or *Vl. schrenckii*. The Eurasian *Dolichovespula sylvestris* with its northern limit at 58° N is endemic to Eurasia at the subgenus level.
4. The Nearctic *Vl. maculifrons* with its northern limit at 52° N has no Eurasian vicariant. The Nearctic *Vl. vidua* and the *Vl. squamosa* species group, confined below 50° N, and the eastern Asiatic *Rugovespula* of the genus *Vespula* and the *Vl. flaviceps* species group, also confined below 56° N, are endemic to each continent and are structurally specialized. *Vl. flavopilosa,* confined below 50° N and endemic to North America, is supposed to have differentiated in this realm, having a common ancestor with *Vl. vulgaris* (Jacobson et al. 1978).

These results seem to support the hypothesis that the exchange of vespine faunas between Eurasia and North America was mainly made via the Bering land bridge. The most important exception is the absence in Eurasia of *D. arenaria*, which is one of the most common Nearctic *Dolichovespula* entering tundra. The lack of the vicariants of *Vespa crabro* and *V. simillima* in the New World poses another difficulty, because the northern limits of these species reach very near that of *Vl. germanica,* which has its relative in North America. Sk. Yamane et al. (1980b) attempted to solve this apparent paradox by the supposedly delayed arrival of *Vespa* at higher latitudes from the original center. At least the present distribution pattern shows that the number of hornet species drastically decreases toward the northern limit of its range, while that of yellowjackets increases (Fig. 11.2).

2.3 Geographical Variations in High-Latitude Species

If the recent migration and interchange of vespine species between Eurasia and North America were made through the Bering land bridge during the last glaciation, in species common to both continents the differences between populations of a given species ought to be larger between Europe and North America than between eastern Asia and North America (prediction 2). Available information does not necessarily support this. Here we discuss some interesting aspects of distribution and variation in high-latitude species.

The D. norwegica Group and Vl. rufa Group. In the western parts of Europe *D. norwegica* possesses basic body markings of yellow and gastral segments 1 and 2 marked with reddish brown. East of the River Lena (125°–30°E), however, individuals with pale whitish yellow or ivory-white body markings are frequently encountered, and individuals lacking rufous gastral markings are not rare. Even west of the Lena, for example in Alma Ata (43°20′ N, 76°50′ E), wasps with ivory-white markings are occasionally collected (Birula 1930a). Although the taxonomy of the *D. norwegica* group was still in confusion in Birula's time, so that his accounts are not entirely reliable, recent detailed work by Eck (1981) has shown that wasps (especially workers and males) with whitish body markings have been collected in eastern Siberia and Kamchatka. In the two queen specimens from Sakhalin, preserved in the Collection of Hokkaido University, the body markings were ivory-white, with rufous gastral markings completely absent (Sk. Yamane 1975). The Nearctic population, generally called *D. albida*, does not differ from the Eurasian populations in measurements of body parts (Eck 1981), and it is even indistinguishable from some specimens from eastern Siberia in coloration as well. Since only color differences exist between the populations of the two continents, current taxonomists (Sk. Yamane 1975; Sk. Yamane et al. 1980b; Eck 1981; also see Akre 1982) are inclined to regard them as subspecies (geographical races) of one species, *D. norwegica*. Taking this view, it is also possible to include the eastern Siberian population in the Nearctic subspecies *albida*. But, since both yellow and ivory-white types coexist in Alma Ata and eastern Siberia, there is the possibility that this group consists of two closely related species, *D. norwegica* widely spread over Eurasia, and *D. albida* occurring in Siberia and North America, and that both the species coexist in Siberia (R.E. Wagner pers. comm.). At any rate, the distribution of the *D. norwegica* group suggests a close faunal relationship between Siberia and North America.

A similar case involves the *Vespula rufa* group, in which the distribution pattern is more complicated. The so called *Vl. rufa* in Europe has, like *D. norwegica,* yellow basic body markings and rufous gastral markings. But, going eastward beyond the River Lena, body markings become more tinged with white, and become almost white in Japan and Kamchatka, where rufous gastral markings nearly completely disappear (these specimens were called *Vl. sibirica).* In Siberia east of the Lena, intermediate forms between the typical *rufa* and *sibirica* are also collected. Today these intermediate forms and *sibirica* together are called *Vl. "rufa" schrenckii* (in this book *schrenckii* is treated as a separate species). Around the Lena, *r. rufa* and *"r. schrenckii"* coexist (Birula 1930b). The North American counterparts capable of inhabiting tundra are *Vl. intermedia* (ivory-white) and *Vl. acadica* (yellow). Although there is no apparent structural difference, these two forms are no doubt different species, as their ranges widely overlap, yet no intermediate has been found (Miller 1958, 1961). *Vl. intermedia* is not separable from *schrenckii,* except that the former possesses the rufous gastral markings (Bequaert 1932). Furthermore, as mentioned by Birula (1930b), *schrenckii* also often has the rufous markings (such individuals are rarely found in Japan). On the other hand, *Vl. acadica* specimens captured near the Arctic closely resemble some individuals of *schrenckii,* except for yellow basic body markings (Wagner pers. comm.). The questions are whether the Eurasian "*rufa*" is a single species or consists of two sibling species, and what are their relationships to the two Nearctic species. At any rate, the "*rufa*" case seems to suggest again the close faunal connection between eastern Asia and North America.

For both these examples, the body color pattern changes around the River Lena, and the gap here seems more profound than that seen between the east and west sides of the Bering sea. This suggests that during the last glaciation, when the land bridge connected Siberia with North America, there was a barrier somewhere in Eurasia preventing wasps from dispersing from west to east or vice versa. The view that, in the present interglacial, the European population meets the Asian one around the River Lena, with some areas of sympatry, is worth considering. In order to clarify this point, intensive ecological surveys are needed in the areas of sympatry.

The eastern Asian–North American faunal similarity can be explained, especially for the species pairs with slight differentiation, by migration via the Bering land bridge during the last glaciation. An alternative explanation that does not assume the existence of the land bridge is, however, possible. At present the shortest distance between the Cape of Dezhneva on the Chukotskii Peninsula (USSR) and Cape Prince of Wales (USA) is approximately 80 km, which is far shorter than the longest migration distance of yellowjacket spring queens on record (150 km; Mikkola 1978). According to Mikkola, among the migrating Vespinae seen in spring in northern Europe, *Vl. rufa* and *D. saxonica* are the majority. They usually fly en masse along the coast, and cross the sea only when the opposite shore is in sight (at least for human eyes). On the small island of Mellum, which is in the North Sea, 7 km from the European continent, one vespine species (*D. sylvestris*) was collected 100 years after the commencement of colonization by life (Haeseler 1978). The Krakatau islands, about 40 km away from large neighboring islands, have been colonized by three *Vespa* species only 100 years after the destructive eruption (Sk. Yamane 1983). The significance of the Bering sea as a barrier to vespine wasps should be reconsidered.

The D. norvegiioides Group and the D. saxonica Group. D. norvegicoides is a species with the longest oculo-malar space among *Dolichovespula* (Sk. Yamane 1975; Wagner 1978; Eck 1980). Although three "species" have been described from northern Europe (*loekenae*), eastern Siberia and Japan (*pacifica*), and North America (*norvegicoides*), structural differences among them are slight, and they are completely allopatric; thus, we have treated them as subspecies of the same species, *norvegicoides*. Specimens of *loekenae* and *norvegicoides* that have yellow markings differ from each other in the measurements of some body parts (Eck 1980). Body markings of *pacifica* are ivory-white, and a detailed biometric study has not been done for this form. In color pattern at least, the Nearctic *norvegicoides* is more similar to the European *loekenae* than to the eastern Asiatic *pacifica*.

Until recently, the records of *D. saxonica* have been confined to the western parts of Eurasia. Guiglia's (1972) treatment of the Nearctic "*D. norvegicoides*" as a variety of *D. saxonica* seems incorrect. Sk. Yamane (1975) first confirmed the occurrence of *D. saxonica* in eastern Asia (Japan and Sakhalin). Later, a form closely allied to it (not yet named) was found to have been confused with *D. norvegicoides* in North America (Wagner 1978). Eck (1983) mentioned that the population in North America is clearly separated from the Eurasian populations, in the shape of its clypeus and other characters. In this case the gap lies in the Bering sea. The body markings are, on the other hand, yellow in most parts of Eurasia (nominate subspecies), Sakhalin (ssp. *nigrescens*) and North America (unnamed form), and ivory-white or white in Kamchatka and northern Kuriles (ssp. *kamtchatkensis*) and Japan (ssp. *nipponica*).

The reason for the formation of these distribution patterns is not yet clear.

Chapter 12

The Vespines and Man

Economically, vespine wasps can be both beneficial and harmful. They are beneficial as predators of agricultural, forest and hygienic pests. Unfortunately, we still do not know much of this aspect of vespidology. The larvae and pupae of vespine wasps are utilized as foods by man in some parts of the world, such as Asia, but these wasps do not contribute at all to "apiculture", because they do not have honey-storing behavior, as seen in the Apinae and some Neotropical Polybiini. When viewed as harmful insects, vespines have three major aspects: they are nuisance pests, natural enemies of honeybees and horticultural pests. The relative importance of these three aspects depends on the human culture, industrial style and species composition of Vespinae in the regions concerned. People of some countries may coexist with vespines with practically no trouble, while in other countries a large amount of money is used for the control of wasps at recreation parks.

As explained in detail in the previous chapter, the vespines are quite widely distributed, spreading over most parts of the Old World (except the Ethiopian region), and over nearly the whole North American continent, and some species extend onto islands of the South Pacific. Among another vespid subfamily (Polistinae), the Polistini are a cosmopolitan group. the Polybiini are highly differentiated in the New and Old World tropics, and the Ropalidiini are dominant in the subtropics and tropics, except in the New World. Thus, people all over the world have ample opportunity to come into contact with stinging social wasps. In spite of this fact, few reports are available which discuss the relation of man with Polybiini and Ropalidiini prevailing in the tropical and subtropical regions (Akre and Davis 1978). Looking upon the vespines as nuisance insects is at present restricted to such industrialized countries as the United States of America, the European countries and Japan. Asia stands out in the wasp-man confrontation, because most of the large, ferocious species of *Vespa* are concentrated in this region. Therefore, control measures for wasps in Japan differ from those in the United States and Europe. In some countries in Southeast Asia, e.g., Taiwan and Indonesia, several species of *Provespa* and *Vespa* seem to obstruct cultivation and farming. Judging from the scale of nesting of tropical *Vespa,* the peoples of Southeast Asia should suffer from stings, but at present we have no means to know to what extent the hornets have negative effects on human lives and the local economy.

1 Vespine Wasps as Nuisance Pests

Vespine wasps usually inhabit natural environments such as forests, but often occur in human environments such as residential areas, parks and farms (Fig. 12.1). In the United States and Europe, the most synanthropic group is the subgenus *Paravespula* of the genus *Vespula*, since species of *Paravespula* are often general hunters or scavengers, and often build large colonies in structures. As will be mentioned later, in Japan some species of hornets (especially *Vespa simillima*) are more important than *Vespula*. In this section we briefly explain the damage and trouble caused by vespine wasps in the United States and Europe, then point out the particular situation in Japan.

1.1 Trouble Caused by Vespines in the USA and Europe

In these regions, wasps in *Paravespula* are the principal harmful species. In addition to them, *Vl. squamosa* and some *Dolichovespula* species are sometimes troublesome (Edwards 1980; Akre et al. 1981). In Israel and other countries in the Middle East, the Oriental hornet *Vespa orientalis* prefers the human environment for nesting, and causes serious damage (Ishay et al. 1968; Spradbery 1973a). There are three reasons why vespines are regarded as nuisance pests (a) they sting

Fig. 12.1. *Vespula flaviceps* nest constructed in a wall void in the urban area in Hokkaido

people and domesticated animals, (b) they carry disease organisms, and (c) they contaminate human food (Edwards 1980). All of these may bring about economic losses. Among them, stings are the most serious threat. A feeling of annoyance and a terror of being stung when being pursued by wasps in trivial cases, and dying from stings in serious cases should all be included in this danger category. The nature of vespine venoms will be briefly discussed at the end of this section.

There are virtually no statistics for the number of people stung by vespines. Only the records of deaths by stings are occasionally available. For example, in England and Wales there were 70 deaths from wasp (probably vespine) stings in the 21 years from 1949 to 1969, with an average of only 3.3 per year (Edwards 1980). In the United States, on the other hand, 15–20 persons die each year from stings of vespids, including paper wasps (Polistinae). However, as pointed out by several authors, the actual numbers may be somewhat more, because some deaths may have been incorrectly diagnosed as "heart attack", etc. But, at any rate, those killed constitute only a portion of those who were stung by vespine wasps.

People may be stung in various environments, e.g., parks, recreation grounds, residential areas, zoological gardens and farms, in the United States, for example (Wagner 1961; Akre et al. 1981). In these environments, there are many potential sites for wasps to nest, and furthermore they have a great deal of food, such as empty juice cans, fruit and candies, so many yellowjackets fly in from adjacent areas. In Europe, yellowjackets fly in large numbers to bakeries and food stores. The 2300 cases of control service carried out by the English pest control company Rentokil during the six years from 1967 to 1972 were at the following sites: private houses (33.6%), bakeries (17.5%), food manufacturers (13.2%), food shops (11.4%), public places (10.0%) and others (14.4%) (Edwards 1980).

Damage caused by yellowjackets (*Vespula* + *Dolichovespula*) may fluctuate considerably from year to year, with high levels in "wasp years". A 29-year survey in Surrey, England showed that the number of colonies fluctuated between two and 84 in a horticultural garden of 40.5 ha (Archer 1980b). Tallying yellowjacket nests destroyed by 25 local authorities in England in the 15 years, 1963–1977, Edwards (1979) recognized 1974–1976 as wasp years. The destroyed colonies amounted to a total of 49 000. Data from the North District shows that approximately one-half of the nests destroyed in the 15 years were concentrated in the three years from 1974 to 1976. No statistics of this sort are available in North America. As a matter of course, in any year wasps become conspicuous from late summer into fall, when colony populations attain their maximum size.

The vespines, especially yellowjackets, are known to carry disease organisms. As mentioned earlier, wasps of the subgenus *Paravespula* have a strong tendency to scavenge, looking for corpses of mammals, snakes and fish, as well as hunting for live filth-breeding flies. There may be a fair chance, therefore, for these wasps to swallow or transport disease organisms which are later scattered by visiting wasps, on bread, candies and fruit in stores. Thus, in this sense, the yellowjackets play the same role as flies, and may better be called hygienic pests rather than nuisances. In Europe, yellowjackets are reported to carry *Salmonella* bacteria that cause food-poisoning, and a *Escherichia* bacterium which causes gastroenteritis in man (Wellman 1952; Döhring 1960).

1.2 Trouble Caused by Vespines in Japan

In Japan, vespine wasps have not been studied as nuisances until quite recently. According to Ban (1980b), who obtained information through questionnaires on a nationwide scale, of 6298 cases in which a colony (sometimes individual wasps and bees, or a swarm of honeybees) was killed or removed by local authorities, 44% were honeybees, 42% vespines and 14% others (mainly *Polistes*) (Table 12.1). Honeybee cases were more frequent in southwestern Japan, while vespine cases were especially abundant in Hokkaido, making up 95.5% of all cases in the statistics of the three years from 1975 to 1977. Among vespine nuisances the Japanese hornet *Vespa simillima* stands first, and has been paid most attention to by staffs of local public health centers (e.g., Ban 1980a,b,c; Makino et al. 1981). Thus, the situation apparently differs from that in Europe and North America, where instead the yellowjackets are major pests. There are statistics on deaths from hymenopteran stings in Japan since 1979, though the species and groups of stinging insects are not distinguished. According to these data, 170 people died of hymenopteran stings in the 5 years from 1979 to 1983. Most of these deaths seem to have been caused by vespines, for virtually no deaths from polistine or honeybee stings have been reported in newspapers. In Japan, vespines are not significant disease vectors and contaminants of food. In the following section the nesting behavior of, and troubles caused by, vespines (especially *Vespa simillima*) in urban areas are described and discussed.

Table 12.1 Number of destroyed nests[a] of wasps and bees by local authorities[b] during 1975–1977 (After Ban 1980b)

Year	Honeybees	Vespines	Others[c]	Total
1975	620	521	218	1359
1976	1082	1101	292	2475
1977	1081	1018	365	2464
Total	2783 (44.2%)	2640 (41.9%)	875 (13.9%)	6298

[a] A few cases are included where individual insects (rather than nests) were killed.
[b] Statistics cover the following regions: Hokkaido, Akita Pref., Mie Pref., Kanagawa Pref., Hyogo Pref., Aichi Pref., Osaka Pref. and 19 big cities (Kyoto, Kitakyushu, Kamakura etc.).
[c] Mainly paper wasps (*Polistes*).

Nesting Sites of V. simillima in Urban Areas. This species nests in both protected and open sites (cf. Chap. 3, Sect. 1). This flexible selection of nesting site is one of the factors responsible for the success of this species in urban areas. Table 12.2 shows the nesting sites of *V. simillima* in human environments, such as residential regions in Otaru and Sapporo, Hokkaido (subsp. *simillima*) and Kyoto, Honshu (ssp. *xanthoptera*). In Otaru, more than 80% of the nests destroyed by the Otaru Public Health Center were constructed in protected sites, and approximately 59% were in wall voids of wooden buildings. In the Botanical Garden of Hokkaido

Table 12.2. Nesting sites of *Vespa simillima* in some localities of Japan (After Makino et al. 1981)

	Protected sites						Open sites (%)			
	above the ground (%)					In the soil	On twigs	Under the eaves	Others	Total
	Wall voids	Attics	Under the floor	Tree hollows	Others					
Otaru[a]	304 (50.7)	96 (16.0)	27 (4.5)		62 (10.3)	28 (4.7)	23 (3.8)	52 (8.7)	8 (1.3)	600
Sapporo[b]	11 (23.9)	10 (21.7)		3 (6.5)	7 (15.2)	2 (4.3)	7 (15.2)	6 (13.2)		46
Kyoto[c]		107 (68.2)		1 (0.6)	21 (13.4)	1 (0.6)		26 (16.6)	1 (0.6)	157
Wakayama Pref.[d]		15 (9.0)		3 (1.8)	3 (1.8)	11 (6.6)	4 (2.4)	113 (68.1)	17 (10.3)	166

[a] Nests destroyed by the Otaru Public Health Center from 1974 through 1980 in residential areas.
[b] Nests found in Sapporo and its neighboring localities from 1969 through 1980.
[c] Nests destroyed by the Kyoto Epidemic Prevent. Center in 1978.
[d] Nests observed by Matsuura from 1954 through 1969.

University (Sapporo), where natural environments are relatively well preserved, 11 of the 22 nests observed (50%) were constructed in open sites (Ban and Sk. Yamane 1980). It is evident that in human environments buildings are used by this species as the preferred nesting sites in Hokkaido. Also in Kyoto, which is dominated by human environments, protected-site nesting exceeded 80%, in contrast to Wakayama (not far from Kyoto, but dominated by agricultural and horticultural environments) where approximately 80% of the nests were built in open sites. The type of structural nesting clearly differed between Otaru and Kyoto, i.e., in Otaru, wall voids were most preferred, whereas in Kyoto most structural nests were made in attics. The colony size finally attained and differences in the architecture of buildings between these localities may be responsible for their nesting types (Makino et al. 1981).

Stings in Residential Areas. Little information is available on the actual state of people stung by vespine wasps. Here (Table 12.3) we cite the results by Ban (1980a,b) and Makino et al. (1981), who intensively investigated those stung in Otaru, Hokkaido. In this city, the number of people who were stung by vespines

Table 12.3. Situations in which victims were stung by hornets (Otaru, Hokkaido) (After Ban 1980a)

Year	Touching nests	Working around nests	Passing close to nests	Others	Total No. people stung
1977	28	92	23	12	155
1978	18	19	17	9	63
Total	46 (21%)	111 (51%)	40 (18%)	21 (10%)	218

and reported to the public health center amounted to 272 during the period from 1977 to 1980. In most cases, people were stung when they somehow disturbed nests in the close vicinity of their residences, i.e., when working around buildings (51%), touching nests (21%), and passing close to nests (18%). This situation differs from that in the United States and Europe where, people are often stung by foraging individual yellowjackets at recreation grounds and confectionaries. More than half were stung on the arm or hands (44%), and faces (18%) (Table 12.4) as they tried to slap or brush off the approaching hornets. Table 12.5 shows the relationship between nest sites and the number of people stung by *V. simillima*. A large proportion was stung by hornets nesting in wall voids (73%) and under the floor (10%). As wall-void nesting makes up 50.7% of all the nestings in Otaru, the figure (73%) suggests the greater danger from colonies constructed in wall voids. This may be because the colonies of this type are hard to locate in early stages and escape control, and after attaining a large size they are easily aroused by people working in gardens and passing near nests. The large nests in wall voids are especially difficult for residents to remove. Forestry workers are also stung when mowing weeds and bushes on the forest floor. Stings in farms, parks and sight-seeing areas are also reported in newspapers, however we have no reliable statistics concerning people stung in these situations. The number of requests by residents to the public health center for nest destruction varied among months and localities (Table 12.6). The difference between Otaru and Kyoto is likely to be caused by differences in hornet life cycles, particularly colony life span (Ban 1980a).

Table 12.4. Parts of body stung by hornet(s) (Otaru, Hokkaido) (After Ban 1980a)

Year	Whole body	Head	Face	Belly	Back	Arms (Hands)	Legs	Un-known	Total No. people stung
1977	3	19	14	6	9	55	12	37	155
1978	2	4	4	3	2	41	7	0	63
Total	5 (2%)	23 (11%)	18 (8%)	9 (4%)	11 (5%)	96 (44%)	19 (9%)	37 (17%)	218

Table 12.5. Relationship between people stung and type of nesting sites of *V. simillima* in residential areas in Otaru (After Makino et al. 1981)

Year	Number of people stung by hornets nesting (%)					Total
	in wall voids	under the floor	in attics or cabin	under the eaves	at other sites	
1977[a]	109 (70.8)	18 (11.7)	6 (3.9)	2 (1.3)	19 (12.3)	154
1978[a]	51 (81.0)	3 (4.8)	3 (4.8)	0 (0.0)	6 (9.4)	63
1979[a]	10 (45.5)	1 (4.5)	3 (13.6)	0 (0.0)	8 (36.4)	22
1980	21 (63.6)	0 (0.0)	2 (6.1)	3 (9.1)	7 (21.2)	33
Total	191 (70.2)	22 (8.1)	14 (5.1)	5 (1.8)	40 (14.8)	272

[a] After Ban (1980a).

Table 12.6. Number of nests destroyed per month by the Otaru Public Health Center (*V. s. simillima*) and the Kyoto Epidemic Prevent. Center (*V. s. xanthoptera*) in 1977 (After Ban 1980a)

	Number of nests destroyed per month (%)								
	Jan.–May	June	July	Aug.	Sept.	Oct.	Nov.	Dec.	Total
Otaru	0	0	16 (9%)	124 (67%)	43 (22%)	3 (2%)	0	0	186
Kyoto	1 (1%)	2 (1%)	21 (14%)	28 (18%)	65 (41%)	28 (18%)	11 (7%)	0	156

Significance of Vespinae as Nuisances. As mentioned above, among the Japanese Vespinae *Vespa simillima* is becoming a significant nuisance pest and is a target of control throughout most of the country. Of course, most of the other vespines are potential nuisances. For example, four *Vespa*, three *Vespula* and two *Dolichovespula* species occur in residential areas in Hokkaido. Nests in houses and on garden trees are reported for most of these species. Mainly because they are not so frequently encountered, people pay less attention to them than to *Vespa simillima*. In Honshu, Shikoku and Kyushu, yellowjackets are generally confined to mountainous regions so that people, except in mountain villages, have few chances of encountering them. Their much smaller body size also does not arouse much fear in man. The world-famous pest *Vespula germanica* does not occur in Japan. Some species of paper wasp (*Polistes*) are common in residential areas, but these wasps are not so feared by people, since they construct small colonies and are much milder in behavior than hornets and yellowjackets. For these reasons, it is important to reduce the instances of stings by hornets, especially by *V. simillima*, which has such flexible nesting behavior and constructs huge colonies inhabited by numerous aggressive workers.

It should be stressed here that vespine stings sometimes cause anaphylactic shock to people who are allergic to the venom and who may die unless they receive immediate medical treatment. In this sense they are never mere nuisances.

1.3 Vespid Venoms

Vespid venoms have been studied not only from the applied view, but also by medical scientists because of their ability to cause severe allergic reactions. The study of vespid venoms has made rapid progress since the 1960s.

Vespid venoms are not acid-like substances, as believed until recently by most people in Japan. They are a complex consisting of low-molecular weight amines, low-molecular weight peptides and high-molecular weight enzymes (Nakajima 1983). The substances responsible for the cutting pain during a sting are thought to be amines such as histamine and serotonin, and kinins (wasp kinins) involved in low-molecular weight peptides (Edery et al. 1978), but the evaluation of the relative importance of these differs among scientists. Among the components peculiar to vespid venoms are serotonin, kinins and protease. In the vespid venoms there are components that effectively act against mammals through, for example, contributing to lower blood pressure (kinins). There may be an evolutionary interaction between the feeding habits of mammals and the extraordinary aggressiveness of some vespine species.

The effect of vespid venoms greatly varies among victims. In people with normal constitutions, sharp pain and discomfort disappear within a day. Although there are deaths from just one sting inside the mouth (a swollen larynx or throat causes suffocation), or from numerous stings that result in massive intoxication, such cases are quite rare. However, in hypersensitive people, only one or two stings can cause systemic symptoms (for cases in Japan, see Ôri and Hiyama 1977; Ban 1980c). The systemic symptoms may abate in some patients without further aggravation, while others may die very quickly, within only one hour in many cases. The reaction leading to death by wasp venom proceeds much more quickly than those by spider venom (more than 18 h) and snake venom (6–48 h). Today, this systemic symptom is regarded as an allergic reaction called anaphylactic shock that occurs in hypersensitive persons. Since the allergic shock is an abnormal reaction by antibodies to an antigen to a second (or following) injection of the same antigen, hypersensitive individuals should not be stung twice or more, especially by the same or related species. Although the ratio of sensitive people has not been investigated on a large scale in Japan, Shioda et al. (1974) showed that of 384 school children investigated in a mountain village in Kanagawa Prefecture (Honshu) 24 (6%) had experienced generalized (systemic) symptoms after being stung by wasps. This figure suggests tne potential danger of severe allergic reactions like anaphylactic shock. The relationship between allergies to vespid venoms and other allergens (penicillin, flower pollens, etc.) has been studied by various workers (e.g., Miyachi 1980).

Though wounds from the sting of *Vespa mandarinia* often remain as scars for several years, normal people generally recover in two or three days after being stung. Antihistamines sprayed onto the affected area or taken by mouth in the form of tablets or capsules are said to reduce pain (Edery et al. 1978; Edwards 1980). Hypersensitive people must receive medical treatment very soon (preferably within 30 min of being stung). Even normal people should consult a doctor when stung by a number of wasps, especially when generalized symptoms appear. To patients in serious condition, deep subcutaneous injections of a 1% solution of adrenalin (epinephrine) in high doses are effective (Edery et al. 1978; Miyachi 1981). Since people may be stung in areas not accessible to professional medical help, in the United States and Europe hypersensitive persons are recommended to carry epinephrine syringe kits (emergency kits) with them (Akre et al. 1981). The massive venous injection of antihistamines, at one time regarded as an effective treatment, is not recommended because it brings on difficulty in breathing (Miyachi 1981).

Some articles are available for readers who are interested in or need precise knowledge about vespid venoms and medical treatment: Edery et al. (1978); Edwards (1980); Akre et al. (1981); MacDonald (1980); Miyachi (1981).

2 The Wasp as Enemy of the Honeybee

2.1 Damage-Inflicting Species

In Japanese apiculture, a native species, *Apis cerana*, had been reared mainly in the southwestern region until the introduction of *A. mellifera* in 1876. *Vespa mandarinia* had often been reported as an enemy of the former species (e.g., "Hachimitsu Ichiran", 1876). Specific details of the damage caused, however, were not clear. Among the wasps attacking *A. mellifera* in and around Tokyo, *V. mandarinia, V. simillima* and *V. crabro* have been reported by Okada (1961) and *V. analis* by Matsuura (1967a). Table 12.7 shows the distribution of *Vespa* species found in hornet traps set at the entrances of beehives at an apiary in Kibi-chô, Wakayama Prefecture. *V. mandarinia* accounted for the highest number of catches, followed by *V. simillima*. Numbers of *V. analis* and *V. crabro* by contrast were negligible. Small numbers of *V. tropica* were also trapped. This species, however, is thought to come to the apiary through attraction to the honey, for it does not attack the honeybee colonies themselves (Matsuura and Sakagami 1973). The relationship between *V. mandarinia* and *V. simillima* is of great interest. When over 900 individuals of the former species were trapped in the course of a single year, this represented only a few percent of the total number of captures. By contrast, in years when the numbers of *V. mandarinia* were low, the ratio of *V. simillima* was remarkably high. This relationship can be explained by exclusion of *V. simillima* from the apiary, since any *V. mandarinia* exhibits the remarkable behavior of driving other wasps, even of the same species, away from other hives and attacking other species of *Vespa*. In the northern region of Japan, *V. mandarinia* generally occurs infrequently and the numbers of *V. simillima* visiting apiaries are higher. This fact also reflects the interspecies relationship (Matsuura and Sakagami 1973).

Attacks by *V. crabro* on *A. mellifera* have been reported in Germany (Olberg 1959), England (Mace 1927; Buzzard 1951), South Africa (Mace 1927), Italy

Table 12.7. Number of *Vespa* wasps captured in hornet traps attached to hives at an apiary in Kibi (Matsuura 1984)

Year	No. of wasps captured										Total No.
	mandarinia		*simillima*		*analis*		*crabro*		*tropica*		
	No.	(%)	No.	(%)	No.	(%)	No.	(%)	No.	(%)	
1966	970	93.6	49	4.7	10	1.0	1	0.1	6	0.6	1036
1967	1320	96.8	24	1.8	11	0.8	2	0.1	5	0.4	1362
1968	1324	96.4	19	1.4	21	1.5	2	0.1	8	0.6	1374
1969	943	91.6	68	6.6	11	1.1	2	0.2	5	0.5	1029
1970	700	65.8	340	32.0	16	1.5	1	0.1	7	0.7	1064
1971	834	80.6	182	17.6	7	0.7	1	0.1	11	1.1	1035
1972	1115	93.9	58	4.9	9	0.8	1	0.1	4	0.3	1187
1973	1434	97.8	14	1.0	10	0.7	2	0.1	7	0.5	1467
1974	706	70.0	288	28.6	8	0.8	3	0.3	3	0.3	1008
1975	1375	96.6	30	2.1	12	0.8	1	0.1	6	0.4	1424
Total	10 721	89.4	1072	8.9	115	1.0	16	0.1	62	0.5	11 986

(Alber 1953) and North America (Bromley 1948). *V. orientalis* has been noted to be an important apicultural pest in Egypt (Wafa 1956), India (Singh 1962), Israel (Ishay et al. 1967) and southern Italy (Alber 1953). Also, in the tropical and subtropical zones of Southeast Asia, attacks have been reported on a native honeybee, *Apis cerana* as well as *A. mellifera*, by *V. velutina*, (India: Singh 1962; Indonesia; van der Vecht 1957), *V. basalis* (India: Singh 1962), *V. analis* (Indonesia: van der Vecht 1957) and *V. affinis* (Sumatra, Taiwan: Matsuura unpubl.).

Although *Vespula* is not as important as *Vespa* as a honeybee predator, there have been some reports on *Vl. vulgaris* and *Vl. germanica* (England: Mace 1927; Rawson 1963) and *Vl. maculifrons* (North America: Bromley 1948). *Vl. flaviceps* in Japan is known to steal honey by invading weak hives (Matsuura and Sakagami 1973).

Regarding *Dolichovespula*, there is a report of *Apis mellifera* being preyed upon by *D. sylvestris* (England: Watson 1922).

2.2 Type of Damage

V. mandarinia and *V. simillima* show completely different methods of attack (see Chap. 5). The former species inflicts lethal damage by invading the beehive in a group, killing all bees, occupying the hive and removing all the brood. The degree of damage inflicted by *V. simillima*, as well as by other *Vespa* species, is quite low in comparison with *V. mandarinia*, because this species always preys upon honeybees in front of the hive entrance and does not engage in either invasion or brood removal.

From mid-August to early November, in a *V. mandarinia* habitat, groups of this hornet fly to apiaries located there almost every day and attack persistently, so that populations of *A. mellifera* are completely destroyed unless they are given artificial protection. In the case of two apiaries in a mountain area of Kibi-chô, Wakayama Prefecture, patrolling was performed every day during the *V. mandarinia* attacking season during each year from 1966 to 1975, and every member of the species was captured and killed as soon as it was found. In spite of these efforts, however, 32 of the total of 720 hives suffered group attacks and were completely destroyed; another 48 hives became weakened and were unable to recover in the fall. Since each of the remaining hives also suffered attack to some extent, it was obvious that they all would have been totally destroyed if they had not been protected (Matsuura unpubl.). In addition, when a total of 86 hives were set up in a similar location in Kawage-chô, Mie Prefecture, and observed each year from 1976 to 1983, being without any protection, they were totally destroyed every year by attacks of *V. mandarinia* at the very beginning of the attacking season, from late August to early September (Matsuura 1984).

Although no data are available on general damage in Japan, it can be assumed that at least 10–20% of any bee hive suffers hopeless damage every year, judging from the descriptions given by apiculturists of their experiences in various parts of the country. According to the data reported annually by the Livestock

Industry Bureau, Ministry of Agriculture, Forestry and Fisheries, the number of honeybee hives existing per year in Japan has been about 320 000 for the past 5 years. Consequently, the number of damaged hives can be estimated to be 30 000–60 000 per year. In addition, if bees were attacked by *V. mandarinia* in the fall, which is the season for raising honeybees for hibernation, numbers of worker bees would be reduced and their field activities like flower-visiting would be repressed. This, therefore, would be a great obstacle to bee raising and would inflict considerable damage to the formation of hibernating populations.

On the other hand, some other species, including *V. simillima, V. crabro* and *V. analis*, attack and prey on bees not only in apiaries but also in the field when bees go flower-visiting, thus causing a chronic phenomenon of bee reduction in the fall, the season in which the number of wasps increases (Matsuura and Sakagami 1973).

The most damage-inflicting species of *Vespa* in foreign countries is *V. orientalis*. This wasp captures bees at the entrance of a hive and transfers them to the wasp's own nest and, furthermore, the wasp sometimes breaks into the hive by killing the guards. Forming a group, they prey on the brood and young bees (Rivnay and Bytinski-Salz 1949). The Israeli Apicultural Association has reported that 3000 of 28 000 hives owned by union members were destroyed by this wasp. Such an attack, therefore, is noteworthy from the viewpoint that the damage it causes is equivalent to that by *V. mandarinia* in eastern Asia (de Jong 1978). The actual conditions of its attacks, however, are not yet clear.

Apicultural damage by *Vespula* has not stimulated any public discussion, because of its low predatory ability compared with *Vespa*, owing to its small body size. Although the damage by *Vespula* observed to date has mostly been its use of weak and dead bees at the entrance of a hive as food, honey-stealing behavior (which is quite uncommon in *Vespa*) has also been observed frequently (Thomas 1960; Rawson 1963; Matsuura and Sakagami 1973). In New Zealand, however, a naturalized species, *Vl. germanica*, has been reported as an apicultural pest. During 1974–1975, a period in which outbreaks of this species were most frequent, the total of bees preyed upon by this species was estimated to be equivalent to 3900 of 96 000 hives. Moreover, it exerted an influence upon 10 100 hives and the total economic loss reached 65 000 £ (Walton and Reid 1976).

3 The Wasp as Fruit Pest

3.1 The Wasp in Orchards

A variety of carbohydrate sources are used by the wasp, including tree sap, nectar, ripe fruit, honeydew and the remains of commercially produced juice and soft drinks (Table 5.1). In an orchard a wasp visits fruit just before it is harvested, or that which has become overripe on the trees, is injured by fruit-piercing moths or else ends as fermenting windfalls. It gnaws the sarcocarp, sucks up juice and fermented liquid, stuffs it into its crop and carries it back to its nest.

In reference books on fruit pests in Japan, the wasp, mainly *Vespa*, is described as inflicting various sorts of damage. Such damage can be divided into two types: feeding on fruit just prior to harvest (primary damage) and feeding on overripe fruit first injured by other organisms or else fermenting (secondary damage). In addition, since the wasp itself is a large-sized and poisonous stinging insect, it is of further significance as a nuisance to the cultivators of fruit trees and visitors to farms, especially at times when the insect is building its nest or roaming orchards. Wasps are further known as pests because they sting the fingers and hands of harvesters who accidentally touch them during harvesting.

3.2 Damage-Inflicting Species

All species of *Vespa* in Japan use various types of ripe fruit as food, but most damage is inflicted by *V. mandarinia, V. simillima, V. tropica* and *V. crabro*. Shiraki (1933) reported that in Taiwan, *V. mandarinia nobilis, V. affinis affinis, V. tropica pseudosoror* and *V. velutina flavitarsus* spoiled grapes and *banzakuro* (a kind of pomegranate) by eating the fruit, but that the damage was not so severe. In India, *V. orientalis* has been reported to injure grapes, apples, peaches and other types of fruit (Horne 1870; Butani 1979). This species has also been known to gnaw the bark off orange saplings in Palestine (Bodenheimer 1951). When *V. crabro germana* was introduced to North America, it was observed to feed off apples (Wray 1954).

Various species of *Vespula* also visit ripe fruit. In Japan, *Vl. flaviceps* mostly consumes grapes (Matsuura unpubl.). In Europe, native species including *Vl. germanica* and *Vl. vulgaris* sometimes further inflict severe damage on plums and pears. In North America, there has been a case of damage to grapes eaten by *Vl. pensylvanica* (Edwards 1980).

There are no reports of *Dolichovespula* damaging fruit. However, there is a possibility that this genus, as well as *Vespa* and *Vespula*, uses grapes as a source of carbohydrate because of their soft pericarp and high sugar content.

3.3 Type of Damage

In Japan, wasps frequently damage figs and grapes in addition to pears, peaches, persimmons, pineapples and apples. The amount of damage varies greatly depending on the region and the year in question (Matsuura unpubl.). Such variations seem to be related to the availability of other effective carbohydrate sources, such as tree sap.

In the tropics, ripe non-acidic fruits such as papaya and guava are damaged by the wasp (Iwata 1971a). In England, the plum, pear, grape and raspberry are also damaged, mainly by *Vespula*, as is the apple occasionally. No damage to peaches or apricots is reported (Edwards 1980).

The season of wasp attack on fruit is concentrated in the general ripening season (August-October). Workers inflict damage in most cases, but in the case of *V. tropica* males occasionally visit fruit.

The extent of damage to the main types of fruits in Japan is as follows:

Figs. Since ripe fruit of almost all varieties including "Dolphin" have soft pericarps and sarcocarps, and a high sugar content, they are eaten by various species, not only of *Vespa* but also of other insects, such as beetles, butterflies and fruit-piercing moths. Damage by *Vespa* is characterized by complete consumption of the sarcocarp. In *V. mandarinia* one to two workers start to gnaw the sarcocarp at the open part of the apex of a ripe fruit and suck the juice up into their crops, leaving the remainder of the sarcocarp in situ. With a full stomach, the wasp returns to its nest, disgorges the juice for consumption by larvae and adults and revisits the same fruit to continue eating. Consequently, the fruit is completely eaten, leaving only the peduncle. Usually this will take 3–4 h from the first visit. The wasps then search out other ripe fruit. Their attacks continue from the harvesting season of the first crop in August to the season of the second crop sometime after September. In *V. mandarinia*, workers feeding on the same fruit always belong to the same nest and keep a monopoly on the fruit by launching severe attacks on wasps of the same species from other nests, other species or insects. Since they also show the same threatening and attacking behavior towards humans, it is necessary to be wary of them, especially during harvesting. Such monopoly of a food source is a habit specific to the *V. mandarinia*; thus, other species of *Vespa*, even when they are eating, do not attack humans who are not in the vicinity of their nests.

Grapes. Open-culture varieties which ripen after August, including "Delaware", "Neo-Muscat" and "Kyohô", suffer damage from *V. mandarinia, V. simillima, V. crabro* and *V. tropica*, which frequently and specifically visit grapes of European varieties whose pericarps are especially soft and which dehisce easily. They spoil these grapes by gnawing the sarcocarp from the dehiscent part and sucking up the juice. In addition, once the pericarp of a grape has been gnawed, the juice flows out, sticks to the surface of other healthy grapes in the same bunch and causes gray mold and white-rot disease on the wetted parts. This often spreads to the whole crop. In this way, a wasp attack sometimes causes a greater amount of damage through the secondary spread of disease than through direct damage caused by eating. Once a bunch of grapes is attacked by *V. simillima*, it suffers continuous injury for several days. Sometimes over ten workers of the same nest feed before moving to the next bunch. Late varieties, such as the brewing variety "Cabernet Sauvignon", which is harvested around the middle of October, are particularly badly damaged by the cumulative attacks of wasps, sometimes becoming unharvestable. However, almost nothing has been reported throughout the country on wasp damage to hard-pericarped, dehiscence-resistant varieties like "Kôshû".

Pears. Damage by *V. simillima, V. analis, V. mandarinia* and *V. tropica* has been reported to reddish varieties, including "Shinsui" and "Kôhsui", in addition to the late variety "Hônsui", which are low in acidity but with a sugar content over 12° Brix. They are rich in juice, have a soft sarcocarp and are harvested in mid- to late August. Damage by *V. basalis* has also been confirmed at an orchard of reddish pears in Puli, Taiwan (Matsuura, unpubl.). In both cases, the insects gnaw a hole through part of the pericarp of the ripe fruit and suck juice up from the sarcocarp, thus injuring the surface of the fruit and causing spotted scars to form. Such damage, however, is generally regarded as secondary to a weakened fruit caused by the juice-sucking of the fruit-piercing moth, most fruits having already been injured by some noctuid moths. In Europe, European pears often suffer damage from *Vespula* just before harvest, as it gnaws the pericarp and makes a hole reaching as far as the core, thus destroying the fruit's financial value. Counter-measures have been encouraged, such as advancing the date of harvest as far as possible, and introducing late varieties which can be harvested when wasp activities drop (Edwards 1980).

Apples. Outbreaks of damage vary from year to year. Watanabe (1949), reporting on damage to apples in Hokkaido, noted that among over 200 varieties cultured in one orchard, "Shikishima" suffered cumulative damage from *V. crabro* and *V. mandarinia* starting in mid-September and that 30% of the fruit was completely eaten, leaving only the pericarp and core. This variety has a relatively thin pericarp and a soft sarcocarp, with a slightly strong acidity. It ripens in late September. Late varieties with a thick pericarp and hard sarcocarp, however, survived completely undamaged.

4 The Wasp as Beneficial Insect

In Japan, the wasp has also been regarded as a beneficial insect, both as food (the larvae and pupae) and in Chinese medicine (the wasp's nest). In Europe it is seen as a natural enemy of pests and a pollen-carrying insect. In addition, worldwide studies are now in progress on the value of the toxic component of its sting as a medicine for the treatment of neuralgia.

4.1 The Wasp as Food

Insect-eating customs, although not generally widespread, can be observed in southeast Asia including Japan, and among the native populations of North and South America. In particular, vespine wasps are used in certain countries as a source of protein, since their large nests contain large amounts of larvae and pupae (van der Vecht 1957). In Japan, *Vl. flaviceps* and *Vl. shidai* have been eaten since ancient times under the names of *Sugare, Jibachi* and *Hebo*. This has mainly occurred in Central Japan in Nagano, Gifu, Aichi and Yamanashi Prefectures. In addition to *Vespula*, people in Iida City and the Shimoina district of Nagano Prefecture collect the nests of *V. simillima, V. analis, V. mandarinia* and *V. dybowskii* and use the larvae and pupae in the nests as food. The origins of these customs are not clear. According to "Sôzan Chomon-shû" (Miyoshi 1830), published in the Tenpô era in the Edo Period, samurai belonging to the Gujô clan of Mino (now Aichi Prefecture) but residing in Edo collected nests of *Hebo* wasps in the fall in the garden of their Aoyama villa, seasoned them with soy sauce and ate them either as a relish taken with *sake* or else mixed the larvae with boiled rice. Since the same author also described how people in other prefectures "had no taste superior to those in Shinshu (Nagano Prefecture) and Mino, who had a preference for wasps as food", such a custom must have originated in quite ancient times.

Vespa is abundant in Southeast Asia. The habit of eating the larvae of *V. velutina* and *V. tropica* has been reported in Thailand (van der Vecht 1957) and Burma (Bristowe 1932). A tribe in North Sumatra collects the nests of *V. tropica, Provespa anomala* and *Pv. nocturna* and cooks the larvae they contain by seasoning them with salt, onions and pepper, or by mixing them with sugar, condiments and coconut oil (van der Vecht 1957).

Discovery and Collection of Nests. Collection of *Vespula* nests is generally carried out from August to November. Recently, however, many people have been doing so in June or July, and then raising them. There are two common methods of discovering nests. In one method, wasps in the field are first introduced to the meat of a frog or a cicada; a lump of the meat (3–5 mm in diameter) is tied to a piece of silk spread to a width of 1–2 cm; the meat tied to the silk is given to a wasp and the location of its nest is discovered by following the thread of silk carried by the wasp. This method can be performed individually or by two or three people. The other method is called "Sukashi". A searcher stands still against the sunlight near a bank or the edge of a thicket where wasp nests are frequently seen, and traces wasps flying to their nest by carefully observing them. When a skilled person finds a wasp flying in the field, he can judge whether the wasp is returning to its nest or going out to hunt by observing whether or not the wasp is holding prey in its mandibles. When the observer is aware that the nest lies within a short distance, he begins a thorough search of the immediate vicinity. In the case of *Vespa*, wasps coming and going to their nest congregate around any person coming within 10 m or so of the nest; by threatening in this manner, they make it possible for him to find the nest.

As for the collection method, wasps in a nest are rendered symcopic prior to digging out the nest by igniting a smoke-producing firework, commercially obtained or prepared personally. When the purpose is to take the insects for immediate consumption or to transport them to market, only the nest is collected and the wasps left in situ. For the purpose of raising for future use as food, the queen is taken with as many workers as possible with the nest. Nowadays, not only private individuals but also hotel or restaurant owners keep nests to serve the insects to customers. In the latter case, food such as fish is hung near the nest and the workers are allowed to carry the meat to the nest. The weight of the nest can then reach 3–4 kg in October to November, and the population grows larger than any population could in the field.

The Market for Broods. Wasp broods and their nests have been on sale together in Nagano and Gifu Prefectures since ancient times. This has been so for every collection season of *Vespula* wasps (August - November). It is recorded that in 1912 the market price was 15–16 sen (one sen = 1/100 yen) per 375 g (Okumura 1912). Wasp broods are still sold as food today.

There are many markets which trade the nests and broods of *Vespula* in Nagano and Gifu Prefectures, and so the total number of transactions is difficult to estimate. Many people are worried by the indiscriminate collection of these wasps, as private and professional collectors motivated by the high prices which such broods bring extend their hunting areas throughout the localities of the two prefectures themselves, and even go as far as Hokkaido.

Canned Broods. Canned and bottled broods of *Vl. flaviceps* and *Vl. shidai* are produced in many areas of Nagano and Gifu Prefectures and are shipped to other districts or exported abroad. The product is prepared by pulling out the larvae and pupae from a nest and boiling them with soy sauce, sugar and synthetic seasoning. This type of food has been produced since the Meiji period and is considered high grade. Its price between 1977 and 1983 was around 2200 yen per 200 g tin. This is quite expensive for tinned food. The amount of the product available is limited, since the broods used as raw material are collected from nests of the two above-mentioned species of *Vespula* in the field. Recently, beekeepers have started to mass-produce male broods of *Apis mellifera* and sell them as a substitute for wild *Vespula*. The market price of this substitute in 1981–1983 was around 1800 yen per 200 g tin.

4.2 Application of the Wasp as Predatory Insect

Wasps prey on various insects including unsanitary pests like the house- and horseflies, agricultural pests like Lepidoptera, Hemiptera and Coleoptera and many forest pests (Table 5.1). For the practical application of the wasp as a natural enemy, *Vespa crabro germana* was introduced artificially from Europe to North America (1840–1860) with the aim of controlling forest pests. It has naturalized in the eastern region of North America since its introduction (Shaw and Weidhaas 1956; Akre et al. 1981).

Farmers in Haibara-chô in Shizuoka Prefecture, a tea-cultivating district in Japan, have observed in their tea fields that *Vl. flaviceps* preys frequently on the larvae of important lepidopteran tea pests, including the tea fortix. Prompted by these observations, the whole town has made an effort to protect these wasps under the guidance of the Town Hall and the Agricultural Cooperative Union of Haibara-chô, and the collecting of *Vespula* nests has been prohibited since 1981. Shizuoka Prefectural Agricultural Experiment Station has also started studies on the application of this wasp as a natural enemy of insect pests (Ikeda and Kodomari 1982).

In Canada, an attempt has been made to apply *Vespula* to the control of a forest pest, *Neodiprion swainei* Midd (a species of sawfly which damages *Pinus banksiana*). *Vespula* is used as a carrier of a sawfly virus disease, and hunts larvae when weakened. They are preyed upon by *Vl. consobrina* and *Vl. vulgaris* (Smirnoff 1959). The principal aim of this scheme was that while a wasp is hunting disease-infected larvae and forming them into flesh balls, viruses would be spread onto neighboring needle leaves, infecting healthy larvae when they ate the leaves.

Vespula also preys on flies and horseflies associated with livestock. Once a suitable hunting place is found, the wasp visits the same place frequently (Iwata 1971a). Berrington (1896) reported that, in Ireland, 300–400 flies around two cows were preyed upon by wasps within 20 min. Watson (1922) has also reported that he observed flies being carried into a wasp nest at a rate of 227 flies per hour.

Application of wasps as predatory natural enemies should be possible since, in general, they have a habit of visiting the same place repeatedly and continuing to prey on the same game once they find another suitable hunting place. This will be a subject for further study, as the possibility has so far been little investigated.

5 Control

The study of yellowjacket control intensified in the 1960s, and various measures have been tried in the United States and Europe (Edwards 1980; Akre et al. 1981; Akre and MacDonald 1986). Yellowjacket control in North America may be divided into the following three categories: (1) Destruction of individual colonies (nests), mainly when they are constructed in or near houses, (2) area-wide destruction of colonies using baits with insecticides incorporated, and (3) area-wide abatement by capturing wasps with artificial attractants. North American workers have concentrated on the second and third devices to reduce the danger of being stung and other problems in parks, zoological and botanical gardens, farms and orchards. In Japan, the vespine problems were noticed by the beginning of the 1970s, and some species (especially a few hornets) are recognized as nuisances in urban areas. Since trouble mostly occurs near human habitations, the main control method is the destruction of individual colonies (Ban 1980a,b).

5.1 Destruction or Removal of Colonies

Control should be practiced taking notice of the following: The nest site and the passage(s) through which wasps come in and out must be carefully examined in the daytime. Although nests made on tree limbs generally have only one entrance hole in the envelope, nests in tree hollows, building wall voids and underground cavities may have more than one passage connecting the space accommodating the nest with the outside. At least 30 min after sunset when worker foragings completely cease, an insecticide such as pyrethroid and DDVP may be injected into the nest. Immediately close all the holes with cotton wads that are preferably saturated with an insecticide that will kill any returning wasps the next morning. Some insecticides used for these purposes are listed in Table 12.8. Cotton wads permeated with ether or chloroform may be applied to the entrance and passages, instead of injecting insecticides. After all the wasps are anesthetized, the nest will be removed and destroyed. Most Japanese vespidologists adopt the anesthesia method in order to get population data and live specimens. During the control work fire should never be used. Some pest control companies in Japan use carbondioxide instead of ether. One must practice with caution when destroying

Table 12.8. Chemicals used for controlling vespines

Types of control		Chemicals used
Destroying nests	In the soil	1% Dichlorvos, 1% propoxur, 3% DDVP, 5% carbaryl, gasoline, paraffin
	In wall voids	Pyrethrin, resmethrin, 3% DDVP
	On tree twigs	1% Dichlorvos, 1% propoxur, pyrethrin, 3% DDVP
Insecticidal baiting	Baits	Cat food, dog food, zoo meat, solids with sugar
	Insecticides	Diazinon in capsules, iodofenphos, dieldrin, chlordane, mirex
Attracting individual wasps		Heptyl butyrate, 2,4-hexadienyl butyrate, octyl butyrate

nests of *Vespa crabro*, in which extranidal activities continue at night. Since the species of *Provespa* in Southeast Asia are completely nocturnal, control should be conducted in the daytime when nearly all the wasps are in the nest, except for a few guarding on the envelope. A flashlight covered with red cellophane is usefui to aid work at night. Generally the naked light excites and attracts wasps. If some wasps escape from the nest due to clumsily applied insecticides, a naked flashlight may be placed slightly away from the nest to attract them. They should then be killed by spraying insecticide, otherwise such recklessly flying wasps may sting the exterminators. If diurnal control is unavoidable, the wearing of protective suits is strongly advised (Wagner and Reierson 1975; Sk. Yamane et al. 1979; Ban 1980c; Fig. 12.2). The best way to destroy aerial nests in the daytime is to use a pressure sprayer fitted with a long lance containing quick-knockdown insecticides such as pyrethrines and Dichlorvos (Edwards 1980).

The above-mentioned methods appear easy even for laymen, but in many cases unexpected accidents occur. People had better consult public health centers or pest control companies. Some local public health centers in Japan (e.g., Otaru and Kyoto) practice cost-free control services. In other Japanese cities such services are not available, and pest control companies often demand a large amount of money, because destroying wasp colonies, they claim, is highly dangerous, and an extra-charge must be paid for night work. For example, 20 000 yen is demanded to destroy a single *Polistes* nest, 30 000–40 000 yen for *Vespa* in Nagoya, and 20 000–40 000 yen for *Vespa* in Sapporo. Even on such occasions people are strongly advised not to practice a self-taught control measure. If the control measures fail, more serious problems may result than if the nests had been left as they were. Since colonies of hornets are extraordinarily huge in subtropical and tropical regions such as Taiwan and Indonesia, control measures should be conducted there with extreme care (S. Yamane 1977; Matsuura 1983b).

Fig. 12.2. Officer of Otaru Public Health Center testing the efficiency of '*Vespa*-suits' against *Vespa simillima*

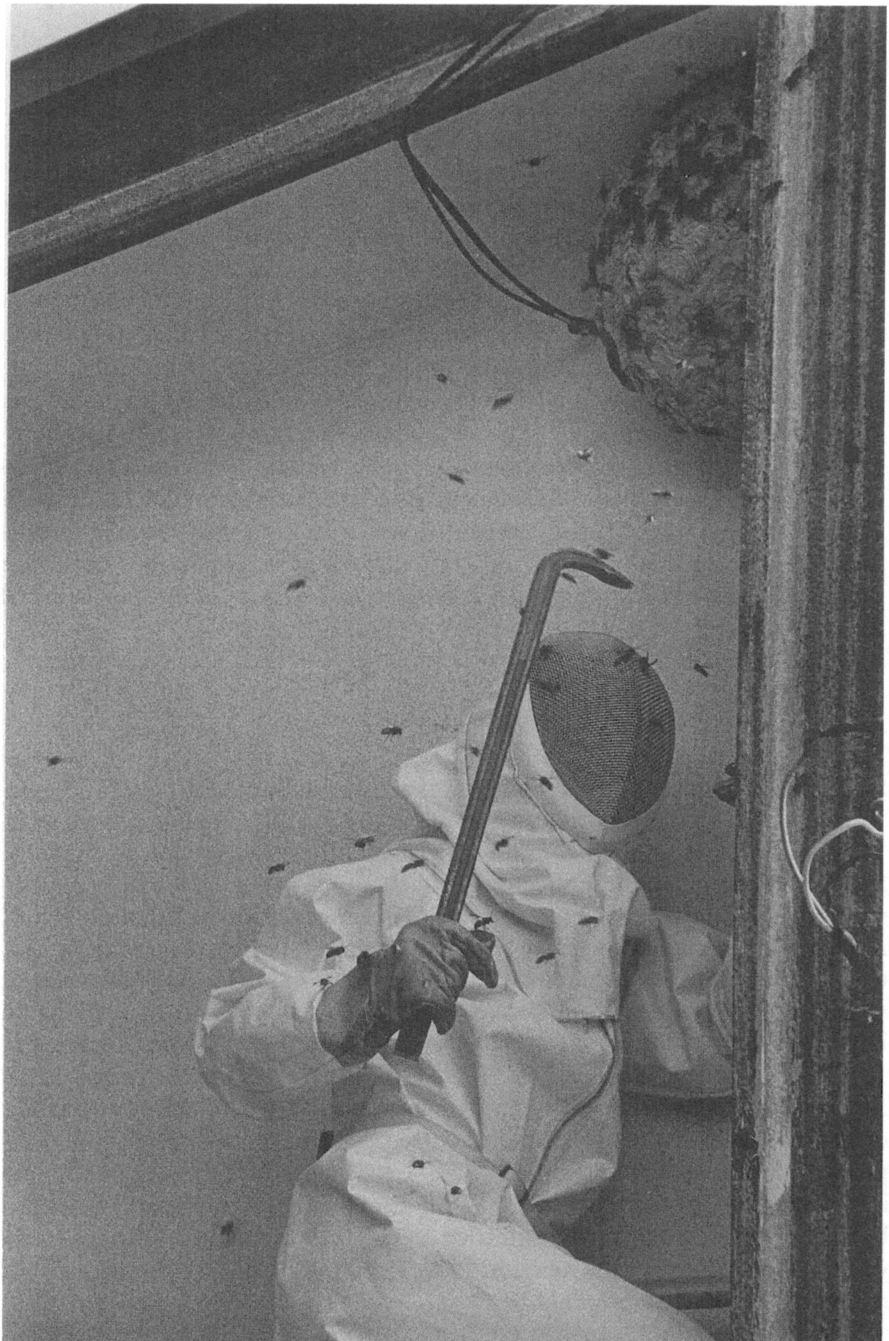

5.2 Control by Insecticide Baits

This control measure, not often employed in Japan, uses baits and insecticides, and has been implemented to exterminate colonies of scavenging yellowjackets of the subgenus *Paravespula*. Meat, fish and cat food are generally used as baits to attract yellowjackets. These baits, incorporated with slow-acting insecticides (diazinon, etc.), are set in places frequently visited by wasps. Returned foragers that fed on the poisonous food may distribute it to nestmates or feed it to larvae. Through this process the insecticide spreads to other colony members, which are finally exterminated. Quick-knockdown insecticides that kill wasps before returning to the nest are not suitable for this purpose. Although this method has been successful to some extent in controlling yellowjackets in parks, etc., there are several weak points (Akre and MacDonald 1986): (1) differences in foraging behavior among species render some baits unattractive (this method is not applicable to *Dolichovespula, Vespa* and *Vespula* s. str., which are hunters of live arthropods); (2) reduced attractiveness of the bait following addition of the insecticide; (3) short "field life" (2–3 days) of the bait due to dehydration and spoilage; and (4) lack of bait competitiveness with discarded foods in infested areas. In spite of these defects, the improvement of this method is hoped for, because it can be powerful against *Paravespula*, the most serious pest in North America.

5.3 Attracting Yellowjackets with Synthetic Chemicals

Since Davis et al. (1967, 1968) reported the attraction of queens and workers of *Vespula* species to some synthetic chemicals (heptyl butyrate and others, Table 12.8), several North American entomologists have been involved with this theme. Since this method was quite effective against *Vl. (Paravespula) pensylvanica* in the western United States, a great success was expected in the whole of North America (these chemicals never attract honeybees). Later investigations, however, revealed that these chemicals have extremely weak attraction for all eastern species of *Paravespula*, while attracting species of *Vespula* (s. str.), which are beneficial as predators of insect pests (Grothaus et al. 1973; MacDonald et al. 1973; Reierson and Wagner 1978; Ross 1983). From our limited experience in Japan, the effectiveness of these chemicals against Asian *Vespula* species is questionable. This control measure, currently being much talked about, has proved applicable only to a portion of the yellowjackets (chiefly *Vl. pensylvanica*) in the western United States. The addition of synthetic chemicals like heptyl butyrate to poison baits has been tried to enhance the effect of attraction for yellowjackets (Wagner and Reierson 1969; Ennik 1973), but the success rate was not high (Akre and MacDonald 1986).

Chapter 13

The Japanese Species

The phylogenetic relationships of vespine groups are discussed at length in Chapter 10. This chapter will present information that may be of help to specialists other than hymenopterists, technicians in applied fields and non-specialists, in understanding Japanese Vespinae. For this practical purpose, body markings rather than structural characteristics will more often be used in the keys to the genera and species.

1 Identification

1.1 Characteristics of the Vespinae

In order to know the correct names of vespine species, a general knowledge of the morphology of Hymenoptera is indispensable. First, we will list a minimal number of characters which are used to separate vespine wasps from other aculeate Hymenoptera which are represented by stinging wasps and bees, and ants.

If the eyes, when seen in front, are deeply concave in their inner margin, most of these hymenopterans belong to the Vespoidea or Scolioidea (Fig. 13.1B,C). In the latter group, there are more than a few species that have a straight inner margin of the eye. The forewings of Vespoidea, with a few non-Japanese exceptions, fold longitudinally at rest, while in the Scolioidea these wings are never folded. In the Vespoidea, the folding often remains even after the wasps are killed and pinned, and the main longitudinal fold is sometimes observed even in spread forewings. The wasps of the subfamily Trypoxyloninae (Sphecidae) also have the eyes with inner concavities, but they are small and very slender wasps, whose forewings are not folded.

Two families are recognized in the Japanese Vespoidea: Vespidae and Eumenidae. In behavioral aspects, it is very easy to separate them, because vespid wasps (in at least Japanese species) all lead a social life, while eumenids are solitary wasps, at most subsocial. The distinguishing of these two based upon adult morphology is, on the other hand, not so easy. The shape of mandibles is the best character in this case. The vespids have short, broad mandibles, at most overlapping mesally to some extent, while eumenids have long mandibles which cross each other (Fig. 13.1E,D). Furthermore, the tarsal claws of the legs are apically simple in the former, but are bifid in the latter (Fig. 13.1G,F). These conditions can be better observed under a dissecting binocular microscope.

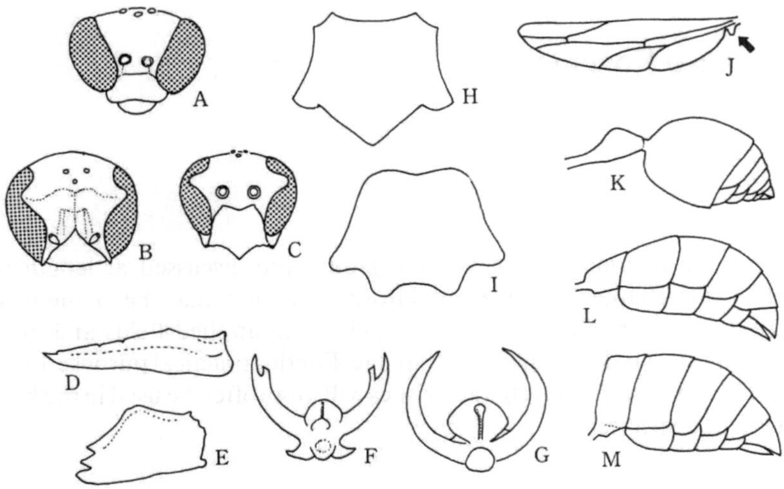

Fig. 13.1. Morphological characters used in vespoid taxonomy. **A-C** Head in front (**A** Sphecidae, **B** Scoliidae, **C** Vespoidea); **D,E** mandible (**D** Eumenidae, **E** Vespidae); **F,G** tarsal claw (**F** Eumenidae, **G** Vespidae); **H,I** clypeus (**H** Polistinae, **I** Vespinae); **J** hind wing in Polistinae (*arrow* indicates anal lobe); **K-M** gaster in profile (**K** *Ropalidia*, **L** *Polistes*, **M** Vespinae)

After determining a given specimen at family level, the following step is to assign it to one of the two subfamilies, Vespinae and Polistinae, found in Japan. They are characterized as follows:

Vespinae. Gastral segment 1 truncate anteriorly (Fig. 13.1M). Clypeus truncate apically; lateral angles more or less produced (these angles are less developed in the male) (Fig. 13.11). Hind wing without anal lobe (Fig. 10.1F). Nests with an envelope.

Polistinae. Gastral segment 1 not truncate, but more or less narrowed anteriorly (Fig. 13.1K,L). Apical margin of clypeus produced medially (sometimes pointed), less in the male (Fig. 13.1H). Hind-wing with an anal lobe (Fig. 13.1J). Nest combs naked (in some tropical genera an envelope is constructed).

1.2 Identification at Genus Level

The Japanese species of Vespinae belong to three genera (*Provespa* is not distributed in Japan). To separate these, more expert knowledge is needed, and utilizing a dissecting binocular microscope will greatly facilitate the work. A more complete list of characters for generic and subgeneric divisions is given in Table 10.1.

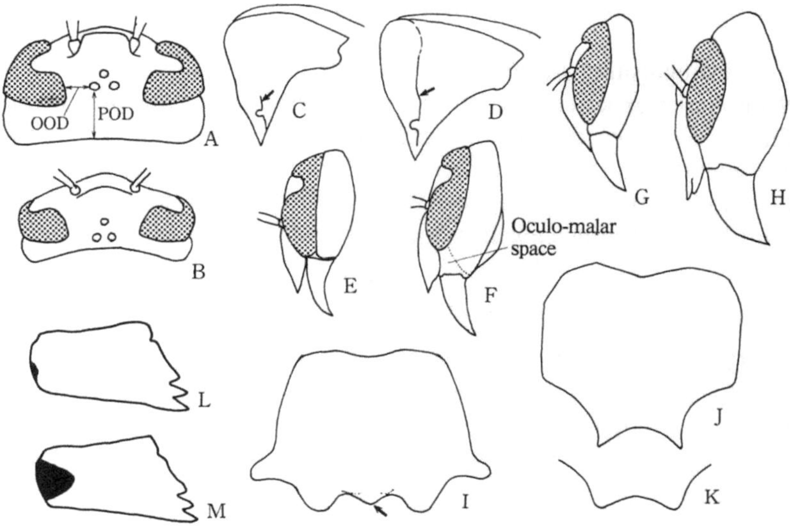

Fig. 13.2. Characters used for separating vespine groups. **A,B** Head from above in *Vespa* (**A**) and *Vespula* and *Dolichovespula* (**B**); **C,D** pronotal carina (*arrow*) in *Vespula* (**C**) and *Dolichovespula* (**D**); **E-H** head in profile in *Vespula* (**E**), *Dolichovespula* (**F**), *Vespa simillima* (**G**), and *V. mandarinia* (**H**); **I-K** female clypeus in *Vespa analis* (**I**; *arrow* indicates median process), *Vespula austriaca* (**J**), and *Vl. schrenckii* (**K**); **L,M** male mandible in *Vl. flaviceps* (**L**) and *Vl. shidai* (**M**)

Key to the Japanese Genera of Vespinae

1. Large forms, usually measuring more than 20 mm in body length. Temple wide, as seen from above, the distance between posterior ocellus to posterior of head longer than the distance between posterior ocellus and eye (Fig. 13.2A). Clypeus usually lacking distinct marking. . . . . *Vespa*

– Small to medium-sized forms, usually not exceeding 20 mm in body length. Temple narrow, as seen from above, the distance between posterior ocellus and posterior of head shorter than the distance between posterior ocellus and eye (Fig. 13.2B). Clypeus often with distinct black marking(s). 2

2. Gena short; lower margin of eye almost touching the mandibular base (Fig. 13.2E). Pronotal carina visible only in its lower portion (Fig. 13.2C). . . . . *Vespula*

– Gena longer; lower margin of eye separated from mandibular base by a wide space (Fig. 13.2F). Pronotal carina more developed (Fig. 13.2D). . . . . *Dolichovespula*

The queen of *Dolichovespula media* is exceptional in body size (often approaching 25 mm) and color pattern; in both respects it greatly resembles a certain species of *Vespa*. Unless the position of posterior ocelli is correctly measured, even professional entomologists may make a mistake.

1.3 Identification of the Japanese Species

For distinguishing between species, still more expert knowledge is needed. Without rich experience, when determining specimens other than *Vespa* it is often difficult to dispense with a dissecting binocular microscope. The following keys are intended to facilitate determination, mainly with color pattern characters. Diagnosis for each species (Sect. 2) includes some additional characters that are useful in separating a given species from the other.

Key to the Japanese Species of *Vespa*

1. Gaster almost wholly blackish brown. .... *V. dybowskii*
– Gaster richly marked with yellow, orange yellow or reddish brown. ... 2
2. Gastral tergites 1 (except for anterior vertical face) and 2 almost wholly orange-yellow. The subsequent tergites wholly blackish brown. Head rufous. .... *V. affinis affinis*
– Tergites 1 and 2 with blackish brown parts, or tergites 3–6(7) with yellowish markings. 3
3. Vertex and gena quite developed; as seen from above the distance between eye and posterior of head approximately two times as long as eye width (slightly shorter in the male). The largest species of the vespines, usually measuring more than 30 mm in body length (sometimes more than 45 mm in the queen). Last tergite of gaster wholly orange yellow. .... *V. mandarinia japonica*
– Vertex and gena moderately developed; as seen from above the distance between eye and posterior of head less than 1.5 times as long as eye width. Body length usually less than 30 mm (in *V. tropica* the female often measures 35 mm). Last tergite of gaster varies in coloration 4
4. Area between the eyes on vertex extensively blackish or with black markings. Pronotum, scutellum and metanotum (postscutellum) often with orange-yellow or rufous markings. 5
– Area between the eyes on vertex wholly yellowish or with ill-developed brownish markings around ocelli. Yellowish or rufous markings on pronotum, scutellum and metanotum usually much reduced or lost. Tibiae rufous to blackish-brown in color. 7
5. Pronotum with rufous markings. Scutellum without marking. Anterior vertical face of tergite 1 rufous; posterior horizontal part with a narrow yellow band at apex. Tibiae rufous. .... *V. crabro flavofasciata*
– Pronotum with yellow or orange-yellow markings. Scutellum often with a yellow marking. Anterior vertical face of tergite 1 wholly orange-yellow or largely blackish-brown (rarely with rufous marking); posterior horizontal part with a wider apical band. Tibiae yellow or yellowish-brown. .... *V. simillima* 6
6. Pronotum, scutellum and metanotum usually extensively orange-yellow. Propodeum (abdominal segment 1) with large yellow markings. Orange-yellow markings on gaster well developed; often gaster almost wholly

orange-yellow. (Honshu – Yaku-shima) .... *V. simillima xanthoptera*

– Orange-yellow markings on pronotum, scutellum and metanotum often reduced or lost. Propodeum wholly blackish-brown. At least basal three tergites of gaster widely blackish-brown at base. (Hokkaido) .... *V. simillima simillima*

7. Last gastral tergite usually yellow or orange-yellow; the preceding tergite with a yellow apical band. Clypeus with a median process (denticle) on the apical margin in the queen and worker (this condition is peculiar to *V. analis*). .... *V. analis insularis*

– Last two segments of gaster usually wholly black (only in the population of Tsushima this part is marked with yellow). Clypeus without median process. Body slender. .... *V. tropica* 8

8. Last two segments of gaster marked with yellow. (Tsushima) .... *V. tropica esakii*

– Last two segments of gaster wholly black. 9

9. Tergites 1 and 2 without blackish-brown part, basally yellowish-brown, with a pale yellow apical band. Pronotum, scutellum and metanotum marked with yellowish brown. (Yaeyama Is). .... *V. tropica loochooensis*

– Tergites 1 and 2 with blackish-brown part at base. Markings on pronotum, scutellum and metanotum tending to be reduced, and rufous in color. (Honshu – Osumi Is.) .... *V. tropica pulchra*

Key to the Japanese Species of *Vespula*

1. Body markings yellow or orange-yellow 2

– Body markings ivory-white or white 3

2. Metanotum (postscutellum) without yellow marking. Yellow markings on scutellum reduced to two widely isolated spots. Clypeus yellow, in the female with 1–3 black spots. Lateral angles of apical margin of clypeus pointed in the female (Fig. 13.2J). Social parasite; lacking worker caste. .. *Vl. austriaca*

– Metanotum usually marked in yellow. Scutellum with a transverse marking that is often interrupted medially. Clypeus always with an anchor-shaped black marking in the queen and worker (in the male this is more or less reduced). Lateral angles of clypeus blunt (Fig. 13.2K). With a worker caste. .... *Vl. vulgaris*

3. Posterior horizontal part of tergite 1 with a white basal band that is interrupted medially. Sinus of eye with a white line in its lower half (Fig. 13.3D). .... *Vl. schrenckii*

– Posterior horizontal part of tergite 1 rarely with a white basal band. Sinus of eye with a white marking extending into its upper half (Fig. 13.3A,B). . 4

4. White area of sinus of eye slightly convex, straight or at most slightly concave on mesal margin (Fig. 13.3A). In the queen and worker, black bar on clypeus usually not reaching the apex of clypeus (Fig. 13.4C). Male mandible without basal black marking (Fig. 13.2L). .... *Vl. flaviceps lewisii*

– White area of sinus of eye with deeply concave mesal margin (Fig. 13.3B). In the queen and worker, black bar on clypeus more developed, usually

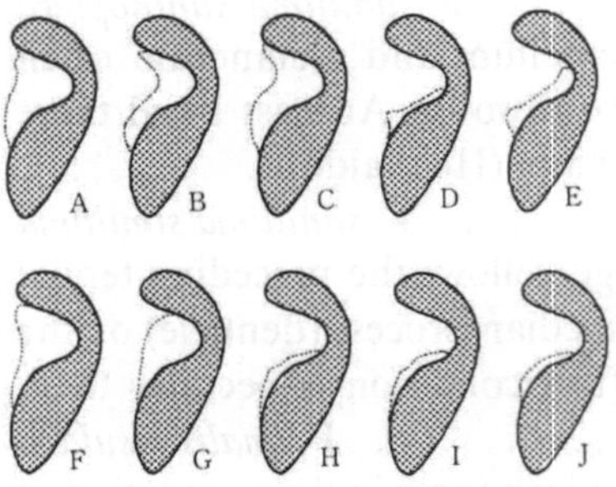

Fig. 13.3. Inner-orbital white (yellow) marking in the Japanese *Vespula* and *Dolichovespula*. **A** *Vl. flaviceps lewisii;* **B** *Vl. shidai;* **C** *Vl. vulgaris;* **D** *Vl. schrenckii;* **E** *Vl. austriaca;* **F** *D. media* (queen); **G** same species (worker); **H** *D. norvegicoides pacifica;* **I** *D. saxonica nipponica;* **J** *D. adulterina montivaga*

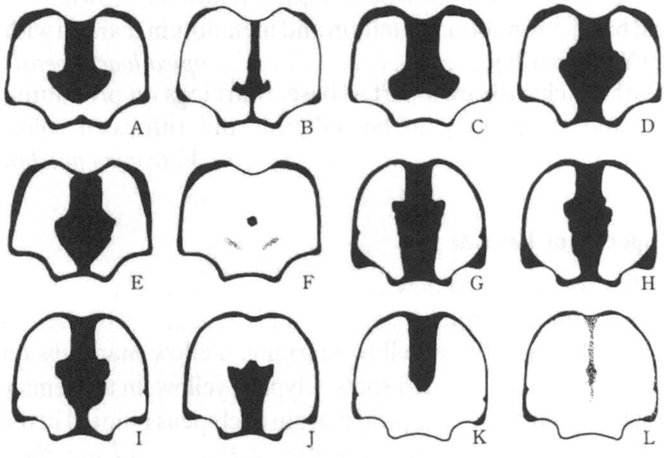

Fig. 13.4. Clypeal markings in the Japanese *Vespula* and *Dolichovespula* (queen and worker). **A,B,** *Vl. vulgaris;* **C** *Vl. flaviceps lewisii;* **D** *Vl. shidai;* **E** *Vl. schrenckii;* **F** *Vl. austriaca;* **G,H** *D. norvegicoides pacifica;* **I** *D. saxonica nipponica;* **J** *D. adulterina montivaga;* **K,L** *D. media.* Note that every species has a certain amount of intraspecific variation

reaching the apex of clypeus (Fig. 13.4D). Male mandible basally with a black triangular marking (Fig. 13.2M). . . . . *Vl. shidai*

Key to the Japanese Species of *Dolichovespula*

1. Body markings yellow or orange-yellow. Sinus of eye almost wholly yellow, especially so in the queen (Fig. 13.3F,G). Clypeal bar ill-developed, not reaching the apex of clypeus (Fig. 13.4K,L). Large species; the queen is often misconceived as *Vespa*. . . . . *D. media . . .2*

– Body markings ivory-white or white (rarely pale yellow in the male). Sinus of eye with a whitish line in its lower half (Fig. 13.3H-J). In the queen and worker, black clypeal bar widened at half-length from the base, usually reaching the apex of clypeus (Fig. 13.4G-J). 3

2. Yellow markings abundant. In the queen, mesoscutum with a V-shaped orange or brownish marking; metapleuron usually with two markings; scutellum almost wholly orange-yellow. Gaster extensively marked with orange-yellow. In the worker and male also, orange yellow markings rather developed, but less than in the queen. (Honshu) .... *D. media sugare*

– Yellowish markings less developed. In the queen, mesoscutum and metapleuron wholly black; at least one-third of scutellum black; apical bands on tergites yellow and generally narrow. In the worker and male, body markings pale yellow, less developed than in subsp, *sugare*. (Hokkaido) .... *D. media media*

3. Lateral angles on the apical margin of clypeus pointed in the female. Black clypeal bar not reaching the base of clypeus (Fig. 13.4J). White marking on the male metanotum reduced into two spots. Hairs on lateral side of thorax black in the male. Social parasite; lacking worker caste. .... *D. adulterina montivaga*

– Lateral angles of clypeus blunt in the queen and worker. Black clypeal bar reaching the base of clypeus. White marking on male metanotum interrupted medially, but each half not reduced into a round spot (rectangle or wedge-shaped). Hairs on lateral side of thorax largely grayish white.......... 4

4. (Applicable to the queen and worker alone) Oculo-malar space very long. Distance between posterior ocelli longer than the distance between posterior ocellus and posterior of head. Black clypeal bar without a distinct constriction just above the apical margin of clypeus (Fig. 13.4G,H). Metanotum often lacking white marking. Males are quite difficult to identify. .... *D. norvegicoides pacifica*

– (Applicable to the queen and worker alone) Oculo-malar space shorter. Distance between the posterior ocelli as long as or shorter than the distance between posterior ocellus and posterior of head. Black clypeal bar with a distinct constriction above the apical margin of clypeus (Fig. 13.4I). Metanotum usually with white marking. .... *D. saxonica nipponica*

2 The Japanese Vespinae

In total, 16 species of Vespinae, seven of *Vespa*, five of *Vespula* and four of *Dolichovespula*, occur in Japan, including the Ryukyu Islands (Nansei-shotô). Although new species are not expected to be found, information about new geographical races (subspecies) and new distribution areas of the known forms will emerge, especially from the Ryûkyûs. In this Section, distribution, diagnostic characters and accounts for general biology are given for all the Japanese species, mainly based upon our observations (also see Ishikawa 1965; S. Yamane 1970). Japanese names will be given in parentheses.

2.1 Vespa mandarinia Smith (Oo-suzumebachi)

The subspecies *japonica* Radoszkowski occurs in Hokkaido (mainly central and western parts), Honshu, Sadoga-shima, Shikoku, Kyushu, Tsushima, Tanega-shima and Yaku-shima.

This is the largest species in this genus, measuring in body length 40–45 mm in ♀, 27–38 mm in ☿, 27–40 mm in ♂. The head is very large, with enormously developed gena which is, in profile, two times wider than the eye. The head has no black markings, being almost wholly orange-yellow. Gastral tergite 1 has an orange-yellow band apically and also at the base. The last tergite is entirely yellow.

V. mandarinia nests underground or in tree hollows. Their nests have a very thin envelope sheet without scallops, not completely covering the lower combs. Mature nests in southwest Japan consist of 4–10 combs and 3000–5000 cells. *V. mandarinia* has the most dominant status among the Japanese Vespinae over feeding sites such as sap-producing tree trunks. Mating behavior is very peculiar in that the males wait at the entrance of other nests for new queens leaving for hibernation sites and copulate with them. Workers are quite aggressive toward humans, their stings causing a sharp pain. *V. mandarinia* preys on sluggish insects including adult scarab beetles (most important food of larvae) and caterpillars. Attacks on colonies of other social wasps and bees are one of the most remarkable behavior traits of this species. After the workers of victimized nests are killed, larvae, pupae and honey are carried away. This species is the most serious enemy of both cultivated and wild honeybees in Japan.

2.2 Vespa tropica (Linnaeus) (Hime-suzumebachi)

At least three subspecies are recognized in Japan. The subspecies *pulchra* Buysson is known from Honshu, Sadoga-shima, Shikoku, Kyushu, Yaku-shima, Kuchinoerabu-jima, Nakano-shima of the Tokara Islands, subspecies *esakii* Sonan from Tsushima, and subspecies *loochooensis* Bequaert from Yaeyama Islands (southern Ryukyus). An unnamed race occurs on Amami-ôshima, Tokuno-shima and Okinawa-jima (central Ryukyus).

This is the second largest species of *Vespa* in Japan, measuring 24–37 mm in length. In body length, queens and workers nearly completely overlap, except for small first brood workers that are easily separable from the queens. Gena is less developed than in the preceding species. This species is relatively easily distinguished from other congeners by its slender body and pointed lateral teeth of anterior margin of clypeus. The apical two tergites (T5, 6) are entirely black except in the subspecies *esakii*; this state is seen among other congeners only in *V. affinis* in which also tergites 3 and 4 are black.

Overwintered queens appear from late May, more than one month later than do the queens of other species. *V. tropica* nests in closed places such as underground cavities, tree hollows and attics. The thin envelope without distinct scallops does not cover the lower nest combs. The colony size of this species is the smallest among the Japanese species of *Vespa*. Mature nests have only 200–300 cells, and 10–50 new queens are produced in late summer. Workers are generally not aggressive toward humans, even when nests are disturbed or excavated. *V. tropica* queens and workers attack polistine nests, pull larvae and pupae out of cells, suck up their body fluid, and regurgitate it in their own nests to feed larvae.

2.3 Vespa analis Fabricius (Kogata-suzumebachi)

The Japanese subspecies *insularis* Dalla Torre is distributed evenly in Hokkaido, Honshu, Sadoga-shima, Shikoku, Kyushu, Tsushima, Tanega-shima and Yaku-shima, but is rare or absent in the mountains. There are two to three unnamed subspecies in the Ryûkyû Islands.

V. analis is smaller than the preceding two species, measuring in body length 25–29 mm in ♀, 22–27 mm in ☿, 23–26 mm in ♂. Gena is moderately developed and oculo-malar space is short. This species differs from all the other congeners in having a small denticle medially on the anterior margin of clypeus between the lateral teeth. In color pattern it resembles *V. mandarinia*, with which it is sympatric, though in the specimens from northern Japan the head often has a dark marking in the ocellar area.

Nests are generally constructed on tree twigs and rarely under the eaves of houses or on electric wires. Completed embryo (queen) nests have an envelope with a long entrance tube produced below, which is always removed after worker emergence by the first workers. Nest combs are completely covered by a thick envelope, with distinct scallops throughout the post-emergence period. Mature nests have two to four combs and 150–450 cells in Hokkaido, and two to four combs and 300–800 cells in southwestern parts of Honshu. Workers are relatively non-aggressive toward humans approaching the nest. *V. analis* preys on winged insects quick in action, such as flies.

2.4 Vespa affinis (Linnaeus) (Tsumaguro-suzumebachi)

In Japan only the nominate subspecies has been recorded from the subtropical areas of the Ryukyus: Okinawa-jima (?), Miyako-jima, Ishigaki-jima, Iriomote-jima, Yonaguni-jima, etc.

V. affinis is a medium-sized species, measuring in body length 24–28 mm in ♀ and around 20 mm in ☿ and ♂. It is distinguished from the other congeners by the distinct punctation on clypeus, genae, temple, scutellum, metanotum and propodeum (♀, ☿). It is also unique in its color pattern: ground color is reddish-brown, with gastral tergites 1 and 2 yellow and 3–6 entirely black.

Little information is available for the biology of this species in Japan. Generally it constructs its nests on twigs of small trees. A thick scalloped envelope completely covers nest combs, as in *V. analis*. According to observations made on Ishigaki-jima, mature nests have four to five combs and 800–1500 cells. Workers are not so aggressive when disturbed by humans.

2.5 *Vespa crabro Linnaeus (Mon-suzumebachi)*

In Japan the range of this species (subsp. *flavofasciata* Cameron) overlaps with that of *V. simillima*, but the former has so far been known only from the four mainlands (Hokkaido, Honshu, Shikoku and Kyushu). Through most of its range *V. crabro* seems to be less dominant than *V. simillima*.

Body length is 28–30 mm in ♀, 21–28 mm in ☿ and 25–27 mm in ♂. In *V. crabro* and the following two species, the ocellar area is extensively blackish. This species is distinguished from the most closely related *V. simillima* in the strongly punctated clypeus. The entirely black scutellum and basally reddish-brown gastral tergite 1 characterize this species in color pattern.

Nests are generally founded in closed spaces including hollow trees, building wall voids and attics. When nesting spaces become filled up, whole colonies often move to open sites, mostly under the eaves (nest relocation). The nest envelope is not typically scalloped but composed of elongated shells or tunnels, and does not cover the lower nest combs. Mature nests have from three to 12 combs with a total of 500–3000 cells (rarely reaching 4500). Fifty to 200 new queens are produced per nest. Although *V. crabro* is principally diurnal, foraging activities continue after sunset; this species is the only hornet in Japan with nocturnal habits. Workers are aggressive in instances of human interference, and colonies have the ability to defend themselves against mass attacks by *V. mandarinia*. *V. crabro* preys on various insects, especially cicadas.

2.6 *Vespa simillima Smith (Kebuka-suzumebachi/Kiiro-suzumebachi)*

In Japan two subspecies are distributed, *simillima* (Hokkaido) and *xanthoptera* Cameron (Honshu, Sadoga-shima, Shikoku, Kyushu, Tsushima, Yaku-shima). In rural areas this species is often called "kame-bachi" (pot wasp) or "aka-bachi" (red wasp).

Slightly smaller than the preceding species, body length is 25–28 mm in ♀, 18–24 mm in ☿ and around 27 mm in ♂. Punctation on clypeus is weak. Scutellum usually has yellow spots. Gastral tergite 1 is only rarely reddish-brown in basal part. In the nominate subspecies orange-yellow maculation is poorly developed, while in *xanthoptera* going from north to south the orange-yellow body markings become more extensive; sometimes the gaster is wholly orange-yellow. In mountainous regions of central Honshu and Tohoku district, however, individuals not clearly separable from the nominate race are not rare.

V. simillima is the most dominant hornet in Japan. It nests in both open (on twigs, on precipices and under eaves) and closed sites (in hollow trees, between building walls and in underground cavities), showing the most expedient nesting habits. When the nests founded in closed spaces become too large, the colonies moves to new locations, as with the aforementioned species. The nest envelope is typically scalloped and thick, completely covering the combs. This species has the largest nesting scale among the Japanese hornets. Mature nests comprise from four to ten combs with a total of 500–2500 cells (rarely reaching 4000), even in a cold temperate region (Hokkaido). In southwestern parts of Honshu, nesting activities last until mid-November; mature nests have from seven to ten combs and 4000–7000 cells (sometimes more than 10 000). Three hundred to 1700 new queens are produced. Workers are quite aggressive, and it is difficult for a man to approach within a 10 m radius of nests at the peak of their activity. *V. simillima* is the representative nuisance pest among the Japanese vespines. It preys on various insects (mainly flies) and spiders, and occasionally visits colonies of *Apis* and *Vespula* to capture workers around the nest entrance.

2.7 Vespa dybowskii André (Chairo-suzumebachi)

In Japan this species (the nominate subspecies) is restricted to certain localities, and generally scarce in number. It is chiefly collected in the mountainous regions of Nagano and Niigata Prefecture. It is also found in Tohoku district, mainly along the Japan Sea. In Hokkaido only one worker from Jôzankei, Sapporo has been found, and constant occurrence of this species in Hokkaido has been doubted.

Body length is about 29 mm in ♀, 18–24 mm in ☿ and around 26 mm in ♂. *V. dybowskii* is easily distinguished from Japanese congeners by its peculiar coloration: head and thorax largely reddish-brown and gaster almost wholly blackish-brown, as in *V. basalis*, which inhabits subtropical and tropical Asia.

V. dybowskii is a facultative, temporary social parasite of *V. crabro* and *V. simillima*. Queens often invade host's nests in the postemergence stage, kill host queens and force host workers to rear their own broods. For some time after the emergence of parasite's workers both the host and parasite workers coexist, but gradually the former are replaced by the latter to the point at which colonies comprise *V. dybowskii* alone. When matured, *V. dybowskii* nests are generally smaller in size than those of host species. Workers respond aggressively to the disturbance of nests by humans. Little is known of ecology and behavior of this rare species.

2.8 Vespula vulgaris (Linnaeus) (Kiobi-kuro-suzumebachi)

This species shows little color pattern variation over its range. In coloration, the Japanese population is not clearly distinguishable from those inhabiting Europe and North America. It occurs almost throughout Hokkaido and in mountainous regions of Honshu, and its range can almost be seen to coincide with coniferous forests. This and two subsequent species belong to the subgenus *Paravespula*.

The species of the genus *Vespula* have quite short oculo-malar spaces, the lower margin of eyes almost touching mandibles. *Vl. vulgaris* measures in body length 17–18 mm in ♀, 10–14 mm in ☿ and 15–18 mm in ♂. Among the Japanese *Vespula* only this species and *Vl. austriaca* possess yellow body markings. In workers there is an arrowhead-shaped black marking on the basal part of gastral tergite 1.

Biology of *Vl. vulgaris* has been studied intensively in Europe, while only a little is known of the Japanese population. A few examples of observation indicate that this species builds its nests in underground cavities and tree hollows. The nest envelope consists of numerous scallops and completely covers the combs, which are brownish in color, as in the following two species. Mature nests have 5–9 combs and as many as 8500 cells. The upper three to four combs are used to rear workers, and the lower ones are for new queens (males may be reared in any comb). Not very aggressive.

2.9 Vespula flaviceps (Smith) (Kuro-suzumebachi)

Unlike the nominate race of continental Asia (with yellow markings), the Japanese race *lewisii* (Camron) possesses white body markings. In Japan *Vl. flaviceps* commonly occurs in the lowlands and lower altitudes of mountains in Hokkaido, Honshu, Sadoga-shima, Shikoku, Kyushu, Tsushima, Yaku-shima and Tanega-shima.

Body length is 15–16 mm in ♀, 10–12 mm in ☿ and 12–14 mm in ♂. Gastral tergites have white apical bands which are narrow and not sinuated. Black bar on clypeus rarely reaches the lower margin of clypeus. White marking situated in ocular sinus is medially not concave.

Vl. flaviceps lewisii nests in closed spaces such as underground cavities, building wall voids, etc., and rarely under eaves. It has the largest nesting scale among the Japanese species of *Vespula*. Mature nests have from five to seven combs and 2500–6000 cells in Hokkaido, and from eight to 12 combs and 8000–12 000 cells in southwestern parts of Honshu. Colonies last long, often into early winter in southwest Japan. Males swarm on the crowns of trees, wait for new queens that fly past, and may copulate with them. Workers show weak aggressiveness; people will be stung only when they approach the nest too close. *Vl. flaviceps* preys on various types of insects and spiders, and in addition to these living arthropods it collects flesh from corpses of frogs, snakes and mammals. Adult wasps have a preference for flower nectar and honeydew secreted by aphids and scale insects. In Nagano and some other prefectures people excavate colonies of this wasp (and *Vl. shidai*) to get larvae and pupae, which are important food for them. Canned wasp larvae and pupae are shipped to other prefectures in Japan.

2.10 Vespula shidai Ishikawa et al. (Shida-kuro-suzumebachi)

This species has a more northerly distribution than the preceding species, but in Japan it extends as far south as Amami-ôshima (the Ryukyus), where the latter is not found. The known range covers Hokkaido, Honshu, Sadoga-shima, Shikoku, Kyushu, Yaku-shima and Amami-ôshima. Except in Hokkaido and Amami-ôshima, it occurs chiefly in mountain regions.

This species, slightly larger than *Vl. flaviceps*, measures in body length 15–19 mm in ♀, 11–14 mm in ☿ and 14–16 mm in ♂. Both species are very alike in structure and color pattern, but in this species the black clypeal bar usually reaches the lower margin of clypeus (♀, ☿) and white marking in ocular sinus is medially strongly concave. Males have a black triangular marking at the base of mandible (absent in *Vl. flaviceps*).

Nesting behavior does not much differ from that of *Vl. flaviceps*. Although in Hokkaido this species has a colony size larger than in *Vl. flaviceps*, in the plains of Honshu the reverse is often the case. Workers are more aggressive than in *Vl. flaviceps*. Food preference is as in *Vl. flaviceps*.

2.11 Vespula schrenckii (Radoszkowskii) (Tsuya-kuro-suzumebachi)

The Japanese form has often been treated as a geographical race of *Vl. rufa* Linnaeus, though in this book we regard it as a separate species. It prefers colder districts than do the two preceding species, and occurs in Hokkaido, Honshu and Shikoku. Except in Hokkaido, it is confined to mountainous regions, and is often collected in alpine zones. This species and *Vl. austriaca* belong to the subgenus *Vespula*.

Body length is 16–17 mm in ♀, 12–14 mm in ☿ and 13–17 mm in ♂. *Vl. schrenckii* resembles in coloration the preceding two species, but differs from the latter in that the white area in the oculo-malar space is quite narrow and absent in the upper portion, and that the gastral tergite 1 has a white basal band which is interrupted medially. In the queen and worker, white genal band is much reduced into a line situated in the upper part of gena behind the eye.

A little information is available concerning the nesting biology of this species. Nests are found underground and between building walls, and never in open sites. Unlike brownish nests of the three *Paravespula* species, the nests of *Vl. schrenckii* are wholly grayish, with an envelope comprising many paper sheets. The envelope is usually not scalloped, but its upper portion sometimes presents a scallop-like structure. In Nagano Prefecture, mature nests have in most cases three combs and 500–1000 cells, being much smaller than in *Paravespula*. Workers are mainly produced in the first comb and the new queens in the lower ones. Aggressiveness varies among nests, and is generally weak. *Vl. schrenckii* preys upon larval Lepidoptera, etc. Adult wasps prefer flower nectar and honeydew.

2.12 Vespula austriaca (Panzer) (Yadori-suzumebachi)

Like *Vl. schrenckii* this is also a northern species, distributed in the Holarctic region. In Japan it is restricted to Hokkaido and Honshu (mainly mountainous regions). The color pattern is fairly stable throughout its range, so that no subspecific division has been attempted.

Body length is 16–18 mm in ♀ and 11–16 mm in ♂ (worker caste is lacking). This yellow-marked species apparently resembles *Vl. vulgaris*, but in the female it is easily distinguished from the latter by the pointed lateral teeth on the anterior margin of clypeus and the clypeal black bar reduced to from one to three small spots. The sting is strongly curved. Male clypeus is almost wholly yellow.

Vl. austriaca is a labor parasite of *Vl. schrenckii*, to which it is closely related. Overwintered queens appear one month later than do other congeners. Queens are thought to invade hosts' nests that have already produced first workers, and usurp them killing host queens. Usurpation may also occur among queens of *Vl. austriaca*.

2.13 Dolichovespula norvegicoides (Sladen) (Shiroobi-hoonaga-suzumebachi)

The eastern Asiatic population has been treated as a separate species, *D. pacifica* (Birula). However, since we could not find important structural differences between it and the Nearctic *D. norvegicoides*, it is considered here to be a subspecies of the latter. In Japan *D. norvegicoides pacifica* is known from Hokkaido, Honshu and Shikoku, chiefly at altitudes more than 500 m above sea level and up to the alpine zones.

The species of *Dolichovespula* have a long oculo-malar space, which is longest in *D. norvegicoides*. Body length is 16–18 mm in ♀, 11–14 mm in ☿ and 13–17 mm in ♂. Body markings are ivory white or white; gastral tergites each have an apical band narrow and weakly sinuate. In ♀ and ☿ the black clypeal bar is well developed.

Little is known of the behavior of this species in Japan. Nests are constructed on tree twigs, in above-ground concealed spaces and in underground cavities. Nests are almost uniformly gray, with an envelope composed of many pliable sheets covering all the nest combs. Suspensoria between combs are ribbon-like (in *Vespula* these are pillars). No data are available on the nesting scale or productivity.

2.14 Dolichovespula saxonica (Fabricius) (Nippon-hoonaga-suzumebachi)

The Japanese subspecies *nipponica* Sk. Yamane occurs in Hokkaido, and only one worker specimen was collected from a mountainous area in Tohoku district (northern Honshu). In Hokkaido it is commonly found in both plains and mountains up to 1000 m in altitude, but disappears at altitudes of more than 1000 m, where the preceding species dominates. The two species coexist at between 500 and 1000 m.

D. saxonica nipponica has long been confused with *D. norvegicoides pacifica*, which is very similar to the former in size, structure and color pattern. It is apparently more richly marked with white than in *D. n. pacifica*, but intermediate specimens are not rare occurrences (see Key, p. 278–279, for distinction).

This wasp constructs its nests in both open (on tree twigs and under eaves) and closed sites (in hollows of fallen trees and between building walls), showing the most expedient nest site preference among the Japanese species of *Dolichovespula*. Nest architecture is the same as the preceding species. Colonies are started in early or mid-May and last until August. Mature nests consists of three to six combs and from several 100 to 1500 cells. Workers are always reared in the uppermost (first) comb, and new queens in the second and consecutive combs. Males can be produced in any of the combs. Aggressiveness toward humans disturbing nests is not so intense. *D. saxonica* preys on insects quick in action such as Diptera. Adult wasps frequently visit flowers to take nectar.

2.15 Dolichovespula adulterina (Buysson) (Yadori-hoonaga-suzumebachi)

The Japanese race *montivaga* Sk. Yamane occurs in mountainous regions of Hokkaido and Honshu. It is generally rare (especially the female), but in alpine zones males are relatively common.

D. adulterina lacks worker caste. In body size, queens and males are the same as the preceding species. Body markings are ivory-white, though males with yellowish markings are sometimes collected in Honshu. In color pattern, this species resembles the preceding two species, but differs from the latter by the poorly developed black clypeal marking in ♀ (usually absent in upper portion), much reduced scutellar spots in ♂ and long black hairs on lateral sides of thorax (♂). Structurally it parallels *Vl. austriaca*: in ♀ lateral teeth on anterior margin of clypeus are sharply pointed and sting is strongly curved.

D. adulterina is a labor parasite of *D. saxonica* in Japan. Its distribution range, however, suggests that *D. norvegicoides* is a principal host.

2.16 Dolichovespula media (Retzius) (Kiobi-hoonaga-suzumebachi)

Two subspecies, *D. media media* (Hokkaido) and *D. m. sugare* Ishikawa (mountainous regions of central and northern Honshu) are known from Japan.

D. media is the largest species of the Japanese *Dolichovespula*, measuring in body length 19–22 mm in ♀, 14–16 mm in ☿ and ca. 19 mm in ♂. This is the only *Dolichovespula* species in Japan with yellow markings. Queens are very similar in color pattern to *Vespa simillima* and *V. crabro*, from which they are easily distinguished by the smaller oculo-occipital distance and clypeus with distinct black markings. In ☿ and ♂ yellow body markings are paler than in ♀.

D. media is a common species in the plains of Hokkaido. It shows an inflexible nest site preference comparable to that by *V. analis*. It usually nests on tree twigs (rarely under eaves), and no subterranean nest has been found. Nest architecture does not differ from that of *D. saxonica*. Mature nests have 2–4 combs, and 300–2000 (in Hokkaido) or 250–1700 cells (in mountainous regions of Nagano Prefecture). Workers are rather aggressive toward humans. *D. media* preys upon insects such as flies. Adult wasps feed on flower nectar and honeydew.

3 Distribution in Japan

The subfamily Vespinae contains large and conspicuous species, but the range for each species has not been thoroughly studied. In this section the distributions of the Japanese forms are briefly discussed.

Japan comprises Hokkaido, Japan proper (Honshu, Shikoku, Kyushu), the Ryûkyû islands (Nansei-shotô), and some other island groups, covering approximately 2400 km from Hokkaido in the north to the Yaeyama islands in the south. Somewhere in the Tokara strait (northern Ryukyus) lies a famous zoogeographical border (Watase's line) by which the Palearctic region is bordered upon the Oriental region. The four Japanese mainlands (Hokkaido and Japan proper) are thought to have connected with the Asiatic continent twice in the ice age. Probably reflecting faunal exchange during these periods, the present-day vespine fauna of Japan closely resembles those of northern China and Siberia. All of the Japanese species of Vespinae also occur in the continent, while Japan lacks the following four species inhabiting the continent: *Vespula (Rugovespula) koreensis, Vl. (Paravespula) germanica, Dolichovespula (Boreovespula) norwegica* and *D. (Metavespula) sylvestris.* It is not clear whether these species became extinct in Japan during the post-glacial period or whether they never entered Japan even in the age of faunal exchange. *D. norwegica* extends to Sakhalin but is not found in Hokkaido. Two species *Vl. germanica* and *D. sylvestris,* prefer drier environments, and may not be adaptable to Japan (except Hokkaido) where wet climate predominates.

Compared with Europe, Japan is rich in hornet fauna (seven vs. two species). Yellowjacket faunas of these two regions are similar in species number but different in species composition. For example, *Vl. vulgaris*, the most common yellowjacket in Europe, is confined to mountainous regions and anywhere not common in Japan. In the lowlands it is replaced by *Vl. flaviceps*, which is peculiar to eastern Asia. The ecological niche of *D. norwegica* seems to be occupied by *D. norvegicoides* in Japan.

Entomofauna of Tsushima is transitional between the Kyushu mainland and Korea (Shirôzu 1980), though in vespid fauna Tsushima has more Kyushu-elements than Korean ones. One remarkable case of subspeciation on Tsushima is of *Vespa tropica* (Ishikawa 1970); the other forms of Vespinae are regarded as belonging to the races occurring in the Kyushu mainland.

The vespine fauna of the Ryûkyû islands has a mixture of the elements of both Japan proper and Taiwan. No *Dolichovespula* species are found there. Islands situated in the northern Ryukyus and the northern part of the central Ryukyus are characterized by the elements of Japan proper. The southern limits of the ranges of *Vespa simillima* and *Vespula flaviceps lewisii* lie in Yakushima, and that of *Vl. shidai* in Amami-ôshima. No *Vespula* is found on any of the islands south of the Amami group, while in Taiwan a peculiar subspecies of *Vl. flaviceps* occurs in the mountains. The southern Ryukyus (Miyako and Yaeyama islands) have received some of the Taiwanese elements. *Vespa affinis*, widespread over the tropical and subtropical areas of Asia and South Pacific, has probably extended through Taiwan into the Ryukyus up to the Miyako islands. The populations of

V. analis and *V. tropica* in the southern Ryukyus (and perhaps in Okinawa-jima) also seem to have no close relationship with those of Japan proper.

Wasps constructing large nests, such as hornets, may not inhabit very small islands. Little information is available relating to area-species relation in the Japanese archipelago. Two species of *Vespa* are recorded from Okushiri-tô (148 km^2), the largest island associated with Hokkaido (Munakata and S. Yamane 1970), four from Sadoga-shima (857 km^2), the largest island associated with Honshu (Sk. Yamane 1982a), and four from Tsushima, large islands close to Kyushu (Shirôzu and Miyata 1976). *Vespa crabro* has so far not been recorded from any island other than the four main islands (Hokkaido and Japan proper). This suggests the occurrence of competition with the more dominant *V. simillima,* though there is no direct evidence.

The islands of the Izu-shotô (the largest is Hachijô-jima: 70 km^2) are poor in vespid fauna. Two *Polistes* species are relatively common on Kôzu-jima (18.6 km^2) and some other islets (Umeya 1956), and *Vespa tropica* completely depending on *Polistes* immatures for prey has been recorded. Other vespine species are not known to establish themselves on these small islands.

Table 13.1 lists all the known hornet species in the Ryukyu islands. Although more intensive surveys are needed, from this list it can be said that there is a tendency for species number to increase with the increase in area of islands. The smallest island where records of *Vespa* exist is Kuro-shima of the Yaeyama group. In this case the constant immigration from the neighboring large islands (Iriomote-jima and Ishigaki-jima) may support the Kuro-shima population. Mention should be made of Nakano-shima of the Tokara group, where two hornet species have been recorded, because this is a very small (27.5 km^2) and solitary island, remote from large islands such as Amami-ôshima (142 km from Nakano-shima) and Yaku-shima (66 km from Nakano-shima). From the other islets (less than 20 km^2 in area) of the Tokara group and three islets of the Mi-shima group (Kuro-shima, Iô-jima and Take-shima), there is no reliable record of hornets. Thus, it is generally concluded that the threshold area for the constant settlement of *Vespa* is around 30 km^2, and that two or three species can inhabit islands more than 200 km^2 in area.

Table 13.1. Distribution of *Vespa* in the Ryûkyû Islands

Islands	Area (km)^2	*affinis*	*tropica*	*analis*	*mandarinia*	*simillima*
Okinawa-jima (Okinawa Is)	1182.52	?	+	+		
Amami-ôshima (Amami Is)	718.74		+	+		
Yaku-shima (Ôsumi Is)	500.59		+	+	+	+
Tanega-shima (Ôsumi Is)	447.42			+	+	
Iriomote-jima (Yaeyama Is)	287.66	+	+	+		
Ishigaki-jima (Yaeyama Is)	223.41	+	+	+		
Miyako-jima (Miyako Is)	159.00	+				
Kuchinoerabu-jima (Ôsumi Is)	38.04		+	+		
Yonaguni-jima (Yaeyama Is)	28.38	+				
Nakano-shima (Tokara Is)	27.50		+	+		
Kuro-shima (Yaeyama Is)	10.04	+				
Akuseki-jima (Tokara Is)	7.03					

References

Adachi J (1979) The wonder of the hornet. Published by the author, Kôbe, 222 pp (in Japanese)

Akima M (1979) A natural enemy of *Vespula lewisii*. Insectarium (Tokyo) 16:19 (in Japanese)

Akre RD (1982) Social wasps. In: Hermann HR (ed) Social insects, vol 4. Academic Press, New York London, pp 1–105

Akre RD, Davis HG (1978) Biology and pest status of venomous wasps. Annu Rev Entomol 23:215–238

Akre RD, MacDonald JF (1986) Biology, economic importance and control of yellowjackets. In: Vinson SB (ed) Economic impact and control of social insects. Praeger, New York, pp 353–412

Akre RD, Reed HC (1981) Population cycles of yellowjackets (Hymenoptera: Vespidae) in the Pacific Northwest. Environ Entomol 10:267–274

Akre RD, Reed HC (1983) Evidence for a queen pheromone in *Vespula* (Hymenoptera: Vespidae). Can Entomol 115:371–377

Akre RD, Hill WB, MacDonald JF (1973) Magnetic recovery of ferrous labels in a capture-recapture system for yellowjackets. J Econ Entomol 66:903–905

Akre RD, Reed HC, Landolt PJ (1982) Nesting biology and behavior of the blackjacket, *Vespula consobrina* (Hymenoptera: Vespidae). J Kans Entomol Soc 55:373–405

Akre RD, Roush CF, Landolt PJ (1977) A *Vespula pensylvanica*/ *Vespula vulgaris* nest (Hymenoptera: Vespidae). Environ Entomol 6:525–526

Akre RD, Hill WB, MacDonald JF, Garnett WB (1975) Foraging distance of *Vespula pensylvanica* workers (Hymenoptera: Vespidae). J Kans Entomol Soc 48:12–16

Akre RD, Garnett WB, MacDonald JF, Greene A, Landolt PJ (1976) Behavior and colony development of *Vespula pensylvanica* and *V. atropilosa* (Hymenoptera: Vespidae). J Kans Entomol Soc 49:63–84

Akre Rd, Greene A, MacDonald JF, Landolt PJ, Davis HG (1981) The yellowjackets of America north of Mexico. USDA, Washington DC, 102 pp

Alber MA (1953) Il calabrone, tigre dellária. O. Apicoltore d'Italis 20:188–189, 192

Alexander RD (1974) The evolution of social behavior. Annu Rev Ecol Syst 5:325–383

Aoi T (1979) Diet of the brown bear in northern Hokkaido. Abstr 26th Annu Meet Ecol Soc Jpn, p 252 (in Japanese)

Aoki S (1979) Aphids with aggressive behaviors. Shokubutsu Bôeki 33:157–162, 1 pl (in Japanese)

Aoki S (1982) Soldiers and altruistic dispersal in aphids. In: Breed MD, Michener CD, Evans HE (eds) The biology of social insects. Westview, Boulder, pp 154–158

Archer ME (1972) The importance of worker behaviour in the colony organization of the social wasps, *Vespula vulgaris* (L.) and *V. germanica* (F.) (Vespidae: Hymenoptera). Insectes Soc 19:227–242

Archer ME (1977) A nest of *Vespula rufa* (L.) – *V. austriaca* (Panzer) (Hymenoptera, Vespidae). Entomol Gaz 28:263–264

Archer ME (1980a) Possible causes of the yearly fluctuations in wasp numbers. Br Isles Bee Breeders' Assoc News 18:26–29

Archer ME (1980b) Population dynamics. In: Edwards R Social wasps: their biology and control. Rentokil Libr, East Grinstead, pp 172–207

Archer ME (1981a) A simulation model for the colonial development of *Paravespula vulgaris* (Linnaeus) and *Dolichovespula sylvestris* (Scopoli) (Hymenoptera: Vespidae). Melanderia 36:1–59

Archer ME (1981b) The Euro-Asian species of the *Vespula rufa* Group (Hymenoptera, Vespidae), with descriptions of two new species and one new subspecies. Kontyû 49:54–64

Arnold TS (1966) Biology of social wasps: comparative ecology of the British species of social wasps belonging to the family Vespidae. M Sc Thesis, University of London (cited in Spradbery 1973a)

Ban T (1980a) Control of social wasps as a public service. Otaru, 44 pp (in Japanese)
Ban T (1980b) The present state of controlling nuisance Hymenoptera by local authorities. Seikatsu to Kankyô 25(11):1-14 (in Japanese)
Ban T (1980c) Injury by the Japanese hornet *Vespa simillima*, and its control. Seikatsu to Kankyô 25(11):15-22
Ban T, Yamane Sk (1980) Nesting sites of *Vespa simillima* in urban areas (abstr). Jpn J Sanit Zool 31:150
Barrington RM (1896) Wasps catching flies on cattle. Ir Nat 5:272
de Beaumont J (1944) Les guêpes (*Vespa* L. *s.l.*) de la Suisse. Bull Soc Vaud Sci Nat 62:329-362
de Beaumont J (1958) Le parasitisme social chez les guêpes et les bourdons. Mitt Schweiz Entomol Ges 31:168-176
de Beaumont J, Matthey R (1945) Observations sur les *Polistes* parasites de la Suisse. Bull Soc Vaud Sci Nat 62:439-454
Beirne BP (1944) The causes of the occasional abundance or scarcity of wasps (*Vespula* spp.). Entomol Mon Mag 80:121-124
Bequaert J (1918) A revision of the Vespidae of the Belgian Congo based on the collection of the American Museum Congo Expedition, with a list of Ethiopian Diplopterous wasps. Bull Am Mus Nat Hist 39:1-384
Bequaert J (1930) On the generic and subgeneric divisions of the Vespinae (Hymenoptera). Bull Brooklyn Entomol Soc 25:59-70
Bequaert J (1932) A tentative synopsis of the hornets and yellowjackets (Vespinae; Hymenoptera) of America. Entomol Am 12:71-138
Bequaert J (1936) The common oriental hornets, *Vespa tropica* and *Vespa affinis*, and their color forms. Treubia 15:329-351
Billups TR (1882) Early appearance of *Vespa germanica*, Fab. Entomol Mon Mag 18:234
Birula A (1930a) Über die russischen Wespen und ihre geographische Verbreitung. III. Ann Mus Zool Acad Sci URSS 31:291-340
Birula A (1930b) Über die russischen Wespen und ihre geographische Verbreitung. IV. Zool Anz 87:127-143
Bischoff H (1931) Zur Kenntnis der Gattung *Pseudovespa*. Sber Ges Naturfreunde Berlin 1930:329-346
Bock WJ (1959) Preadaptation and multiple evolutionary pathways. Evolution 13:194-211
Bock WJ, von Wahlert G (1965) Adaptation and the form-function complex. Evolution 19:269-299
Bodenheimer FS (1951) Citrus entomology in the Middle East. Junk, The Hague, 663 pp
Bohart GE (1970) The evolution of parasitism among bees. 41st Honor Lect, Utah State Univ, Fac Assoc (Logan), 30 pp
Bohart RM, Bechtel RC (1957) The social wasps of California (Vespinae, Polistinae, Polybiinae). Bull Cal Insect Surv 4:73-101
Brian AD (1951) Brood development in *Bombus agrorum*. Entomol Mon Mag 87:207-212
Brian MV (1951) Summer population changes in colonies of the ant *Myrmica*. Physiol Comp Oecol 2:248-262
Brian MV, Brian AD (1948) Nest construction by queens of *Vespula sylvestris* Scop. (Hym., Vespidae). Entomol Mon Mag 84:193-198
Brian MV, Brian AD (1952) The wasp, *Vespula sylvestris* Scopoli: feeding, foraging and colony development. Trans R Entomol Soc London 103:1-26
Bristowe WS (1932) Insects and other invertebrates for human consumption in Siam. Trans Entomol Soc London 80:387-404
Broekhuizen S, Hordijk C (1968) Untersuchungen über die Beute von *Paravespula vulgaris* L. (Hym., Vespidae) und ihre Abhangigkeit von der Beutetierdichte. Z Angew Zool 62:68-77
Bromley SW (1948) Honey-bee predators. J NY Entomol Soc 56:195-199
Brothers DJ (1975) Phylogeny and classification of the aculeate Hymenoptera, with special reference to Mutillidae. Univ Kans Sci Bull 50:483-648
Bunn DS (1982) Notes on the nesting cycle of *Dolichovespula sylvestris* Scop. (Hym., Vespidae). Entomol Mon Mag 118:213-218
Butani DK (1979) Insects and fruits. Khosla & Pragati, Delhi, 415 pp

du Buysson R (1905) Monographie des guêpes ou *Vespa*. Ann Soc Entomol Fr 73 (1904): 485–556, 565–634

Buzzard CN (1951) Ways of a hornet. Country Life (London) 110:344–345

Carmean D, Akre RD, Zack RS, Reed HC (1981) Notes on the yellowjackets parasite *Bareogonalos canadensis* (Hymenoptera: Trigonalidae). Entomol News 92:23–26

Carpenter JM (1982) The phylogenetic relationships and natural classification of the Vespoidea (Hymenoptera). Syst Entomol 7:11–38

Chan KL (1972) The hornets of Singapore: their identification, biology and control. Singapore Med J 13:178–187

Chapman TA (1870) Some facts towards a life-history of *Rhipiphorus paradoxus*. Ann Mag Nat Hist 6(4S): 314–326

Chapman TA (1891) On the oviposition of *Metoecus (Rhipiphorus) paradoxus*. Entomol Mon Mag 27:18–20

Charnley HW (1973) The value of the propodeal orifice and the phallic capsule in vespid taxonomy (Hymenoptera, Vespidae). Bull Buff Soc Nat Sci 26:1–79

Cox CB, Healey IN, Moore PD (1976) Biogeography. An ecological and evolutionary approach. Blackwell, Oxford, London, ix + 194 pp

Craig R (1979) Parental manipulation, kin selection, and the evolution of altruism. Evolution 33:319–334

Crozier RH (1982) On insects and insects: twists and turns in our understanding of the evolution of eusociality. In: Breed MD, Michener CD, Evans HE (eds) The biology of social insects. Westview, Boulder, pp 4–9

Cumber RA (1949) The biology of bumblebees with special reference to production of the worker caste. Trans R Entomol Soc London 100:1–45

Dammerman KW (1948) The fauna of Krakatau 1883–1933. Kon Ned Akad Wet Verh Wet Afdel Natuurk (2)44:1–594

Darlington PJ (1957) Zoogeography: the geographical distribution of animals. John Wiley & Sons, New York, xiv + 675 pp

Davis HG, Eddy GW, McGovern TP, Beroza M (1967) 2–4 Hexadienyl butyrate and related compound highly attractive to yellowjackets (*Vespula* spp.). J Econ Entomol 61:459–462

Davis HG, McGovern TP, Eddy GW, Nelson TE, Bertun KM, Beroza M, Ingangi JC (1968) New chemical attractants for yellowjackets (*Vespula* spp.). J Econ Entomol 61:459–462

Davis WT (1925) *Cicada tibicen*, a South American species, with records and descriptions of North American cicadas. J NY Entomol Soc 33:35–51

Deevey ES (1947) Life tables for natural populations of animals. Q Rev Biol 22:283–314

Deleurance EP (1957) Contribution à l'étude biologique des *Polistes* (Hyménoptèrés, Vespidés) I. L'activité de construction. Ann Sci Nat Zool Biol Anim 11:91–222

Döhring E (1960) Zur Häufigkeit, hygienischen Bedeutung und zum Fang sozialer Faltenwespen in einer Großstadt. Z Angew Entomol 47:69–79

Duncan CD (1939) A contribution to the biology of North American vespine wasps. Stanford Univ Publ Biol Sci 8:1–272

Eck R (1980) *Dolichovespula loekenae* n. sp., eine neue soziale Faltenwespe aus Skandinavien (Hymenoptera, Vespidae). Reichenbackia 18:213–217

Eck R (1981) Zur Verbreitung und Variabilität von *Dolichovespula norwegica* (Hymenoptera, Vespidae). Entomol Abh Mus Tierkd Dresden 44:133–152

Eck R (1983) Zur Verbreitung und Variabilität von *Dolichovespula saxonica* (Hymenoptera, Vespidae). Entomol Abh Mus Tierkd Dresden 46:151–176

Edery H, Ishay J, Gitter S, Joshua H (1978) Venoms of Vespidae. In: Bettini S (ed) Arthropod venoms. Springer, Berlin Heidelberg New York, pp 691–771

Edwards R (1979) The number of wasp's nests destroyed by local authorities. 15 year report. Environ Health 87:100, 105

Edwards R (1980) Social wasps. Their biology and control. Rentokil Libr, East Grinstead, 398 pp

Eickwort GC (1981) Presocial insects. In: Hermann HR (ed) Social insects, vol 2. Academic Press, New York London, pp 199–280

Ennik F (1973) Abatement of yellowjackets using encapsulated formulations of diazinon and Radon. J Econ Entomol 66:1097–1098

Evans HE (1975) Social parasitism of a common yellowjacket. Insect World Digest 2(1):6–13
Evans HE, West-Eberhard MJ (1970) The wasps. David & Charles, Newton Abbot, vi + 265 pp
Evans W (1903) *Vespa austriaca* and *V. rufa* in Scotland. Entomol Mon Mag 39:299–300
Fox-Wilson G (1946) Factors affecting populations of social wasps, *Vespula* species in England (Hymenoptera). Proc R Entomol Soc London Ser A 21:17–27
Free JB, Butler CG (1959) *Bumblebees*. Collins, London, 208 pp
Fukuda H, Ohtani T (1977) Survival and life span of drone honeybees. Res Popul Ecol 19:51–68
Fukuda H, Sakagami SF (1968a) Worker brood survival in honeybees. Res Popul Ecol 10:31–39
Fukuda H, Sakagami SF (1968b) Life tables for worker honeybees. Res Popul Ecol 10:127–139
Gaul AT (1942) Additions to vespine biology. I. Notes on mating and brood rearing. Bull Brooklyn Entomol Soc 37:57–61
Gaul AT (1948) Additions to vespine biology. IV. Notes on *Vespula rufa* var. *consobrina* Saussure. Bull Brooklyn Entomol Soc 43:160–162
Gaul AT (1952) The awakening and diurnal flight activities of vespine wasps. Proc R Entomol Soc London Ser A 27:33–38
Gauss R (1968) Über Parasiten bei sozialen Wespen (Vespidae). Z Angew Entomol 61:453–454
Gibo DL, Temporale A, Lamarre TP, Soutar BM, Dew HE (1977) Thermoregulation in colonies of *Vespula arenaria* and *Vespula maculata* (Hymenoptera: Vespidae) III. Heat production in queen nests. Can Entomol 109:615–620
Grant J (1959) Hummingbirds attacked by wasps. Can Field Nat 73:174
Greene A (1979) Behavioral characters as indicators of yellowjacket phylogeny (Hymenoptera: Vespidae). Ann Entomol Soc Am 72:614–619
Greene A, Akre RD, Landolt PJ (1976) The aerial yellowjacket, *Dolichovespula arenaria* (Fab.): nesting biology, reproductive production and behavior (Hymenoptera: Vespidae). Melanderia 26:1–34
Greene A, Akre RD, Landolt PJ (1978) Behavior of the yellowjacket social parasite, *Dolichovespula arctica* (Rohwer) (Hymenoptera: Vespidae). Melanderia 29:1–28
Grogan DE, Hunt JH (1977) Digestive proteases of wasps of the genus *Vespula*. Insect Biochem 7:191–196
Grothaus RH, Davis HG, Rogoff WM, Fluno JA, Hirst JM (1973) Baits and attractants for east coast yellowjackets, *Vespula* spp. Environ Entomol 2:717–718
Guiglia D (1972) Les guêpes sociales (Hymenoptera, Vespidae) d'Europe occidentale et septentrionale. Faun Eur Bassin Medit 6:181 pp
Haeseler V (1978) Zur Fauna der aculeaten Hymenopteren der Nordseeinsel Mellem. Ein Beitrag zur Besiedlung küstennaher Inseln. Zool Jahrb (Syst) 105:368–385
Haeseler V (1981) Ein von *Dolichovespula adulterina* (du Buysson) infiziertes *Dolichovespula saxonica* (Fabricius) – Nest aus Norddeutschland (Hymenoptera: Vespidae: Vespinae). Z Angew Zool 68:437–452
Hamilton WD (1964a) The genetical evolution of social behaviour. I. J Theor Biol 7:1–16
Hamilton WD (1964b) The genetical evolution of social behaviour. II. J Theor Biol 7:17–52
Harrington WH (1896) A Canadian *Trigonalys*. Can Entomol 28:108
Hattori T, Yamane Sk (1975) Notes on *Metoecus paradoxus* and *M. vespae* parasitic on the *Vespula* species in northern Japan (Coleoptera, Rhipiphoridae; Hymenoptera, Vespidae) 1. New Entomol (Ueda) 24:1–7 (in Japanese with English summary)
Hennig W (1966) Phylogenetic Systematics. Univ Ill Press, Urbana, xiii + 263 pp
Hermann HR, Blum MS (1981) Defensive mechanisms in the social Hymenoptera. In: Hermann HR (ed) Social insects, vol 2. Academic Press, New York London, pp 77–197
Himmer A (1932) Die Temperaturverhältniss bei den sozialen Hymenopteren. Biol Rev 7:224–253
Hoffman DR, MacDonald CA (1982) Allergens in Hymenoptera venom VIII. Immunological comparison of venoms from six species of *Vespula* (yellowjacket). Ann Allerg 48:78–81
Hopkins DM (1967) The Cenozoic history of Beringia – a synthesis. In: Hopkins DM (ed) The Bering Land Bridge. Univ Press, Stanford, pp 451–483
Horne C (1870) Notes on the habits of some Hymenopterous insects from the North-West Provinces of India. Trans Zool Soc London 7:161–185
Ikan R, Bergmann ED, Ishay J, Gitter S (1968) Proteolytic enzyme activity in the various colony members of the Oriental hornet, *Vespa orientalis* F. Life Sci 7:929–934

Ikan R, Gottlieb R, Bergmann ED, Ishay J (1969) The pheromone of the queen of the Oriental hornet, *Vespa orientalis*. J Insect Physiol 15:1709–1712
Ikeda F, Kodomari S (1982) *Biology of Vespula lewisii*. Sizuoka Pref Agric Exp Stn, 18 pp (in Japanese)
Ikezaki Y, Ariga F (1975) Puparium of *Volucella nitobei* Matsumura. Nature and Insects (Tokyo) 10(14):27 (in Japanese)
Ishay J (1964) Observations sur la biologie de la guêpe orientale *Vespa orientalis* F. Insectes Soc 11:193–206
Ishay J (1973) Thermoregulation by social wasps: behavior and pheromones. Trans NY Acad Sci (2)35:447–462
Ishay J (1975) Frequencies of the sounds produced by the Oriental hornet, *Vespa orientalis*. J Insect Physiol 21:1737–1740
Ishay J, Brown MB (1975) Patterns in the sounds produced by *Paravespula germanica* wasps. J Acoust Soc Am 57:1521–1525
Ishay J, Ikan R (1968) Food exchange between adults and larvae in *Vespa orientalis* F. Anim Behav 16:298–303
Ishay J, Landau EM (1972) *Vespa* larvae send out rhythmic hunger signals. Nature (London) 237:286–287
Ishay J, Schwartz A (1973) Acoustic communication between the members of the Oriental hornet (*Vespa orientalis*) colony. J Acoust Soc Am 53:640–649
Ishay J, Bytinski-Salz H, Shulov A (1968) Contributions to the bionomics of the Oriental hornet (*Vespa orientalis* Fab.). Isr J Entomol 2:45–106
Ishikawa R (1965) Vespidae. In: Asahina S, Ishihara T, Yasumatsu K (eds) Iconographia insectorum aponicorum colore naturali edita, vol. 3. Hokuryükan, Tokyo, pp 291–294, pls 146, 147 (in Japanese)
Ishikawa R (1970) On some noteworthy vespid wasps of Tsushima. Mem Nat Sci Mus Tokyo 3:269–272, pl 22 (in Japanese)
Itô Y (1978) Comparative ecology, 2nd edn. Iwanami-shoten, Tokyo, xvii + 421 pp (in Japanese) [English Transl by Kikkawa J (1980) Comparative ecology. Univ Press, Cambridge, xii + 436 pp]
Itô Y (1982) Introduction to social ecology. Univ Press, Tokyo, vi + 210 pp (in Japanese)
Iwasa Y (1981) Adaptation strategies in living things. Saiensusha, Tokyo, v + 229 pp (in Japanese)
Iwata K (1953) Biology of *Eumenes* in Japan (Hym., Vespidae). Mushi 25:25–46
Iwata K (1971a) Evolution of instinct. Comparative ethology of Hymenoptera. Mano-shoten, Kanagawa, vi + 503 pp [English Transl by Smithsonian Inst (1976) Evolution of instinct. Amerind, New Delhi, xii + 535 pp]
Iwata K (1971b) Ethological notes on four Japanese species of *Vespa* (Hymenoptera). Entomol essays to commemorate the retirement of Prof. K. Yasumatsu. Hokuryûkan, Tokyo, pp 219–223
Jacobson RS, Matthews RW, MacDonald JF (1978) A systematic study of the *Vespula vulgaris* group with a description of a new yellowjacket species in eastern North America (Hymenoptera: Vespidae). Ann Entomol Soc Am 71:299–312
Janet C (1895) Études sur les fourmis, les guêpes et les abeilles. 9th note. Sur *Vespa crabro* L.—histoire d'un nid depuis son origine. Mém Soc Zool Fr 8:1–140
Janet C (1903) Observations sur les Guêpes (études sur les fourmis, les guépes et les abeilles. 23rd note). Naud, Paris, 85 pp
Jeanne RL (1972) Social biology of the Neotropical wasp *Mischocyttarus drewseni*. Bull Mus Comp Zool Harvard 144 (3):63–105
Jeanne RL (1975) The adaptiveness of social wasp nest architecture. Q Rev Biol 50:267–287
Jeanne RL (1977a) A specialization in nest petiole construction by queens of *Vespula* spp. (Hymenoptera: Vespidae). J NY Entomol Soc 85:127–129
Jeanne RL (1977b) Behavior of the obligate social parasite *Vespula arctica* (Hymenoptera: Vespidae). J Kans Entomol Soc 50:541–557
Jeanne RL (1980) Evolution of social behavior in the Vespidae. Annu Rev Entomol 25:371–396
de Jong D (1978) Insects: Hymenoptera (ants, wasps, and bees). In: Morse RA (ed) Honey bee pests, predators, and diseases. Cornell Univ Press, New York, pp 138–157
Kaibara E (1709) Yamato Honzô XIV. Edo (Tokyo), 405 pp (in Japanese)
Kasuya E (1980) Simultaneous maintenance of two nests by a single foundress of the Japanese paper wasp, *Polistes chinensis antennalis* (Hymenoptera: Vespidae). Appl Entomol Zool 15:188–189

Kasuya E (1981) Why did the social insects evolve? Critical review of recent theoretical trends. I. Seibutsu Kagaku 33:209–217 (in Japanese)
Kasuya E (1982a) Why did the social insects evolve? Critical review of recent theoretical trends. II. Seibutsu Kagaku 34:7–14 (in Japanese)
Kasuya E (1982b) Variation and evolution of behavior in the social insects. Seibutsu Kagaku 34:169–174 (in Japanese)
Katô M (1950) A hover fly (*Volucella tabanoides*) parasitic on *Vespula lewisii* Cameron. Shin-konchû 3(2):30 (in Japanese)
Kemper H, Döhring E (1962) Untersuchungen über die Ernährung sozialer Faltenwespen Deutschlands, insbesondere von *P. germanica* und *P. vulgaris*. Z Angew Zool 49:227–280
Kemper H, Döhring E (1967) Die Sozialen Faltenwespen Mitteleuropas. Parey, Berlin Hamburg, 180 pp
Kifune T (1971) What are the Strepsiptera? Gekkan Mushi 8:6–11 (in Japanese)
Kifune T, Maeta Y (1975) A new subgenus and new species of the genus *Xenos* (Strepsiptera, Stylopidae) from Japan. Kontyû 43:446–455
Kojima J (1982) The genus *Vespa* in the Philippines (Hymenoptera, Vespidae). Kontyû 50:434–444
Kojima J, Yamane Sk (1980) Biological notes on *Vespa luctuosa luzonensis* from Leyte Island, the Philippines, with descriptions of adults and larvae (Hymenoptera: Vespidae). Insect Matsum NS 19:79–87
Krebs JR, Davies NB (1981) An introduction to behavioural ecology. Blackwell, Oxford London, x + 292 pp
Kugler K, Orion T, Ishay J (1976) The number of ovarioles in the Vespinae (Hymenoptera). Insectes Soc 23:525–533
Kuse T (1791) Kahô Chikuyô-ki. Edo (Tokyo), 220 pp (in Japanese)
Laidlaw WBR (1930) Notes on some humble bees and wasps in Scotland. Scott Nat 185:135–136
Landolt PJ, Akre RD, Greene A (1977) Effects of colony division on *Vespula atropilosa* (Sladen) (Hymenoptera: Vespidae). J Kans Entomol Soc 50:135–147
Latter OH (1895) Wasps and weather. Nat Sci (London) 6:178–179
Latter OH (1915) Remarkable nest of *Vespa norwegica* and fertility of workers of this species. Nature (London) 96:59
Lee T-S (1982) Hornets from agricultural regions of China (Hymenoptera: Vespoidea). Agric Publ House, Beijing, 255 pp, 6 tabs (in Chinese)
Lindauer M (1952) Ein Beitrag zur Frage der Arbeitsteilung im Bienenstaat. Z Vergl Physiol 34:299–345
van Lith JP (1956) Merkwaardige Nesten van *Dolichovespula saxonica* (F.). Entomol Ber Amsterdam 16:33–35
Løken A (1964) Social wasps in Norway (Hymenoptera, Vespidae). Norsk Entomol Tidssk 12:195–218, 2 pls
MacDonald JF (1977) Comparative and adaptive aspects of vespine nest construction. Proc 8th Int Congr IUSSI, Wageningen 1977, pp 169–172
MacDonald JF (1980) Biology, recognition, medical importance and control of Indiana social wasps. Coop Ext Serv Purdue Univ Bull E-19: 24 pp
MacDonald JF, Matthews RW (1975) *Vespula squamosa*: a yellowjacket wasp evolving towards parasitism. Science 190:1003–1004
MacDonald JF, Matthews RW (1976) Nest structure and colony composition of *Vespula vidua* and *V. consobrina* (Hymenoptera: Vespidae). Ann Entomol Soc Am 69:471–475
MacDonald JF, Matthews RW (1981) Nesting biology of the eastern yellowjacket, *Vespula maculifrons* (Hymenoptera: Vespidae). J Kans Entomol Soc 54:433–457
MacDonald JF, Akre RD, Hill WB (1973) Attraction of yellowjackets (*Vespula* spp.) to heptyl butyrate in Washington State (Hymenoptera: Vespidae). Environ Entomol 2:375–379
MacDonald JF, Akre RD, Hill WB (1974) Comparative biology and behavior of *Vespula atropilosa* and *V. pensylvanica* (Hymenoptera, Vespidae). Melanderia 18:1–66
MacDonald JF, Akre RD, Hill WB (1975) Locations and structure of nests of *Vespula atropilosa* and *V. acadica* (Hymenoptera: Vespidae). J Kans Entomol Soc 48:114–121
MacDonald JF, Matthews RW, Jacobson RS (1980) Nesting biology of the yellowjacket, *Vespula flavopilosa* (Hymenoptera: Vespidae). J Kans Entomol Soc 53:448–458

Mace H (1927) Modern beekeeping. Wyman & Sons, London, 340 pp
Madden JL (1981) Factors influencing the abundance of the European wasp (*Paravespula germanica* [F.]). J Aust Entomol Soc 20:59–65
Maeta Y (1963) Some biological notes on *Pseudoxenos iwatai* Esaki (I). Its biology and life history. Kontyû 31:1–15 (in Japanese with English summary)
Makino S (1980) Bionomics of *Dolichovespula media media* and *D. saxonica nipponica*: behaviour of the foundress, embryo and mature nest structure, and colony productivity (Hymenoptera: Vespidae). MS Thesis, Hokkaido Univ, 36 pp, 18 figs, 6 tabs
Makino S (1982) Nest structure, colony composition and productivity of *Dolichovespula media media* and *D. saxonica nipponica* in Japan (Hymenoptera, Vespidae). Kontyû 50:212–224
Makino S (1983a) Polistine wasps as hosts. Kotaigun-seitaigakkai Kaihô 37:53–66 (in Japanese)
Makino S (1983b) Biology of *Latibulus argiolus* (Hymenoptera, Ichneumonidae), a parasitoid of the paper wasp *Polistes biglumis* (Hymenoptera, Vespidae). Kontyû 51:426–434
Makino S, Yamane Sk (1979) An example of simultaneous maintenance of two nests by a single foundress of *Dolichovespula saxonica nipponica* (Hymenoptera: Vespidae). Kontyû 47:78–84 (in Japanese with English summary)
Makino S, Yamane Sk (1980) Heat production by the foundress of *Vespa simillima*, with description of its embryo nest (Hymenoptera: Vespidae). Insect Matsum NS 19:89–101
Makino S, Yamane Sk, Ban T, Kunou I (1981) The Japanese hornet *Vespa simillima* Smith, an important nuisance pest in urban areas (Hymenoptera: Vespidae). Jpn J Sanit Zool 32:203–213
Marchal P (1896) La reproduction et l'évolution des guêpes sociales. Arch Zool Exp Gén 4(3S):1–100
Maschwitz U (1966) Das Speichelsekret der Wespenlarven und seine biologische Bedeutung. Z Vergl Physiol 53:228–252
Matsuura M (1966a) On the early emergence of *Vespa analis* male. Insectarium (Tokyo) 3:12 (in Japanese)
Matsuura M (1966b) Notes on the hibernating habits of the genus *Vespa* in Japan. Kontyû 34:52–67 (in Japanese with English summary)
Matsuura M (1967a) Life of hornets, pt 1. Jpn Bee J 20:6–9 (in Japanese)
Matsuura M (1967b) Stylopized workers of *Vespa crabro flavofasciata*. Insectarium (Tokyo) 4:13 (in Japanese)
Matsuura M (1967c) Life of hornets, pt 10. Jpn Bee J 20:385–388 (in Japanese)
Matsuura M (1968a) Life of hornets, pt 11. Jpn Bee J 21:21–25 (in Japanese)
Matsuura M (1968b) Life of hornets, pt 17. Jpn Bee J 21:319–322 (in Japanese)
Matsuura M (1969) Behaviour of post-hibernating female hornets, *Vespa*, in the pre-nesting stage, with special reference to intra- and interspecific dominance relationships. Jpn J Ecol 19:196–203
Matsuura M (1970a) A journey to Formosa to collect the nests of social Vespidae. Life Stud 14:16–20 (in Japanese)
Matsuura M (1970b) Intraspecific invasion behaviour of *Vespa crabro flavofasciata*. Life Stud 14:21–26 (in Japanese with English summary)
Matsuura M (1971a) Nesting sites of the Japanese *Vespa* species (Hymenoptera, Vespidae). Kontyû 39:43–54 (in Japanese with English summary)
Matsuura M (1971b) Nest foundation by the female wasps of the genus *Vespa* (Hymenoptera, Vespidae). Kontyû 39:99–105
Matsuura M (1973a) Colony development of *Vespa crabro flavofasciata* Cameron in early stage of nesting. Life Stud 17:1–12 (in Japanese with English summary)
Matsuura M (1973b) Intracolonial polyethism in *Vespa* 3. Foraging activities. Life Stud 17:81–99 (in Japanese with English summary)
Matsuura M (1973c) Nesting habits of several species of the genus *Vespa* in Formosa. Kontyû 41:286–293
Matsuura M (1974a) The life of hornets. Shizen 29:62–71 (in Japanese)
Matsuura M (1974b) Intracolonial polyethism in *Vespa* 1. Behaviour and its change of the foundress of *Vespa crabro flavofasciata* in early nesting stage in relation to worker emergence. Kontyû 42:333–350
Matsuura M (1974c) *Volucella nitobei* found in a *Vespa simillima* nest. Insectarium (Tokyo) 11:282 (in Japanese)

Matsuura M (1975a) Notes on the number of eggs laid by *Vespaxenos*. Insectarium (Tokyo) 12:18 (in Japanese)

Matsuura M (1975b) The mass-attack of *Vespa mandarinia*. Shizen 30:1–5, 68–69 (in Japanese)

Matsuura M (1975c) The life of social wasps. Anima 29:68–72 (in Japanese)

Matsuura M (1976a) The effects of stylopization on *Vespa* workers. Abstr 36th Annu Meet Entomol Soc Jpn, p 43 (in Japanese)

Matsuura M (1976b) Notes on the social wasps and bees in Wakayama. Nanki-Seibutsu 18:5–11 (in Japanese)

Matsuura M (1977a) The life of the paper wasps. Shizen 32:26–36 (in Japanese)

Matsuura M (1977b) Observations on the hornets hunting the cicadas. Rostria 27:216–218 (in Japanese with English summary)

Matsuura M (1980a) Behaviour of the giant hornet, *Vespa mandarinia*, at tree sap sources with special reference to intra- and inter-specific dominance relationship. Honeybee Sci 1:111–118 (in Japanese with English summary)

Matsuura M (1980b) Hibernation of the social wasps. Insectarium (Tokyo) 17:344–349 (in Japanese)

Matsuura M (1983a) Preliminary report on the polygynous colonies of *Vespa affinis indosinensis* (Hymenoptera, Vespidae) in Sumatra. Kontyû 51:80–82

Matsuura M (1983b) Tropical hornets and honeybees. Insectarium (Tokyo) 20:4–9 (in Japanese)

Matsuura M (1984) Comparative biology of the five Japanese species of the genus *Vespa* (Hymenoptera, Vespidae). Bull Fac Agr Mie Univ 69:1–131

Matsuura M, Sakagami SF (1973) A bionomic sketch of the giant hornet, *Vespa mandarinia*, a serious pest for Japanese apiculture. J Fac Sci Hokkaido Univ (VI Zool) 9:125–162

Matthews RW (1982) Social parasitism in yellowjacket (*Vespula*). In: Jaisson P (ed) Social insects in the tropics, vol 1. Univ Press, Paris-Nord, pp 193–202

Matthews RW, Matthews JR (1979) War of the yellowjacket queens. Nat Hist 88(8):56–65

Matthews RW, Ross KG, Morse RA (1982) Comparative development of queen nest of four species of yellowjackets (Hymenoptera: Vespidae) reared under identical conditions. Ann Entomol Soc Am 75:123–129

Menke AS, Snelling R (1975) *Vespula germanica* (Fabricius), an adventive yellowjacket in the northeastern United States (Hymenoptera: Vespidae). US Dep Agr Coop Econ Insect Rep 25(11):193–200

Michener CD (1958) The evolution of social behavior in bees. Proc 10th Int Congr Entomology, Montreal, vol 2, pp 441–448

Michener CD (1969) Comparative social behavior of bees. Annu Rev Entomol 14:299–342

Michener CD (1974) The social behavior of bees. A comparative study. Harvard Univ Press, Cambridge, xii + 404 pp

Mikkola K (1978) Spring migrations of wasps and bumble bees on the southern coast of Finland (Hymenoptera, Vespidae and Apidae). Ann Entomol Fenn 44:10–26

Miller CDF (1958) Distribution and nomenclatorial problems in some forms of *Vespula* in North America (Hymenoptera: Vespidae). Proc 10th Int Congr Entomology, Montreal, vol 1, pp 257–264

Miller CDF (1961) Taxonomy and distribution of Nearctic *Vespula*. Can Entomol 93 (Suppl 22): 1–52

Miyachi S (1980) Bee sting allergy. Mod Media 26:191–199 (in Japanese)

Miyachi S (1981) Insect allergy. In: Shin-shôni-igaku-taikei 21B:148–153 (in Japanese)

Miyano S (1980) Life tables of colonies and workers in a paper wasp, *Polistes chinensis antennalis*, in central Japan (Hymenoptera: Vespidae). Res Popul Ecol 22:69–88

Miyazaki M (1981) Eagles hawks falcons. Heibon-sha, Tokyo, 149 pp (in Japanese)

Miyoshi S (1830) Sôzan Chomon-shû. Edo (Tokyo), 112 pp (in Japanese)

Morse RA, Eickwort GC, Jacobson RS (1977) The economic status of an immigrant yellowjacket, *Vespula germanica* (Hymenoptera: Vespidae), in northeastern United States. Environ Entomol 6:109–110

Munakata M, Yamane S (1970) Social vespid wasps from the southern part of the Oshima Peninsula and the Okushiri Island, northern Japan (Hymenoptera, Vespidae). Kontyû 38:281–291

Nakajima T (1983) Chemistry of hymenopteran venom. Honeybee Sci 4:9–14 (in Japanese with English summary)

Nixon GEJ (1934) Notes on wasps 2. On the adoption of an orphaned brood of *Vespa vulgaris* L. by a queen of *Vespa germanica* Fab. Entomol Mon Mag 70:87–89

Nogueira-Neto P (1970) Behavior problems related to the pillages made by some parasitic stigless bees (Meliponinae, Apidae). In: Development and evolution of behavior. Freeman, San Francisco, pp 416–434

Nozaka C (1976) Observations on the egg-laying behavior of *Bareogonalos jezoensis* Uchida. Hymenopterists' Commun 3:1–4 (in Japanese)

Okada I (1961) Notes on the habits of a giant hornet (*Vespa mandarinia* Smith) as a noxious insect against honeybee. Bull Fac Agr Tamagawa Univ 2 C:73–89 (in Japanese with English summary)

Okumura T (1912) Wasp brood for food. Zool Mag (Tokyo) 289:645–650 (in Japanese)

Okuno T (1970) Immature stages of two species of the genus *Volucella* Geoffroy (Diptera, Syrphidae). Kontyû 38:268–270

Okutani T (1952) Notes on the life-history of *Metoecus paradoxus,* nest-parasite of *Vespula lewisii* (Some notes on *Vespula lewisii.* III). Seitai-konchû 4:71–74, pls 9, 10 (in Japanese)

Okutani T, Tomomatsu Y (1951) Some knowledgement of *Vespula lewisii.* I. The preys of the wasps. I. Seitai-konchû 3:95–97 (in Japanese)

Olberg G (1959) Das Verhalten der Solitären Wespen Mitteleuropas (Vespidae, Pompilidae, Sphecidae). Deutscher Verlag der Wissenschaften, Berlin, xiv + 402 pp

Ono R (1805) Honzô Kômoku Keimô XXVII. Edo (Tokyo), 204 pp (in Japanese)

Ôri M, Hiyama O (1977) A case of anaphylactic shock caused by *Vespa tropica*. Jpn J Sanit Zool 28:281–284 (in Japanese)

Oster GF, Wilson EO (1978) Caste and ecology in the social insects. Univ Press, Princeton, xv + 352 pp

Pardi L, Marino Piccioli MT (1981) Studies on the biology of *Belonogaster* (Hymenoptera, Vespidae) 4. On caste differences in *Belonogaster griceus* (Fab.) and the position of this genus among social wasps. Monit Zool It NS Suppl 9:131–146

Parrish MD, Roberts RB (1982) Successful establishment of the German yellowjacket in urban New Jersey indicated by relative abundance of *Vespula germanica* and *V. maculifrons* (Hymenoptera: Vespidae). J Kans Entomol Soc 55:272–276

Payne JA, Mason WRM (1971) Hymenoptera associated with pig carrion. Proc Entomol Soc Washington DC 73:132–141

Perrott DCF (1975) Factors affecting use of mirex-poisoned protein baits for control of European wasp (*Paravespula germanica*) in New Zealand. NZ J Zool 2:491–508

de Portugal-Araüjo V (1963) Subterranean nests of two African stingless bees. J NY Entomol Soc 71: 130–141

Potter NB (1964) A study of the biology of the common wasp, *Vespula S. Vulgaris* L., with special reference to the foraging behaviour. PhD Thesis, Bristol Univ

Rau P (1929) The nesting habits of the bald-faced hornet, *Vespa maculata*. Ann Entomol Soc Am 22:659–675

Rawson JH (1963) Wasps. Bee Craft 45:14

Reed HC, Akre RD (1982) Divergent strategies of colony usurpation among two obligate vespine parasites (Hymenoptera: Vespidae). In: Breed MD, Michener CD, Evans HE (eds) The biology of social insects. Westview, Boulder, p 291

Reed HC, Akre RD (1983) Comparative colony behavior of the forest yellowjacket, *Vespula acadica* (Sladen) (Hymenoptera: Vespidae). J Kans Entomol Soc 56:581–606

Reed HC, Akre RD, Garnett WB (1979) A North American host of the yellowjacket social parasite *Vespula austriaca* (Panzer) (Hymenoptera: Vespidae) Entomol News 90:110–113

Reid JA (1942) On the classification of the larvae of the Vespidae (Hymenoptera). Trans R Entomol Soc London 92:285–331

Reierson DA, Wagner RE (1978) Trapping to determine the sympatry and seasonal abundance of various yellowjackets. Environ Entomol 7:418–422

Richards OW (1927) The specific characters of the British humblebees (Hymenoptera). Trans Entomol Soc London 75:233–268

Richards OW (1937) The mating-habits of species of *Vespa* (Hymen.). Proc R Entomol Soc London Ser A 12:27–29

Richards OW (1962) A revisional study of the masarid wasps (Hymenoptera, Vespoidea). Br Mus NH London, v + 294 pp

Richards OW (1971) The biology of the social wasps (Hymenoptera, Vespidae). Biol Rev 46:483–528
Richards OW (1978) The social wasps of the Americas excluding the Vespinae. Br Mus NH London, 580 pp, 4 pls
Richards OW, Richards MJ (1951) Observations on the social wasps of South America (Hymenoptera, Vespidae). Trans R Entomol Soc London 102:1–170, 4 pls
Ritchie J (1915) Some observations and deductions regarding the habits and biology of the common wasp. Scott Nat 46:318–331
Rivnay E, Bytinski-Salz H (1949) The oriental hornet (*Vespa orientalis* L.); its biology in Israel. Bull Agr Res St Rehovot, Isr 52:1–32 (in Hebrew)
Robson JE (1898) *Vespa austriaca*, a cuckoo-wasp. Sci Gossip 5:69–73
Rösch GA (1925) Untersuchungen über die Arbeitsteilung im Bienenstaat 1. Z Vergl Physiol 2:571–631
Ross D (1983) The role of volatiles and social facilitation in bait discovery by yellowjackets. MS Thesis, Purdue Univ, 96 pp
Ross KG (1982) Atypical yellowjacket nests initiated in the laboratory (Hymenoptera: Vespidae). J Kans Entomol Soc 55:563–567
Ross KG, Matthews RW (1982) Two polygynous overwintered *Vespula squamosa* colonies from the southeastern U.S. (Hymenoptera: Vespidae). Fl Entomol 65:176–184
Ross KG, Matthews RW, Morse RA (1981) Laboratory culture of four species of yellowjackets, *Vespula* spp., foundress nest initiation. Ann Entomol Soc Am 74:247–254
Rothney GA (1877) Notes on the habits of *Chlorin lobotum* and two species of wasps in India. Entomol Mon Mag 14:91–92
Roubaud E (1916) Recherches biologiques sur les guêpes solitaires et sociales d'Afrique. La genèse de la vie sociale et l'évolution de l'instinct maternel chez les vespides. Ann Sci Nat Zool Ser 10/1:1–160
Roush CF, Akre RD (1978) Nesting biologies and seasonal occurrence of yellowjackets in northeastern Oregon forests (Hymenoptera: Vespidae). Melanderia 30:57–94
Rudebeck G (1965) On a migratory movement of wasps, mainly *Vespula rufa* (L.) at Falsterbo, Sweden. Proc R Entomol Soc London 40:1–8
de Ruitter BA (1916) De *Vespa cincta* als poppenroover. Trop Nat 5:190–191
Sakagami SF (1953) Arbeitsteilung der Arbeiterinen in einem Zwergvolk, bestehend aus gleichaltrigen Volksgenossen. J Fac Sci Hokkaido Univ (VI Zool) 11:343–400
Sakagami SF (1959) Some interspecific relations between Japanese and European honeybees. J Anim Ecol 28:51–68
Sakagami SF (1968) Comparative bee sociology I. Origin of communal and caste systems. Zool Mag 77:35–43 (in Japanese)
Sakagami SF (1970) Mitsubachi no tadotta michi (Comparative sociology and evolution of bees). Shisaku-sha, Tokyo, 327 + x pp (in Japanese)
Sakagami SF (1975a) Evolution of caste system in the Hymenoptera. Kagaku 45:138–144 (in Japanese)
Sakagami SF (1975b) My Brasil and its bees. Shisaku-sha, Tokyo, 328 pp (in Japanese)
Sakagami SF (1976) Specific differences in the bionomic characters of bumblebees. A comparative review. J Fac Sci Hokkaido Univ (VI Zool) 20:390–447
Sakagami SF (1977) Seasonal change of nest survival and related aspects in an aggregation of *Lasioglossum duplex* (Dalla Torre), an eusocial halictine bee (Hymenoptera: Halictidae). Res Popul Ecol 19:69–86
Sakagami SF, Fukuda H (1968) Life tables for worker honeybees. Res Popul Ecol 10:127–139
Sakagami SF, Fukushima K (1957a) *Vespa dybowskii* André as a facultative temporary social parasite. Insectes Soc 4:1–12
Sakagami SF, Fukushima K (1957b) Some biological observations on a hornet, *Vespa tropica* var. *pulchra* (du Buysson), with special reference to its dependence on *Polistes* wasps (Hymenoptera). Treubia 24:73–82
Salt G, Bequaert J (1929) Stylopized Vespidae. Psyche 36:249–282
Sanderman RG (1936) Notes on the habits of *Vespula vulgaris* and *V. germanica* (Hymenoptera). Proc R Entomol Soc London Ser A 11:88–90

Sanderman RG (1940) An experiment and some observations on *Vespa crabro* L. (Hymenoptera, Vespidae). Proc R Entomol Soc London Ser A 15:29-32
de Saussure H (1852-1858) Études sur la famille des Vespides. I-III. Masson, Genève Paris
Schremmer F (1962) Wespen und Hornissen. Die Neue Brehm-Bücherei, H 298. Ziemsen, Wittenburg
Schulz-Langer E (1954) Beobachtungen zur Copulation gefangen gehaltener Wespen. Zool Anz 152:39-42
Scott H (1944) Notes on the season of 1943. Entomol Mon Mag 80:1-4
Sekiguchi K, Sakagami SF (1966) Structure of foraging population and related problems in honeybee, with considerations on the division of labour in bee colonies. Hokkaido Agr Exp Stn Rep 69:1-65
Sekijima M, Sugiura M, Matsuura M (1980) Nesting habits and brood development of *Parapolybia indica* Saussure (Hymenoptera: Vespidae). Bull Fac Agr Mie Univ 61:11-23
Sewerzoff AN (1931) Morphologische Gesetzmäßigkeiten der Evolution. Fischer, Jena, xi + 371 pp
Shaw FR, Weidhaas J, Jr (1956) Distribution and habits of the giant hornet in North America. J Econ Entomol 49:275
Shida T (1959) *Vespula lewisii* in Musashino. Nippon-Konchuki, vol 1. Kôdan-sha, Tokyo, pp 77-145 (in Japanese)
Shida T (1963) Prey list of *Vespula lewisii* in the suburbs of Tokyo. Kontyû 31:198-199 (in Japanese)
Shioda H, Mishima T, Nakajima O, Ikeda K (1974) A study on insect allergy (abstr). Allergy 23:99 (in Japanese)
Shiraki T (1933) Crop pests in Taiwan, pt 1. Taiwan Agr Exp Stn, 183 pp, 12 pls (in Japanese)
Shirôzu T (1980) Outline of the insect fauna of Tsushima. Gekkan Mushi 117:3-12 (in Japanese)
Shirôzu T, Miyata A (1976) List of insects of Tsushima. In: Biol Soc Nagasaki (ed) Fauna and flora of Tsushima, pp 567-763 (in Japanese)
Singh S (1962) Beekeeping in India. Indian Council of Agr Res, New Delhi, 214 pp
Smirnoff WA (1959) Predators of *Neodiprion swainei* Midd. (Hymenoptera: Tenthredinidae). Larval vectors of virus diseases. Can Entomol 91:246-248
Snelling RR (1981) Systematics of social Hymenoptera. In: Hermann HR (ed) Social insects, vol 2. Academic Press, New York London, pp 369-453
Sonan J (1929) On *Vespa* from Formosa (1). Trans Nat Hist Soc Formosa 19:136-149 (in Japanese)
Southwood TRE (1967) The interpretation of population change. J Anim Ecol 36:519-529
Spradbery JP (1965) The social organization of wasp communities. Symp Zool Soc London 14:61-96
Spradbery JP (1971) Seasonal changes in the population structure of wasp colonies (Hymenoptera: (Vespidae). J Anim Ecol 40:501-523
Spradbery JP (1973a) Wasps. An account of the biology and natural history of social and solitary wasps. Sidgwick & Jackson, London, xvi + 408 pp
Spradbery JP (1973b) The European social wasp, *Paravespula germanica* (F.) (Hymenoptera: Vespidae) in Tasmania, Australia. Proc 7th Int Congr IUSSI, London, pp 375-380
Stage GI, Slobodchikoff CN (1962) New distribution and host record of *Bareogonalos canadensis* (Harrington) (Hemenoptera: Trigonalidae and Vespidae). Pan Pac Entomol 38:97-98
Starr CK (1979) Origin and evolution of insect societies: a review of modern theory. In: Hermann HR (ed) Social insects, vol 1, Academic Press, New York London, pp 35-79
Sudd JH (1967) An introduction to the behaviour of ants. Arnold, London, 200 pp
Sugiura M, Sekijima M, Matsuura M (1983) Intracolonial polyethism in *Parapolybia indica* (Hymenoptera, Vespidae) 1. Behaviour and its change in the foundress and worker wasps in relation to colony development. Bull Fac Agr Mie Univ 66:11-25 (in Japanese with English summary)
Suzuki H (1975) The glacial epoch. Kôdan-sha, Tokyo, 189 pp (in Japanese)
Tagawa Y (1981) Asilinae in Shikoku, Japan (Diptera: Asilidae). Trans Shikoku Entomol Soc 15:187-213
Takamatsu Y (1952) Studies on the mechanism of the sexual function in the *Vespula lewisii* (Cameron). J Fac Agr Shinshu Univ 2:1-46 (in Japanese with English summary)
Takamizawa K (1981a) Some biological notes on *Vespula rufa* in Nagano. New Entomol (Ueda) 30:55-61 (in Japanese)
Takamizawa K (1981b) Colony size of *Vespula shidai*. New Entomol (Ueda) 30:62-63 (in Japanese)

Takamizawa K (1981c) Notes on the nesting habits of *Dolichovespula media sugare*. New Entomol (Ueda) 30:89–93 (in Japanese)

Taylor GW (1898) Notes on *Trigonalys canadensis* Hargt. Can Entomol 30:14–15

Taylor LH (1939) Observations on social parasitism in the genus *Vespula* Thomson. Ann Entomol Soc Am 32:304–315

Taylor LR (1963) Analysis of the effects of temperature on insects in flight. J Anim Ecol 32:99–117

Terayama M (1980) Research on the fauna of Yayeyama Islands (2nd report), with the bees of Miyako Islands. Sci Rep Exp Club Meiji Univ 2(1):1–19 (in Japanese)

Thomas CR (1960) The European wasp (*Vespula germanica* Fab.) in New Zealand. NZ Dep Sci Ind Res Inf Ser 27:1–74

Tissot AN, Robinson FA (1954) Some unusual insect nests. Fla Entomol 37:73–92

Tokuda Y (1924) Studies on the honeybee, with special reference to the Japanese honeybee. Trans Sapporo Nat Hist Soc 9:1–27

Trap-Lind I (1962) Observations on a honey buzzard digging out a wasp's nest. Br Birds 55:36

Tsuneki K (1973) On Trigonalidae of Japan. Life Stud 17:128–130 (in Japanese)

Umeya K (1956) Insect fauna of the Izu Islands 4. Kôzu-jima. Shin-konchû (Tokyo) 9(9):34–38 (in Japanese)

van der Vecht J (1933) Notes on *Nippogonalos jezoensis* Uchida. Tijdshr Entomol 76:79–82

van der Vecht J (1957) The Vespinae of the Indo-Malayan and Papuan areas (Hymenoptera, Vespidae). Zool Verh Leiden 34:1–83, 6 pls

van der Vecht J (1959) Notes on Oriental Vespinae, including some species from China and Japan (Hymenoptera, Vespidae). Zool Meded Leiden 16:205–232

van der Vecht J (1968) The terminal gastral sternite of female and worker social wasps (Hymenoptera, Vespidae). Proc Koninkl Ned Akad Wetensch Amsterdam Ser C 71:411–422

Verlaine L (1932) La spécialisation et la division du travail chez les guêpes. Bull Soc Sci Liége 1:186–191

Wafa AK (1956) Ecological investigations on the activity of the oriental hornet, *Vespa orientalis*. Bull Fac Agr Cairo Univ 103:1–35

Wagner RE (1961) Control of the yellowjacket, *Vespula pensylvanica*, in public parks. J Econ Entomol 54:628–630

Wagner RE (1978) The genus *Dolichovespula* and an addition to its known species of North America. Pan Pac Entomol 54:131–142

Wagner RE, Reierson DA (1969) Yellowjacket control by baiting 1. Influence of toxicants and attractants on bait acceptance. J Econ Entomol 62:1192–1197

Wagner RE, Reierson DA (1975) Clothing for protection against venomous Hymenoptera. J Econ Entomol 68:126–128

Walton GM, Reid GM (1976) The 1975 New Zealand European wasp survey. N Z Beekeeper 38:26–30

Ward P (1965) A raid on a *Stenogaster* nest by a hornet *Vespa tropica*. Malay Nat J 19:152

Watanabe C (1949) *Vespa* wasps on apples. Appl Entomol 5:109 (in Japanese)

Watson EB (1922) The food habits of wasps. Bull Chamber Hort 1:26–31

Welch HE (1958) *Agamomermis pachysoma* (Linstow 1905), a parasite of social wasps. Insectes Soc 5:353–355

Wellman G (1952) Über die Notwendigkeit der Wespenbekämpfung. Berlin Münchener Tierärtzl Wochenschr 65:11–12

West-Eberhard MJ (1969) The social biology of polistine wasps. Misc Publ Mus Zool Univ Mich 140:1–101

Weyrauch W (1935) *Dolichovespula* und *Vespa*. Vergleichende Übersicht über zwei wesentliche Lebenstypen bei sozialen Wespen. Mit Bezugnahme auf die Frage nach der Fortschrittlichkeit tierischer Organisation. I. Biol Zentralbl 55:484–524

Weyrauch W (1936) *Dolichovespula* und *Vespa*. Vergleichende Übersicht über zwei wesentliche Lebenstypen bei sozialen Wespen. Mit Bezugnahme auf die Frage nach der Fortschrittlichkeit tierischer Organisation. II. Biol Zentralbl 56:287–301

Weyrauch W (1937) Zur Systematik und Biologie der Kuckuswespen *Pseudovespa, Pseudovespula* und *Pseudopolistes*. Zool Jahrb (Syst) 70:243–290

Wheeler WM (1928) The social insects: their origin and evolution. Harcourt Brace, New York, 378 pp
Wheeler WM, Taylor LH (1921) *Vespa arctica* Rohwer, a parasite of *Vespa diabolica* de Saussure. Psyche 28:135–144
Wild OH (1927) Wasps destroying young birds. Br Birds 20:254
Williams FX (1919) Philippine wasp studies II. Description of new species and life history studies. Bull Hawaii Sugar Pl Assoc (Ser Entomol) 14:19–186
Wilson EO (1971) The insect societies. Harvard Univ Press, Cambridge, x + 548 pp
Wilson EO (1975) Sociobiology. The new synthesis. Harvard Univ Press, Cambridge, ix + 697 pp
Wilson MVH (1977) New records of insect families from the freshwater Middle Eocene of British Columbia. Can J Earth Sci 14:1139–1155
Wray DL (1954) Nesting notes on the European hornet in N. Carolina. Entomol News 65:1–5
Yamane S (1969) Preliminary observations on the life history of two polistine wasps, *Polistes snelleni* and *P. biglumis* in Sapporo, northern Japan. J Fac Sci Hokkaido Univ (VI Zool) 17:78–105
Yamane S (1970) Classification, life cycle and observations of social wasps of Hokkaido. Seibutsu Kyôzai (Kikonai) 7:20–42 (in Japanese)
Yamane S (1971) Daily activities of the founding queens of two *Polistes* wasps, *P. snelleni* and *P. biglumis*, in the solitary stage (Hymenoptera, Vespidae). Kontyû 39:203–217
Yamane S (1976) Societies seen in the Vespoidea. Iden 1976(4):34–44 (in Japanese)
Yamane S (1977) On the collection technique of vespine nests, based chiefly on practices through a survey in Taiwan from 1972 to 1974 (Hymenoptera, Vespidae). Seibutsu Kyôzai (Kikonai) 12:42–59 (in Japanese)
Yamane S. Kubo T (1970) A brief note on a labor-parasitic wasp, *Vespula austriaca*, in association with *V. rufa schrenckii*. Kontyû 38:171–175
Yamane S, Yamane Sk (1975) Investigating methods of dead vespine nests (Hymenoptera, Vespidae). Seibutsu Kyôzai (Kikonai) 10:18–39 (in Japanese)
Yamane S, Sakagami SF, Ohgushi R, Abbas ND, Matsuura M (1983) Life history and social organization of *Parischnogaster mellyi*. Ecological study on social insects in central Sumatra with special reference to wasps and bees. Rep Overseas Sci Surv (SNS-Entomol) 1981, pp 18–24
Yamane Sk (1973) Descriptions of the second to final instar larvae of *Bareogonalos jezoensis* with some notes on its biology (Hymenoptera, Trigonalidae). Kontyû 41:194–202
Yamane Sk (1974) Observations on an orphan nest of *Vespa simillima* Smith (Hymenoptera, Vespidae). Kontyû 42:404–415
Yamane Sk (1975) Taxonomic notes on the subgenus *Boreovespula* Blüthgen (Hymenoptera, Vespidae) of Japan, with notes on specimens from Sakhalin. Kontyû 43:343–355
Yamane Sk (1976) Morphological and taxonomic studies on vespine larvae, with reference to the phylogeny of the subfamily Vespinae (Hymenoptera: Vespidae). Insect Matsum NS 8:1–45
Yamane Sk (1978) Evolution of social parasitism among the Vespinae. Panmixia 3:1–9 (in Japanese)
Yamane Sk (1982a) Vespoid wasps from Niigata Pref., Honshu, Japan. Trans Essa Entomol Soc (Niigata) 53:3–14 (in Japanese)
Yamane Sk (1982b) Comparison of vespine fauna between eastern Asia and North America. Ecol Soc Jpn Kyushu-chikukai Kaihô 25:12–13 (in Japanese)
Yamane Sk (1982c) A new subspecies of *Dolichovespula adulterina* Buysson from Japan, with a brief note on its host (Hymenoptera: Vespidae). Trans Shikoku Entomol Soc 16:109–115
Yamane Sk (1983) The aculeate fauna of the Krakatau Islands (Insecta, Hymenoptera). Rep Fac Sci Kagoshima Univ (Earth Sci Biol) 16:75–107
Yamane Sk, Kamijo K (1976) Social wasps visiting conifer plantations in Hokkaido, northern Japan (Hymenoptera: Vespidae). Insect Matsum NS 8:59–71
Yamane Sk, Kanda E (1979) Notes on the hibernation of some vespine wasps in northern Japan (Hymenoptera, Vespidae). Kontyû 47:44–47
Yamane Sk, Makino S (1977) Bionomics of *Vespa analis insularis* and *V. mandarinia latilineata* in Hokkaido, northern Japan, with notes on vespine embryo nests (Hymenoptera: Vespidae). Insect Matsum NS 12:1–33
Yamane Sk, Yamane S (1975) A new trigonalid parasite (Hymenoptera, Trigonalidae) obtained from *Vespula* nests in Taiwan. Kontyû 43:456–462
Yamane Sk, Makino S, Macfarlane RP (1981) Embryo nest architecture in three *Vespula* species (Hymenoptera, Vespidae). Kontyû 49:491–497

Yamane Sk, Makino S, Toda MJ (1980a) Nests of *Dolichovespula albida* from the Arctic Canada (Hymenoptera: Vespidae). Low Temp Sci Ser B 38:61–68
Yamane Sk, Wagner RE, Yamane S (1980b) A tentative revision of the subgenus *Paravespula* of eastern Asia (Hymenoptera: Vespidae). Insect Matsum NS 19:1–46
Yamane Sk, Makino S, Ban T, Sato Y (1979) Protective suits against hornets. Kontyû 47:429–430 (in Japanese)
Yoshikawa K (1959) Superindividual organization of polistine wasps. In: Imanishi K (ed) Societies and individuals in the animals. Iwanami-shoten, Tokyo, pp 78–89 (in Japanese)
Yoshikawa K (1963) Comparative sociological studies of hunting wasps in temperate Asia. Proc 9th Pac Sci Congr, Tokyo, 1957, 9:136–140
Yoshikawa K (1973) Social insects, with special reference to the life of wasps. Kyôritsu-shuppan, Tokyo, 130 pp (in Japanese)

Author Index

Taxonomic Index

Subject Index

MIX
Papier aus verantwortungsvollen Quellen
Paper from responsible sources
FSC® C105338

If you have any concerns about our products,
you can contact us on
ProductSafety@springernature.com

In case Publisher is established outside the EU,
the EU authorized representative is:
Springer Nature Customer Service Center GmbH
Europaplatz 3, 69115 Heidelberg, Germany

Printed by Libri Plureos GmbH
in Hamburg, Germany